METROS IN INDIA

VOLUME 2

SHAPING URBAN MOBILITY: METRO RAIL SYSTEMS ACROSS INDIA

DR. BC ROY

notionpress.com

INDIA • SINGAPORE • MALAYSIA

ISBN 979-8-90207-627-8

Contents

List of Figures

Part 4

List of Tables

Part 4

Foreword

Prof SS Chakraborty, FNAE (I), FICE (UK)-Gold Medalist, FIStructE (UK)
Co-Founder & Chairman Emeritus - CES, & Chairman-Burdwan Health Care
Metros in India

It has been my pleasure to know Dr. BC Roy since 1978, when his association with CES began. Over time, what started as a professional acquaintance was transformed into a collaboration that has continued for more than forty-five years. During this period, his steady growth as a versatile engineer has been observed, with contributions made to high-rise buildings, bridges, environmentally sustainable facilities, sports complexes, and more importantly, major transit projects in India and abroad. His commitment to hard work, and constant pursuit of deeper knowledge have consistently been demonstrated.

The foundation of Dr. Roy's expertise in metro systems was laid with his involvement in India's first metro project in Kolkata. Germany was the source of his first metro lesson. For the young engineers, he was considering consolidating all his Metro-related experience in one location and authored a book on "Indian Metros." This early experience was further enriched through exposure to international practices and decades of professional engagement across the country. A distinctive ability to integrate innovative design with practical execution was displayed in projects of national importance, including the development of metro systems in several Indian cities. It is from this wealth of experience of "Metros in India: Evolution of Indian Metros, Kolkata and Beyond" has been authored, with the intention of sharing valuable knowledge with younger generations of engineers and practitioners.

In this volume, a comprehensive account of India's metro journey has been provided by Indian engineers—from the pioneering work in Kolkata to the extensive networks now seen in Delhi, Mumbai, Bengaluru, Chennai, Hyderabad, Pune, and beyond. The technical dimensions of metro development, such as tunnelling, rolling stock, and signalling systems, have been brought out with clarity, while attention has been given to the social, environmental, and policy challenges accompanying such transformative undertakings. Emphasis has been placed on the Kolkata Metro, which served as a crucible of innovation and a source of vital lessons for subsequent projects across the nation.

The expansion of metro networks has been presented not merely as an infrastructure initiative, but as a redefinition of urban mobility in response to rapid population growth and the demands of sustainable development. The broader impact of metro projects on the quality of urban life has been underscored, bearing in mind that these systems are as much about people as they are about engineering.

A distinctive presentation style has been adopted, where practical examples, project experiences, and visual illustrations have been integrated to enhance accessibility. This approach has ensured that the book is not only a reference for policymakers and professionals but also a valuable guide for students and young engineers.

This work will be regarded as a standard reference in the field of urban transport. Inspiration will be drawn by future engineers and planners, who will be encouraged to view metro development as a mission and an absolute essential, aimed at improving urban mobility in Indian cities.

Prof SS Chakraborty FNAE

Preface

With rapid global urbanisation, over half the world's population now lives in cities. In India, nearly 1.46 billion people live, of which 37% (542 million) are urban residents. This scale of growth demands integrated planning and robust Mass Rapid Transit Systems (MRTS) to enhance mobility, living standards, and economic productivity. Metro rail, operating on independent rights-of-way, is central to sustainable urban mobility by offering fast, safe, reliable, and low-emission connectivity. India's metro journey began in Kolkata (construction started in 1970, operations commenced in 1984) and has since expanded to 17 cities, highlighting both technical, institutional, and financial challenges, as well as indigenous innovations and global collaborations. Drawing on the author's direct involvement in projects with CES, AECOM, and Chatterjee & Polk, this book covers planning, design, execution, and operations, supplemented by DPR excerpts, tender samples, case studies, and design details. While financial returns remain challenging, metros are indispensable in addressing congestion and supporting economic growth. This book draws on the author's experience and collaborations across Kolkata, Delhi, Mumbai, Bengaluru, and other cities. It covers the planning, design, construction, and operation of metro projects, with case studies ranging from Kolkata's pioneering North-South and East-West Metro lines - India's first underground line and first river-crossing metro—along with metro developments in Delhi, Mumbai, Bengaluru, Chennai, Jaipur, Hyderabad, and Pune in the PPP models. Insights from DPR preparation, tendering practices, depot & station design, tunnelling technologies, and green building concepts are included.

Structured in two volumes, the book combines technical depth with practical case studies to serve as a comprehensive resource for students, engineers, and practitioners engaged in metro development within India's evolving urban landscape.

Author's Note

A

The rapid pace of urbanisation in India, like in the rest of the world, has transformed cities into densely populated spaces with limited room for expansion. Narrow roads and heavy congestion posed enormous challenges for planners, particularly when designing for future needs. The demand for an efficient urban transport solution capable of carrying large numbers of passengers across cities led to the adoption of Rail based Mass Rapid Transit Systems (MRTS). The metro emerged as the most appropriate choice, offering high capacity, safety, comfort, and short headways of 90–150 seconds.

The foundation of India's metro development was laid in 1969 with the establishment of the Metropolitan Transport Project (MTP) under the Ministry of Railways. With the support of Soviet specialists and East German engineers, MTP prepared the first master plan for Kolkata. The North–South corridor of the Kolkata Metro became India's first metro rail project, designed, engineered, and constructed by Indian engineers. Funded through the Ministry of Railways under budgetary provisions, this metro marked the beginning of India's journey into modern urban mobility. The Kolkata Metro relied on broad-gauge technology, familiar to Indian Railways. This avoided the need to design or import new bogies, saving time and resources on validation and durability. India already had a strong tradition of coach manufacturing for broad, meter, and narrow gauges, with ICF Perambur serving as the principal manufacturing hub. Except for advanced features such as automatic door closure, most requirements were indigenously met. The signalling system was based on interlocking technology developed by Indian Railways, and today the North–South Metro is being upgraded to modern signalling standards.

However, a few experts considered the Kolkata Metro obsolete compared to global standards, but history has proven otherwise. Karnataka Government prepared a Project Report for Bangalore Metro, with an idea to use Indian System in bringing down the cost, in early nineties. In 1994 BMRTL prepared feasibility report for Bangalore Metro. Delhi Metro Rail Corporation (DMRC) was established in 1995, and the report of first phase was approved by Union Cabinet in 1996 and began operations in 2002. Delhi Metro initially used broad gauge with imported coaches before transitioning to standard gauge. International firms such as Hyundai Rotem, Mitsubishi, Bombardier, and CAF supplied the early stock, after which Bharat Earth Movers Limited (BEML), in collaboration with Hyundai Rotem, began indigenous manufacturing. Today, BEML supplies metro coaches for ongoing and planned projects, with other Indian firms, such as Tikamgarh Rail Systems, also entering the coach manufacturing sector. Delhi now operates the country's largest metro network, spanning over 350 km with 256 stations, and has ambitious expansion plans under Phase V. Broad-gauge lines primarily use indigenously manufactured coaches, while standard-gauge corridors employ a mix of imported and locally built stock. Except for Kolkata's East–West Metro, which runs on standard gauge, most new corridors under construction or commissioning continue to adopt broad gauge with Indian-built coaches.

A significant hurdle in the early stages of metro development was the absence of a dedicated legal framework for urban rail transport. In Kolkata, the outdated Kolkata Tramways Act was modified to accommodate metro operations, though its provisions were limited to a single municipal jurisdiction. Similar challenges were faced in Mumbai, governed by the Bombay Tramway Act. Delhi Metro necessitated the creation of an entirely new legal framework to allow operations across and beyond the city limits. This was the groundwork for future expansion across the country. Metros in India have expanded rapidly. Today, seventeen Indian cities operate metro systems, with nine more in different stages of construction and planning. Kolkata's North–South Metro, largely constructed underground through the cut-and-cover method and partly with Tunnel Boring Machines (TBMs), set the standard for engineering practices in subsequent projects. Russian engineers initially recommended sheet piles for deep excavation, but when locally available piles proved inadequate, Indian teams introduced diaphragm wall construction with German technical support, a Milestone innovation at that time. Later, advanced tunnelling techniques such as precast concrete segments and TBMs were introduced, particularly in the East–West Metro, which achieved the distinction of making India's first under-river tunnel beneath the Hooghly.

The East–West Metro also pioneered modern construction practices with extensive use of instrumentation, top–down station construction, and experiences of international consultants. However, frequent alignment changes and delays demonstrated the challenges of integrating metro systems within complex urban environments. Parallel to these developments, other cities started planning and implementing their metro projects. Mumbai, Bangalore, Chennai, Hyderabad, Jaipur, and Pune followed, with Hyderabad and Pune's Line 3 adopting the Public–Private Partnership (PPP) model. Hyderabad Metro, executed by the L&T-LED consortium under HMRL, became India's largest PPP-based metro project, whereas Rapid Light Metro system in Gurgaon was the first fully owned by IL&FS in India. Similarly, Pune Metro's Line 3 was entrusted to a Tata Realty–led consortium under Maha-Metro.

In Mumbai, the Metro Planning Group was established in the mid-1990s to prepare feasibility studies. Alternative systems, such as the Monorail and the Konkan Railway's Sky Bus Metro, were also evaluated as feeder services. These experiences contributed to the broadening of India's urban transport strategies beyond conventional metro systems.

The development of metro systems in India has always been closely related to innovations in design, construction, and project management. From diaphragm walls and precast tunnel segments to advanced TBM tunnelling and modern depot design, each project has introduced new practices that were later adopted by various cities. Building condition surveys, an integral part of the East–West Metro, became SOP for assessing urban infrastructure before construction. Condition of neighbouring structures including Heritage Structures along a Metro Corridor and remedial measures are included in section for Mumbai Metro Line 3 in Volume 2.

Over the years, the metro has transformed from a pilot project in Kolkata to a national urban ecosystem. Supported by domestic manufacturing, evolving legal frameworks, and continuous learning from both global and local experiences, the development of metros in India reflects the resilience and ingenuity of its engineers and planners. Indian Engineers carried out lot of Research work on Hanging Metro, under the leadership of Mr Rajaram, former MD, KRCL at KRCL which was not successful in first attempt, though a few cases of such metros are in operation outside India.

B

The Presentation of the book is a bit different from the general book format. The book is presented using a combination of texts, bullet points, and figures. The figures include design sketches, execution plans and many more photos, as the Author's experience is that this approach will provide a better understanding.

While reading, one may experience repetition, which is done to facilitate comprehension. For comprehensive details on the Maintenance Depot or Systems, one may refer to the details discussed for other metros included in Parts 2, 3 and 4. The systems for the Kolkata North-South Metro, the first metro, and the East-West Metro differ substantially. While metros in other Cities primarily follow the Delhi Metro with variations. Author believes, all most all aspects of an Urban Metro planning, design and construction aspects have been included, may not be in one place but distributed among the Metros included. It is particularly for Depot and System designs. The author included most of the references for both texts and photographs. However, a few references might have been inadvertently missed and may be excused.

Safety for construction employees, important issue, in Metro Construction, has been included, although the author's name could not be located.

Metros in India

Volume 2

PART 4

Part 4 features metros in Thane, Delhi, Mumbai, Bangalore, Chennai, Jaipur, Hyderabad, and Pune. The author has focused on metros in which his team was directly involved. However, the Hyderabad and Pune (Line 3) Metros have been included as they were designed, executed, and are operating under the Public-Private Partnership (PPP) model. The Hyderabad Metro Rail Limited (HMRL) commissioned the L&T-led consortium for the Hyderabad Metro, which was planned, designed, executed, and is now operational. Similarly, Maha-Metro assigned Line 3 of the Pune Metro Network to the Tata Realty-led consortium.

The DPR presented here was prepared by a team led by CES in 2001 based on the RITES report, for the Thane Municipal Corporation and appointed by MSRDC. This report was comprehensive and included all aspects of implementing the PPP mode. However, at present, the Thane Metro is being implemented as a part of the Mumbai Metro Network Plan.

MSRDC was responsible for planning the Metro systems for three cities in Maharashtra —Pune, Thane, and Nagpur. DPRs for the Thane and Nagpur MRTS systems were prepared sometime in 2001-03. However, the projects were not sanctioned for various reasons, primarily due to financial viability. Currently, work on Metro corridors in all these cities is being done by Maha Metro and MMRDA.

The Thane Metro Phase 1 project, with 1 circular line and 22 stations, is a 29 km mass rapid transit system (MRTS) approved for construction in Thane, Maharashtra, by the Maharashtra Metro Rail Corporation Limited (MMRC or Maha Metro).

The project, officially known as the Thane Integral Ring Metro, was approved by India's Central Government in August 2024. Thane's mass transit project was initially planned as a regular (heavy) metro line, had its foundation stone laid by the Prime Minister in December 2018, was scrapped in 2019 due to cost concerns, and then resurrected as a metro lite (light rail transit, LRT) system in early 2020.

In 2023, the project was redesigned as a regular metro line. The Thane Metro will loop around Thane city and consist of 22 stations (20 elevated and 2 underground) to link Mumbai Metro's 32.32 km Line-4 (Green Line), 24.9 km Line-5 (Orange Line), and the Central Railways' Thane Station. Thane Metro's train maintenance depot (car shed) is planned to be built at Wadavli.

CHAPTER 8

Thane MRTS – Detailed Project Report (DPR) - 2001

8.1 Background

In 1989, a feasibility study was conducted for a high-capacity Mass Rapid Transit System proposed in the Thane Development Plan (1987). The study focused solely on assessing the feasibility along the route specified in the plan and the reserved right-of-way for the Corridor. On behalf of the Thane Municipal Corporation, CIDCO conducted a cost-benefit and financial analysis of the Light Rapid Transit (LRT) System for Thane, which was included in the same development plan. The LRT System is designed in a circular route extending over 20 km (as identified in 1989) and consists of 11 stations.

The Thane Municipal Corporation area spans 128.23 sq. km, with a population of 0.80 million in 1991, increasing to 1.4 million by 2000, resulting in a multi-fold increase in traffic —both vehicular and pedestrian —which has caused congestion and traffic jams in the city. The MSRDC appointed a consortium led by Consulting Engineering Services (India) Limited, with PBK&D (UK) and LPT (Malaysia) as associates, and UTI-SEL and ANS Architects as consortium members to prepare a Detailed Project Report (DPR) for the Thane MRTS, planned to be implemented on a BOT/BOOT basis. The consortium was led by the Author.

The DPR was submitted in **seven Volumes** and consisted of Study Area Characteristics, Master Plan, and Ridership, inter alia, including Surveys & Investigations, Traffic, Travel &Travel Characteristics, Alignment, and Travel Demand Modelling & Forecast (Vol 1), System Design (Vol 2), Corridor and Station Area Planning, Architecture & Engineering (Vol 3), Resource Mobilisation, Legal and other Issues, inter Alia including Environmental and Social Impact, Economic Analysis, Fare Policy, Financial Aspects and Viability Analysis, Risk Analysis and Management, Legal Frame work and Institutional Framework for Project Implementation (Vol 4), Alignment and Existing Utilities (Vol 5), General Arrangement Drawings (Vol 6), Station Area Planning and Architectural Drawings (Vol 7). Work commenced in 1998, and the report was submitted in 2001 following MSRDC's approval.

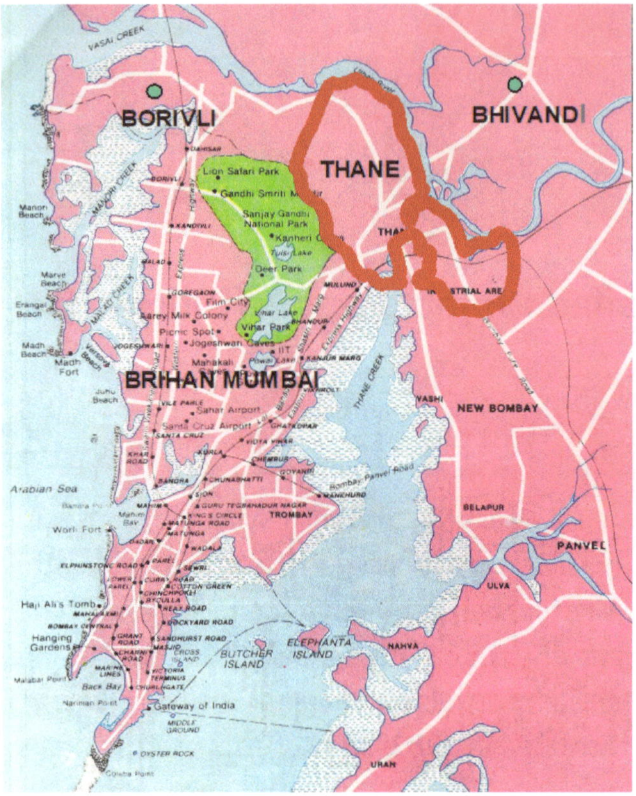

Fig.8.1: Location of Thane

Fig.8.2: Existing Railway Station

The Section includes Background, Existing Transport Facilities, Study Area: Traffic and Travel, Traffic Characteristics, Master Plan, Thane MRTS Network Plan, and Briefs of Three Corridors, Ridership, Track Gauge, Alignment, Stations, Engineering, and System.

8.2 Existing Transport Facility and Need for MRTS

The main railway station serving Thane is south of the Thane Municipal area, and passengers travelling to and from the northern parts. The surrounding roads are narrow and congested. There is inadequate parking near the station, and the large number of taxis and auto-rickshaws operating around the station exacerbates the congestion. This situation is exacerbated by the lack of a road crossing of the railway between the Eastern Express Highway and the narrow road that passes under the railway to the east. The demand for public transport has been increasing due to population growth and the associated rise in employment. The result is more congestion with large numbers of slow-moving vehicles steadily polluting the environment with exhaust emissions and noise.

Detailed surveys and Investigations, which included Topographical, Traffic, and willingness-to-pay surveys, as well as existing road systems, travel characteristics, Geotechnical, and Environmental surveys, were carried out. The collected data were studied in detail, and an alignment was selected by examining several possible alignments, assigning appropriate values to land acquisition, preliminary cost, economic viability, and financial viability, and using a matrix to evaluate. Thane has developed circularly, expanding outwards from the initial Central Business District adjacent to the main railway station, Central Railway, providing commuter services to Mumbai and long-distance services. Three major highways, NH-3, NH-4 and the Eastern Express Highway run through Thane. Several new industries have developed in Thane, concentrating in the Belapur industrial area, the Wagle, Kalwa industrial estates, and the Kolshet and Balkum areas.

Fig.8.3: Thane Station Area

The demand for public transport has been increasing in recent years due to population and employment growth, as well as inadequate existing bus services. This has led to an increase in taxis and auto rickshaws, further straining the road network. The result of this is increasingly congested roads with large numbers of slow-moving vehicles, steadily polluting the environment with exhaust emissions and noise.

To solve the problem, the Government of Maharashtra decided to implement a Public Transport System with higher capacity and a rail-based Mass Rapid Transit System. Rail-based Transit systems carry many more commuters at any given time and transfer passengers more quickly from one location to another. The MRTS is expensive in terms of capital and operating costs, but the benefits include increased employment, environmental benefits, economic expansion, increased revenue earnings, enhanced state tax revenues, and passenger benefits such as reduced travel time, frequent service, and comfort. To achieve higher revenue, the fare structure should be affordable to ensure the desired ridership and to make optimal use of the capacity provided by the MRTS, thereby providing socio-economic benefits.

8.3 Surveys and Studies

To appreciate the study area characteristics, identify issues, constraints, and opportunities, and develop a database for system planning, design, and operation, extensive surveys and studies have been carried out. Extensive data and information on the physical, demographic, economic, and other characteristics of Thane have been collated from several sources, viz. TMC, Maharashtra State Electricity Board, Regional Transport Office, Census of India, Indian Railways, etc.

Fig.8.4: View During O-D Survey.

Extensive primary surveys have been carried out to elicit socio-economic, network traffic and travel, environmental, and geo-technical characteristics of the study area. Important surveys conducted for the study relate to Classified Traffic Volume Counts at the outer cordon (seven count stations). The data and information generated have been used in constructing transport demand models to forecast travel demand, finalise the alignment of the system, plan and design the system components, estimate costs, assess environmental impacts and develop a program of action and Screen Lines (nine count stations); Origin and Destination survey at Outer Cordon (seven count stations); Road Network Inventory (186 km); Speed and Delay Survey (450 links of length 186 km); Household Survey (3507 Households); Willingness to Pay Survey (350); Terminal Surveys (Road and Rail); Topographic Surveys: Air Quality Monitoring surveys, Noise survey, etc. To assess soil strengths and characteristics along the proposed MRTS corridor, geotechnical studies were carried out under the Consultants' supervision.

8.3.1 Study Area: Traffic and Travel

Appreciation of traffic and travel characteristics of the study area is important to assess future intensity and pattern of traffic on the transport network system; evaluate the network system in terms of capacity, constraints and potentials; understand the needs of the people as regards their travel pattern in terms of size, purpose, desire, modal preference, willingness to pay for an improved system and service etc. They also provide a rational basis for policy formulation for system development and operation, including fare policy. Several surveys and studies were carried out for the present study.

The study area encompasses the Thane Municipal Corporation (TMC) territory, spanning 128.23 sq. km and comprising 11 sectors and 95 wards.

Study Area Traffic and Travel Characteristics

The salient traffic and travel characteristics of Thane, as revealed by the surveys, were:

- There were about 1.3 lakh motorised vehicles registered in Thane City, out of which 47% are 2 wheelers, 17% are auto rickshaws, and 16% are cars.
- A large volume of traffic used to move within, to, from and through Thane. A total of 1,15,049 vehicles, carrying 4,73,533 passengers, enter and exit Thane City on average per day.
- 'Through' (i.e. by-passable) traffic was 22.48% by vehicles and 14.22% by passengers.
- At the Screen Line Count Stations, the Average Daily Traffic (ADT) ranged between 16,017 vehicles (22,435 PCUs) and 66,997 vehicles (57,914 PCUs).
- The number of two axle trucks at Bhiwandi Bypass was 9084, accounting for 35.4% of total traffic.
- The average household size in Thane was 3.81.
- 37% of households possess a vehicle. Excluding cycles, 18.2% households possessed one vehicle, and only 3.7% households possessed two vehicles.
- On an average day, 5.58 lakh people used the railway stations at Thane, Kalwa and Mumbra.
- Walking was the predominant (48.50%) access mode of rail passengers.
- More than 1.80 lakh inter-city passenger trips were generated at inter-city bus terminals in Thane.
- A total of 17,78,718 trips were generated in the study area by the residents of Thane Municipal Corporation on an average day. Of the total trips generated, 23.66% were intra-zonal and 76.34% inter-zonal.
- By purpose, Work trips accounted for 51% of total trips. Education trips account for 38%.
- Out of the total trips by vehicular modes, public transport modes, including IPT modes, account for a high share of 77%.
- The journey speeds, during peak periods, on some important roads, far from a low of 8.2 kmph to a high of 45.3 kmph.

8.3.2 Traffic Characteristics

Outer Cordon

A large volume of traffic moves within, from, to and through Thane. A total of 1,15,049 vehicles enter and leave the study area on an average day. In terms of passengers, it totals 4,73,533. The highest traffic was observed at the entry points on Thane-Belapur Road (Vehicles – 24274, passengers – 1,68,378). **Table 8.1** presents the traffic volume at outer cordon count stations. Of the passenger modes, the "through" traffic at the outer cordon was 22.48% by vehicles and 14.22% by passengers. Interestingly, the through traffic by private car/jeep was high at 62.6% by vehicles and 41.05% by passengers. **Table 8.2** presents the passenger mode pattern at the outer cordon.

Table 8.1: Traffic at Outer Cordon

Name of Outer Cordon	No. of Vehicles	Passenger Trips	% Passenger Trips
Ghodbunder Road	7766	31639	6.68%
Old Agra Road (Kasheli)	11416	32480	6.86%
Bhiwandi Bypass Road	12533	42817	9.04%
Mumbra Panvel Road (Shil Phata)	10666	41128	8.69%
Thane Belapur Road	24274	168378	35.56%
Eastern Express Highway	26146	75277	15.90%
Mulund Check Naka (LBS Road)	22248	81814	17.28%
Total	**115049**	**473533**	**100%**

Table 8.2: Passenger Mode Pattern at Outer Cordon

Mode	Intercity Traffic		Through Traffic		Total	
	Vehicles	Passengers	Vehicles	Passengers	Vehicles	Passengers
1. Private Car/Jeep	27153	81126	16197	27632	43350	108759
2. Taxi	1683	4899	672	934	2355	5833
3. 2-Wheeler	31395	47601	5239	4061	36634	51663
4. Auto Rickshaw	22033	58399	1454	1984	23487	60383
5. Bus	6921	214188	2302	32707	9223	246896
Total	**89185**	**406215**	**25864**	**67319**	**115049**	**473533**
	(77.52%)	*(85.78%)*	*(22.48%)*	*(14.22%)*		

Among person trips at the outer cordon, through traffic was highest at Bhiwandi Bypass Road, accounting for 31.73%, while intercity traffic at this location accounted for only 5.28% of the total intercity traffic at all count stations along the outer cordon. **Table 8.3** presents the passenger trip pattern at outer cordon count stations.

Table 8.3: Passenger Trip Pattern at Outer Cordon

Cordon Point	Intercity		Through Traffic		Total	
Old Agra Road (Kasheli)	27210	6.70%	5270	7.83%	32480	6.86%
Bhiwandi Bypass Road	21457	5.28%	21360	31.73%	42817	9.04%
Thane Belapur Road	159630	39.30%	8748	12.99%	168378	35.56%
Mumbra Panvel Road (Shil Phata)	40591	9.99%	538	0.80%	41128	8.69%
LBS Road	74020	18.22%	7793	11.58%	81814	17.28%
Eastern Express Highway	56962	14.02%	18314	27.21%	75277	15.90%
Ghodbunder Road	26344	6.49%	5296	7.87%	31639	6.68%
Total	**406215**		**67319**		**473533**	

The interaction between Thane and other nodes of MMR was very high, accounting for nearly 100% of external (through) trips at count stations on Eastern Express Highway and LBS Marg. On the other hand, it was 2.19% at the Kasheli count station. **Table 8.4** presents the pattern of interaction between Thane and other parts of MMR.

Table 8.4: Traffic Interaction Pattern

	Ghodbunder	Kasheli	Shilphata	Belapur	Bhiwandi Bypass	EEH	LBS
I. Inter City Trips	83.26%	83.77%	98.69%	94.80%	50.11%	75.67%	90.47%
I. External Trips (Through Traffic)	16.74%	16.23%	1.31%	5.20%	49.89%	24.33%	9.53%
II. Thane-MMR Interaction	77.83%	2.19%	72.63%	89.48%	35.15%	98.68%	99.97%
II. Thane-Outside MMR	22.17%	97.81%	27.37%	10.52%	64.85%	0.32%	0.03%

Fig.8.5: Pedestrian Traffic near Thane Rail Station

Fig.8.6: Vehicular Traffic in front of Thane Station (1998)

Along the screen lines 1 to 4, the volume of passenger modes ranged from a low of 11,180 vehicles (9,637 PCUs) at Screen Line 4 to a high of 156,663 vehicles (134,219 PCUs) at Screen Line 2. **Table 8.5** presents the passenger mode traffic at the screen lines.

The ADT at the outer cordon shows a consistently high volume of traffic across all entry count stations. The maximum traffic was at EEH with a total of 36,865 vehicles (52,812 PCUs). **Table 8.6** presents ADT at different count stations along the outer cordon. Truck traffic was consistently high, ranging from 29 to 45%, except on LBS Marg, where it was a low of 9%. **Figure 8.2** shows traffic composition at Outer Cordon points.

Table 8.5: Passenger Mode Traffic at Screen Lines

Vehicle Type	SCR – 1	SCR – 2	SCR – 3	SCR – 4
Two Wheelers	3786	43221	18241	9495
Auto Rickshaw	2700	71216	31375	12628
Car/Jeep/Van	4539	33355	14457	9818
Taxi	37	386	93	31
Bus	1305	8485	5525	3482
Total Vehicles	12367	156663	69691	35454
Total PCU	**12409**	**134219**	**63777**	**34514**

SCR – 1: Ghodbunder Road

SCR – 2: Hutatma Maruti Kumar Marg (Cadbury), Gokhale Marg, Golden Dyes LBS Road, Teen Hath Naka LBS Marg, Mahapalika Marg

SCR – 3: Kalwa Bridge, Bhiwandi Bypass Road

SCR – 4: Mumbra Panvel Road, Thane Belapur Road

Traffic Surveys were carried out at nine screen line points. The ADT ranged from a low of 16,017 vehicles (23,435 PCUs) at Mumbra Panvel road (S–9) and a high of 66,997 vehicles (57,941 PCUs) at Kalwa Bridge (S–5). The number of 2-axle trucks at Bhiwandi Bypass (S-1) reached 9084, accounting for 35.4% of the traffic volume. The ADT at screen line points is given in **Table 8.5. Figure 8.2** shows traffic composition at screen line points.

Travel Characteristics

The household survey reveals the travel characteristics of Thane's residents. This provides important input for developing transport demand models. The salient travel characteristics are presented in the following paragraphs.

Household Size and Vehicle Ownership

The average household size in the TMC area was 3.81. The size ranges from one person to 12 people. The median value was four.

Among households, only 37% have a vehicle. Without bicycles, it is only 23%. Excluding bicycles, 18.2% of households possess one vehicle, and only 3.7% of households own two vehicles. The vehicle ownership pattern indicates a high dependence on public transport.

Person Characteristics

Age Distribution, Literacy and Occupation

Thane is a city of young people. The average age of people at TMC was 31 years. The median age group was 24-40 years. Sector 2 shows a higher concentration of middle-aged (40-58) and older age

(58+) people. The people of Thane are **highly literate** (92%). People with a graduation and higher education accounted for 17.27%. For the **occupation '** Workers', including "Business", accounted for 36.30%. Students accounted for 25.06%. 'Others' include housewives, retired people, etc.

Trip Characteristics

Trip Size: A total of 17,78,178 trips were generated in the study area by the residents of TMC on an average working day.

Trip Rate: Excluding age group 0-5 years, the per capita and per household trip rates worked out to 1.44 and 5.2, respectively (all modes).

Trip Pattern: Of the trips generated, 23.66% were intra-zonal and 76.34% inter-zona

Trip Purpose: Trips for "Work" (including Business) accounted for 51.39%.

"Education" trips accounted for 38.06%.

Trips by Mode and Journey Speeds

"Walk" trips accounted for 46.04% of total trips. Of trips by vehicular modes (i.e. excluding "walk" trips), public transport modes, including IPT modes, accounted for a high share of **77.22**%. Sectors IX and X account for a high share of trips by train, whereas sectors V and VI account for a very low share by train. **Figures 8.4 and 8.5** show modal shares with and without walk, respectively.

Average journey speeds during the peak period on some of Thane's important roads ranged from 8.2 kmph to 45.3 kmph. The generally Low speeds indicate a decline in the level of service on the road network, supporting the policy of developing a rail-based mass transit system.

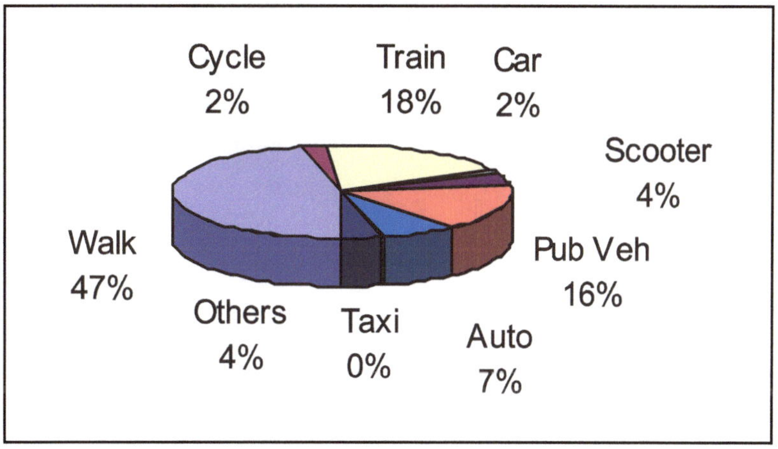

Fig.8.7: Modal Share of Trips with Walk

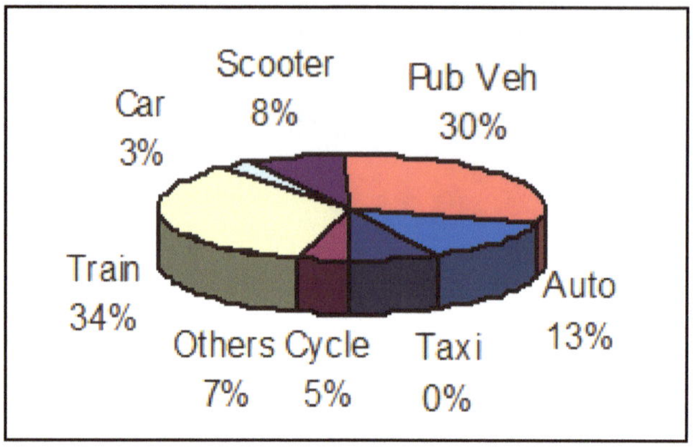

Fig.8.8: Modal Share of Trips without Walk

Rail Commuter Traffic

On an average day, a total of 5.58 lakh people use the railway stations at Thane, Kalwa and Mumbra, within the TMC area. The distribution was as under:

Table 8.6: Rail Commuter Traffic

Station	Entry	Exit	Total
Thane	1,71,097	2,01,283	3,72,380
Kalwa	37,664	49,712	87,376
Mumbra	45,822	52,414	98,236

"Walk" was rail passengers' predominant access mode choice (48.50%). Next in importance was "Bus" mode (34.30%). **Figure 8.9** shows the access mode choice of train passengers.

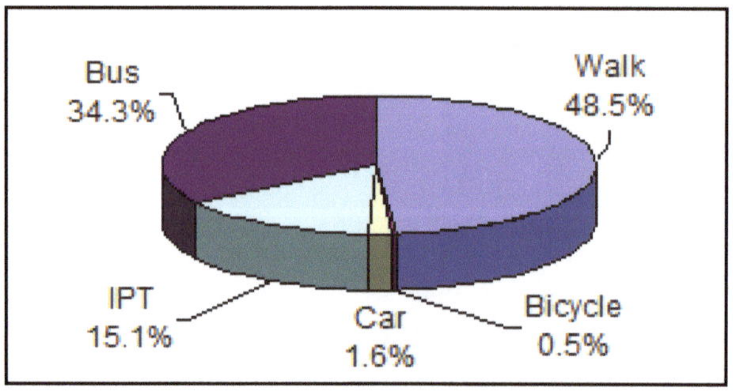

Fig.8.9: Access mode of Train Passenger

Bus Commuter Traffic

Passenger OD and Count surveys were conducted at the three bus terminals, viz. (I) NMMT Bus Stand, Near Railway Station, (ii) State Transport Depot Near Railway Station and (iii) ST Stand near Vandana Theatre. More than 1,80,000 intercity bus passenger trips were generated at the intercity bus terminals in Thane. Their choice of mode for travel to and from the city is shown in Figure 8.10, which shows the access mode choice of bus passengers.

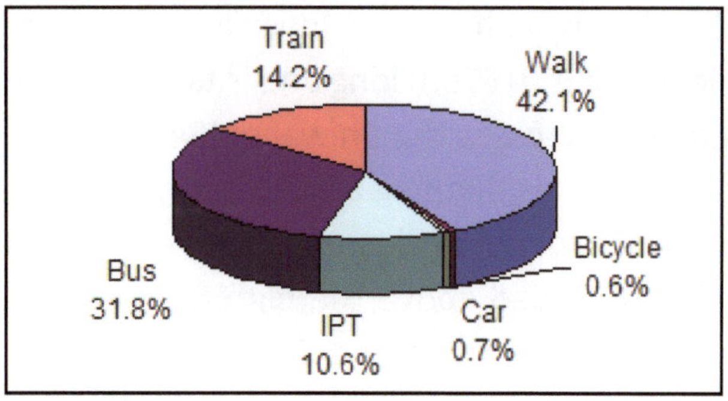

Fig.8.10: Access Mode Choice of Bus Passengers

8.4 Alignment and Basic Features

The Thane Development Plan (TDP-2001) has proposed the development of a mass rapid transit system (MRTS) along the corridor identified in the RITES report. A width of 30 m has been reserved for this use. The corridor, as per the Thane Development Plan (TDP), originally started from Thane Railway Station and ended at Chendani, forming an end-to-end line. As part of subsequent studies (the CIDCO study of 1994), it was decided that, since the two ends of the line at Chendani and Thane Railway Station were only 600 m apart, they should be connected to form a complete ring. As per the TOR and as agreed by MSRDC, this ring (Green Corridor) would be taken as the basis of the present study.

Fig.8.11: Proposed Balkum Station Location

The alignment begins at Balkum Station (km 0+000), which is proposed to be located 650m north of Balkum Road. This station is proposed to be an interchange station for the future Bhiwandi link and a terminal station for operational purposes, given its proximity to the car shed. The alignment will be laid to the west of the major water pipeline. A DP road of 45m was proposed in the Thane Development plan parallel to the water pipeline. Formation levels at Balkum station will be about 2m above the present ground level. The platform level will be reasonably above the high tide level.

Two tracks parallel to the Main line have been provided for trains to go to the car shed. Alignment crosses Balkum Road. The alignment is 0+670m long and is taken at ground level from the alignment after crossing the water mainline. It crosses the Bhiwandi Bypass at Ch. 1+238, the existing service road, and again crosses water main lines, running parallel to the DP road. The height of the elevated structure has been fixed to provide adequate clearance above the proposed D.P. Road, considering the formation of the proposed D.P. Road above the high tide level. Alignment after Ch. 1+800 is shifted over the 30 m wide reserved corridor, and Creek View Station is located at Ch. 1+992 in between the Saket area and Rabodi road to serve both areas. For the Creek View station, connectivity will be from both sides, i.e. from the existing road on the creek side and the proposed DP road on the township side. The alignment then crosses Raboli Road at Ch. 2+420 and continues along DP Road. After crossing Kalwa circle at the ch. 2+824, it follows the existing footpath on the creek side of Dadaji Kandeo Stadium. The existing road width is 15 m with 1.75 m.

Fig.8.12: Creek View Station

The footpath on both sides is lined with several utilities and multi-storied buildings on the landward side of the road. An elevated path is taken along the creek side of the footpath to avoid demolishing buildings, relocating utilities, and retaining the existing road alignment in this stretch. After this, the alignment runs parallel to the existing road at 20 m on the east and passes over the existing slum opposite to Dadaji Kondeo Stadium, where Chendani Station is proposed at Ch. 3+547 m. This station is connected to the stadium via a footbridge. To save the slum as far as possible between the Ch. 3+660 and 3+360, alignment is chosen at the center across the city.

8.5 Master Plan

In addition, as the study horizon is 2031, it was decided to conceptualise the development pattern of Thane up to that year and to cater to development needs and travel demands up to that horizon year. It was also decided to determine if any future extensions to the MRTS network would be needed within this time frame. Accordingly, a Master Plan for the MRTS Network System, up to the year 2031. This identified corridors for possible future extensions,

1. Intra-City System and corridors (within TMC)
2. Inter-City Linkages (between Thane and Bhiwandi, Thane and Borivali

The intra-city network system includes three corridors as follows:

Corridor – 1: The Ring (proposed in TDP with modifications as necessitated)

Corridor – 2: The North–South Axial (with extension to Mira Road and Airport Station)

Corridor – 3: The West-East Axial (with extension to Kopar Khairne)

The **ring corridor** is reserved in TDP. The alignment begins as an independent line at the east end of Thane station, where several goods sidings are located for loading and unloading salt and minerals. It takes a sharp right-handed curve with a turn of almost 135 degrees and then runs straight for over a kilometer and a car shed are proposed in this stretch.

These facilities lie in vacant salt pan lands, adjacent to which new colonies and a school are being built up. At km 2.3, it again turns into a right-hand curve of about 90°. Here, it passes over a truck terminal and crosses Eastern Express Highway by an elevated structure. Its upgradient continues till it crosses the Central Railway lines

Fig.8.13: Alignment crossing EEH at Kopri Bridge

8.6 Thane MRTS Master (Network) Plan

TDP envisages the development of the Thane Municipal Corporation (TMC) area up to 2001. The Government of Maharashtra has sanctioned it. The TMC area covers 128.23 sq. km. It is administratively demarcated into 11 sectors and 95 wards.

Thane is a high-growth city. Its population increased from 0.47 million in 1981 to 0.8 million in 1991, registering a 69% decadal growth rate. TDP projected and planned the city for a size of 1.45 million by 2001. In 2031, the quadratic model gives a size of 3.05 million.

The Thane MRTS network plan was prepared based on Growth (high-growth city), Population in the horizon year, Accessibility Model (with identified mode), Travel time matrix, Population distribution (based on accessibility & Land availability), Accessibility & Land availability, Developable area, Population sizes every five years, Urban structure, TDP, Land use structure, Sector-wise population and Employment by traffic zones. To promote the envisaged urban form and structure of Thane by 2031 and cater to the travel needs of the city's residents, a Master Plan for the MRTS Network has been developed in integration with the land-use plan. The network configuration has been finalised based on traffic estimates, site conditions, engineering feasibility, and detailed stakeholder interactions and discussions. It extends over 46 km in length and comprises three corridors (lines) crisscrossing the TMC area, interconnecting activity areas. The corridors are:

Corridor 1: Green (The Ring)-16.05 Km

Corridor 2: Red (North-South Axial)-11.2 km

Corridor 3: Blue (North-East Axial)- 18.5 Km

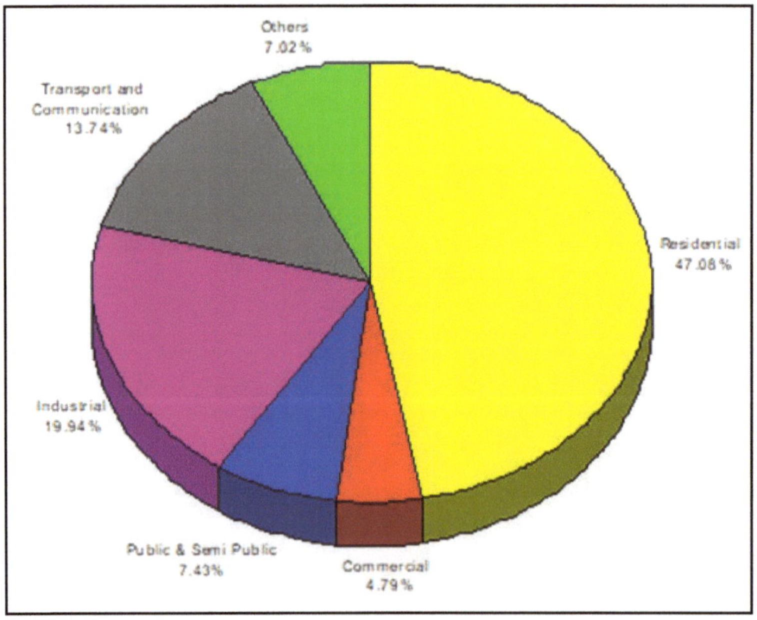

Fig.8.14: Land Use Structure

8.6.1 Regional Linkages

Thane is a crucial node in the MMR. There is intense interaction between Thane and other regional nodes, particularly Navi Mumbai, Kalyan-Dombivli Complex, Bhiwandi, Vasai-Virar, and Mira-Bhayandar. Of the above nodes, Thane and Navi Mumbai will be connected by rail with the development of the Kalwa-Turbhe line and its extension to Belapur. The suburban rail services of the Central Railway connect the Kalyan complex. The rail system connects Diva and Vasai. There is a need to connect Thane with Bhiwandi and Mira Bhayandar with a rail system.

8.6.2 Car Depot and Transfer Stations

The car depot is located at Balkum. Transfer points are Naupada, Golden Dyes, Pokhran, Kolshet, and Creek View (Town Center). Between MRTS and Mumbai Suburban Line, transfer points are Thane, Mumbra, Diva and Koper Khairne. Between MRTS and Regional Links are Naupada and Balkum.

8.7 MRTS Ridership

A total travel demand of 39 lakh trips per day is estimated for the TMC area by 2031. This includes trips generated by the resident population and trips carried out within the study area by people coming to Thane from outside areas. Many modes, including walking, will perform these trips.

According to the modal split model, the estimated travel demand is split between public transport modes at 82%.

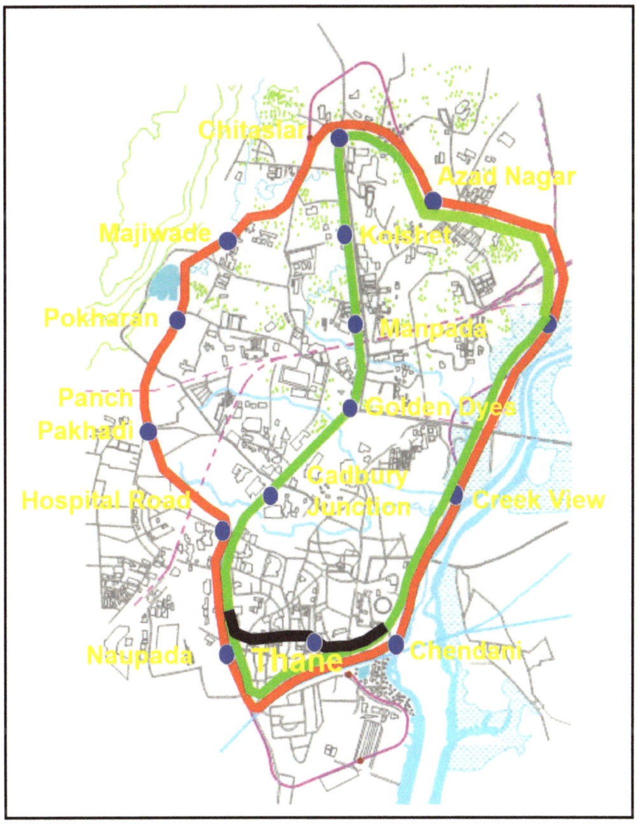

Fig.8.15: Thane Metro Master Plan

This includes road, MRTS, and suburban rail links. The generalised costs by different modes on each link have been estimated and included in the assignment.

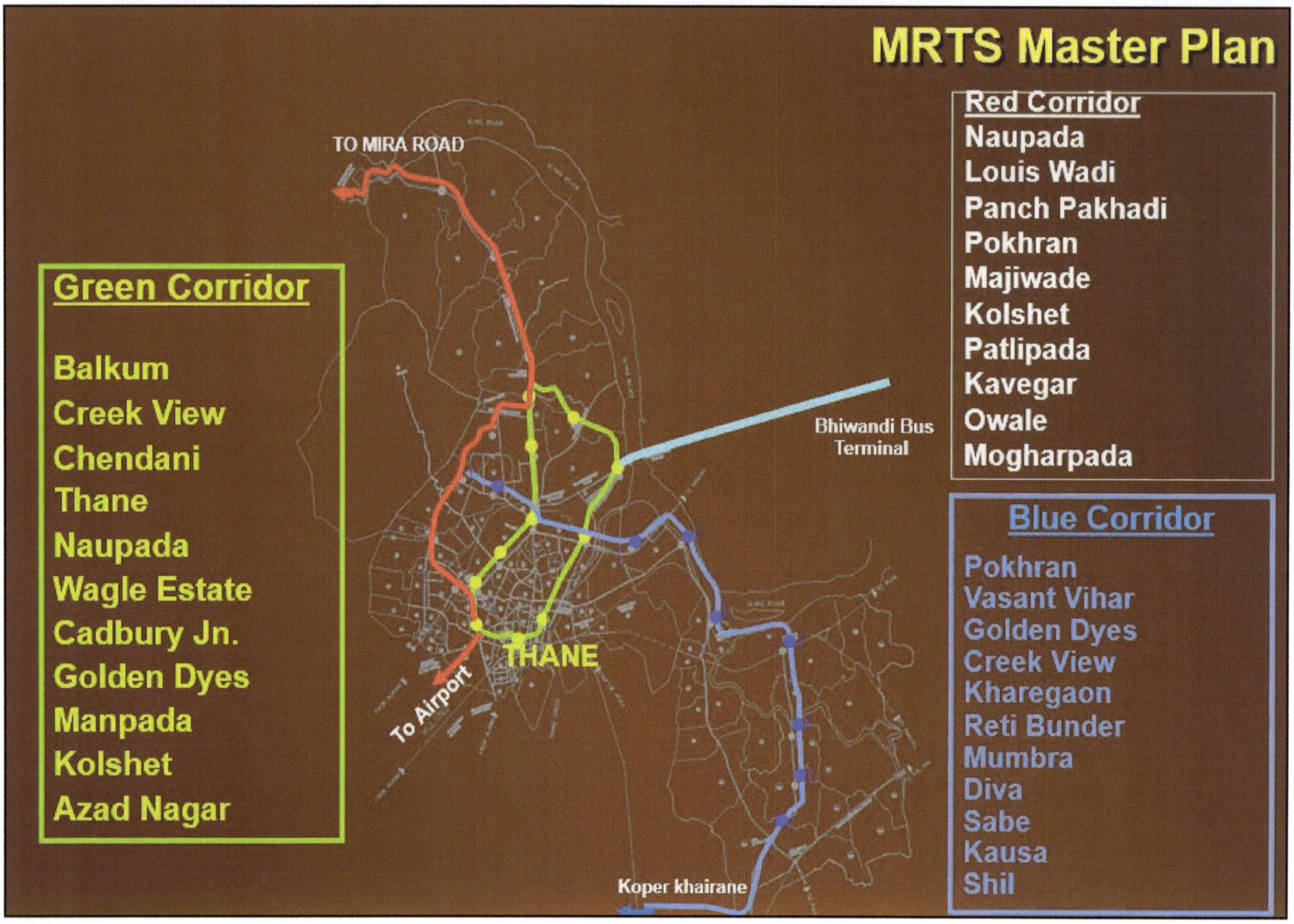

Fig.8.16: Thane Metro Network – Showing Green, Red & Blue Corridor

With a MRTS fare structure of Rs. 4.00 up to three km and Rs. 6.00 for longer trips, and frequency of operation of five minutes on all corridors, the MRTS ridership per day on different corridors is estimated as under:

Corridor	Ridership/day
Green	7,73,404
Red	3,04,126
Blue	2,10,422

MRTS Network Development Phasing

The proposed network system is to be developed in three phases:

Phase I	2001-2011
Phase II	2011-2021

8.8 Track Gauge for Thane MRTS

Technically, there is no difference between the track gauges in question, i.e., the worldwide 1435 mm standard gauge, the Indian 1676 mm broad gauge, and the 1000 mm meter gauge. All gauges allow for safe and comfortable operation, with no significant difference in track stability or riding quality.

From the economic point of view, the 1435 mm gauge is better than the broad gauge, as sleepers can be shorter. Distinct cost savings, however, will not be achieved in track costs. The same applies to meter gauge, but the coach's width is limited for stability.

The commercial conditions within the world market for MRT vehicles and equipment strongly support the adoption of the 1435 mm standard track gauge. In terms of line length worldwide, about 66% of MRT systems and 88% of LRT systems operate on a 1435 mm gauge.

Given these factors and market conditions, the 1435 mm gauge appears to be the most suitable solution for developing the Thane MRTS.

8.9 Design Parameters: Alignment

The track alignment has been designed with the following parameters.

1. Gauge Standard – 1.435 m
2. Distance between two standard gauge tracks – 4.05 (on straight) m.
3. Length of coach about 20 Meters.
4. Width of Coach – 2.65/2.76/3.0 Meters.
5. Distance between bogie centres – 12 Meters.
6. Axle Load – 10 – 14 Tonnes.
7. Sharpest Curve – 55 Meters Radius.
8. All curves are with super elevations and transition curves based on the following

   Equilibrium Can't E = (Ea + D) = 11.82 $\frac{V^2}{R}$ in mm.

   Equilibrium Speed *Ve* = 0.29 $\sqrt{R * E}$ in kmph

   Maximum Speed Vm = 0.29 $\sqrt{R * (E + D)}$ in kmph

   Where

   R = Radius in Meters

   Ve = Speed in Kmph.

   E = Cant in Millimetres.

   D = Cant Deficiency in Millimetres.

Ea = Cant provided in mm.

Vm = Maximum Speed.

9. Transition length (L) in meters is the maximum of the following

L = 0.0082 * Ea * Ve

= 0.0062 * D * Ve

= 0.37 * Ea.

10. Maximum cant deficiency – 110 mm
11. Maximum cant gradient – 1 in 300
12. Minimum transition length – 20 m
13. Gradient (ruling)- 4%
14. At the intersection of two gradient curves of a minimum 750-meter radius has been provided.
15. All stations are planned on straight except Thane Station, where a transition curve extends within the platform area.
16. Maximum Speed – 60 kmph on straights.

Speed on curves is reduced in proportion to the radius of the curve. In some cases, the available space for the transition curve is so restrictive due to ground conditions that the speed had to be further reduced.

All stations are on level and straight, except Thane station, where a small portion at the ends is on a transition curve.

8.9.1 Clearance on the sides

As detailed in the paragraph above, there is a clear space of 650 mm between the dynamic/swept profile and any obstruction. To reduce environmental sound pollution, a 100 mm-thick, 1 m-high sound barrier wall is proposed. Hence, the total width required on each side is 750 mm.

8.9.2 Width of Corridor

For straight alignment, the corridor width is the sum of the Dynamic Profile widths of the two coaches, the clearance between the two coaches, and the two walkways required on each side.

Dynamic/Swept Profile

During the movement, the train lurches due to the spring action of the bogies and occupies more space than the static profile. Figures 2.2 and 2.2 (a) show a typical static profile of an LRT coach.

8.9.3 Clearance between two coaches

The clearance between the kinematic envelopes of two adjacent developments' coaches should be at least 200 mm without centre traction poles and 600 mm with centre traction poles for high-speed Metro. As track maintenance for the elevated tracks of Thane MRTS is done at night, no walkway clearance is required. In cases where trains are disabled in the mid-section, passenger evacuation will be conducted between two stations. To facilitate such evacuation and enable passengers to get down between two coaches, space for a safety margin and a walkway of 650 mm between the two coach profiles has been provided. Moreover, as the Thane MRTS will not be air-conditioned and the windows will be open, this clearance will help prevent accidental protrusion.

8.9.4 Width required for Safety on curves

When a vehicle runs through a curve, there are two accelerations: a lateral acceleration V2/R, where V is the speed and R is the radius of curvature, and a vertical acceleration due to gravity, g.

Fig.8.17: Gap between to Rakes

The resulting acceleration is A. The track may be canted so that the resulting acceleration acts vertically relative to the vehicle, resulting in no lateral acceleration and thus no side force.

Equilibrium Cant is the cant required so that the resultant force from the vehicle's centre of gravity is perpendicular to the rails and halfway between them for a given speed known as the equilibrium speed (V_e).

Equilibrium Cant (E) is given by:

$E = 11.82\ V_e^2/R$ in millimetres, and the equilibrium speed by

$V_e = 0.29\ \sqrt{(R*E)}$ in km/h

where R is the radius in metres.

The actual (applied) Cant E_a is given by

$E_a + D = 11.82\ V_e^2/R$ in millimetres, and the maximum permissible speed V_m by

$V_m = 0.29\ \sqrt{(R*(E + D))}$ in km/h

The maximum applied cant is shown in the following table

Maximum Applied Cants	
Running lines	165 mm
Platforms	Zero
Switches and crossings	Zero
Maximum cant deficiency	110 mm

Maximum design Cant gradient	
Less than 50 km/h	1 in 300
More than 50 km/h	1 in $6V_m$

The minimum transition lengths are shown in the following table.

Minimum transition length	Maximum of $0.0082\ E_a V_m$ $0.0061\ DV_m$ or $0.37\ E_a$

Where the track runs on elevated structures, the potential consequences of a derailment are more severe. Additional precautions are taken to contain the vehicles by arranging to trap the wheels to prevent them from moving too far from the rails.

This can be done on ballasted track by laying a second pair of rails between the running rails. In the event of derailment, the inner wheel is contained between them. Where the track runs on a structure, an upstand concrete beam can perform the same purpose.

8.10 Track Spacing

Table 8.4 shows the required track spacing for a 2.65-meter-wide coach. Similarly, the track spacing for a 3.00-meter-wide coach and a 2.76-meter-wide coach was determined. To facilitate an increase in capacity, without increasing the train/platform length, the spacing required for a 3.0-meter-wide coach has been adopted as the spacing requirement when ridership improves.

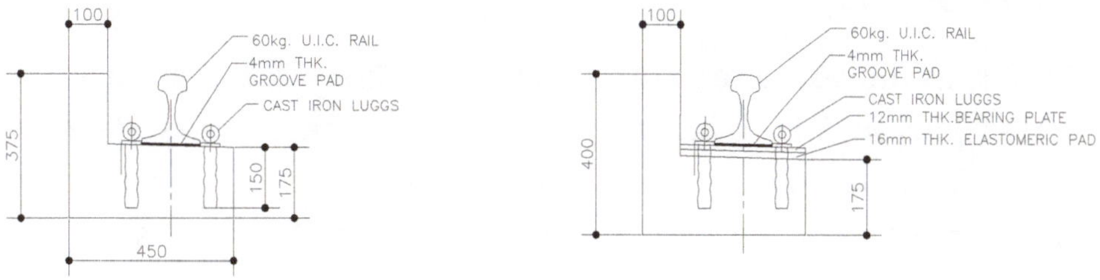

Fig.8.18: Fixing of Rail

8.10.1 Extra Clearance for Superelevation (Cant)

As both tracks of the double line will be provided with the same superelevation, both coaches will be tilted by the same amount, hence the track centers need not be increased on account of this.

Table 8.7: Track Spacing for 2.65 m Coach Width

RADIUS	DYNAMIC	CENTRE	END	SAFETY MARGIN AND	TOTAL OF	TRACK SPACING
(M)	COACH	THROW	THROW	PROVISION FOR	2+3+4+5	
	WIDTH (M)	(M)	(M)	WALKING (M)	(M)	(M)
1	2	3	4	5	6	7
Straight to 300	2.758	0.048	0.055	0.65	3.511	3.65
300 to 250	2.758	0.057	0.066	0.65	3.531	3.65
250 to 200	2.758	0.072	0.082	0.65	3.562	3.75
200 to 150	2.758	0.095	0.109	0.65	3.612	3.75
80 to 75	2.758	0.191		0.65		
75 to 70	2.758	0.205				
70 to 65	2.758	0.221				
65 to 60	2.758	0.239				
			0.214			
			0.228			
			0.245			
			0.265			
					3.813	3.85
					3.841	4.00
				0.65	3.874	4.00
				0.65	3.912	4.00

8.10.2 Corridor Width at Stations

At stations, in addition to the space required for tracks, there will be two platforms, each with a width of 3 to 6 metres, plus a width to accommodate stairs and escalators. Hence, the width of the elevated structure will be equal to the sum of the coach width, track spacing, platform width, and staircase/escalator width, i.e., 2.65 + 4.05 + (3 + 4) × 2 = 20.7 to 26.7 metres.

8.10.3 Width of Corridor in Ballasted Portion

Where the track will be resting on the ground instead of the elevated structure near Balkum Station, the corridor width will include the width required for prestressed concrete sleepers, the ballast slope, and the berm and slope required for the embankment.

8.11 Stations

All stations, except Balkum, have been planned in a three-floor configuration, with the platform approximately 13 meters above ground, a concourse level below, and entry to the station at ground level. This invariably results in large halls at the intermediate concourse level with operational and ticketing facilities. Platforms at the topmost level are covered for protection from the sun and rain. All stations include staircases and Escalators commensurate with Ridership in the coming year. Part of the concourse area is recommended to provide air conditioning. Sufficient provisions have been recommended for people with disabilities. All the elevated station buildings follow this principle.

8.11.1 Traffic Flow

Segregation of vehicular and pedestrian traffic at ground level, within the given physical constraints, has been implemented at each station to ensure a smooth flow and safety for commuters under both normal and emergency conditions. Adequate paved areas have been created. Easy access for 2-wheelers, auto-rickshaws, cars, and taxis has been provided to the station forecourt from the existing road network and vice versa. Wherever necessary, bus stops have also been provided below the station building. Some of the stations have also been linked with large population centers by subway or foot overbridge to overcome the physical constraints for pedestrians at the ground level.

8.11.2 Passenger Movement

The design considers the smooth flow of passengers inside the building. To achieve this, all facilities have been conveniently positioned to avoid conflicts between incoming and outgoing passengers, both under normal and emergency conditions. Accordingly, the stairs, escalators and elevators have been judiciously located on various floors.

8.11.3 Finishes

The selection of finishing materials has been made with consideration for safety, aesthetics, durability, ease of maintenance, and economics.

Signage plays a major role in guiding commuters in normal and emergency situations. Its design and locations are carefully selected to integrate with the interior.

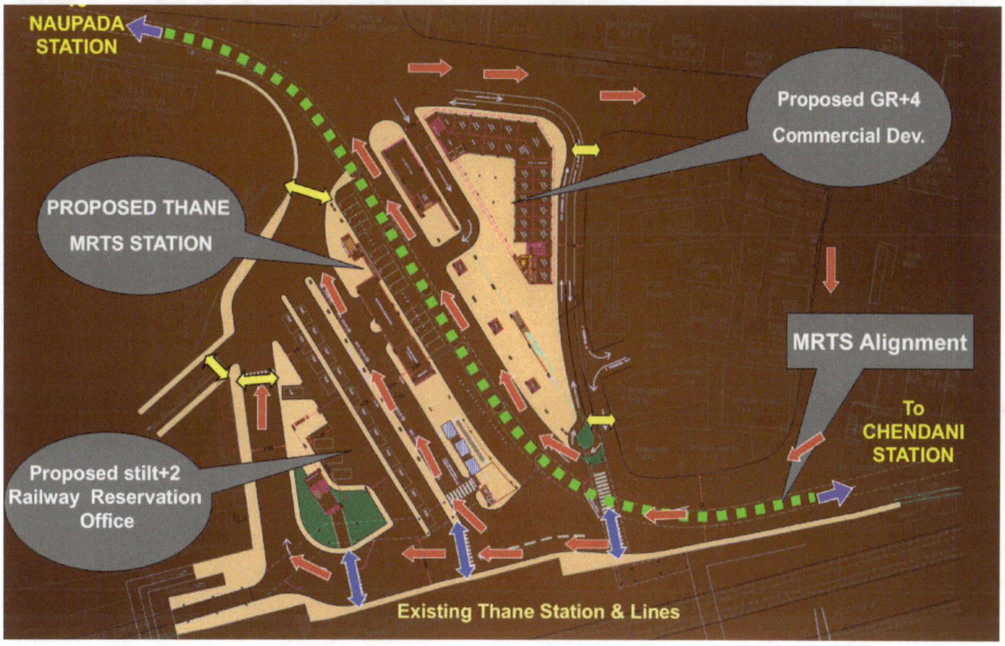

Fig.8.19: Integration of Proposed MRTS Alignment with Thane Station Area

8.12 Planning of Individual Station

DPR covers general planning for all stations. It also covers each station's special features, particularly site planning. This section provides a brief overview of each station.

17. **Thane Station (Intermediate Station with Elevated tracks)**

The site for Thane Station is the forecourt of the existing Main Line Thane Station. The idea is to achieve the quickest and easiest possible transfer of commuters from the mainline station to the proposed MRTS station. This is achieved by connecting the two stations at the ground and concourse levels. The Foot Over Bridge (F.O.B.) on either side of Thane station is proposed to be extended to connect to the concourse level (first floor) of the MRTS station. This direct connection will considerably ease congestion on the ground, both by reducing the number of passengers and autos.

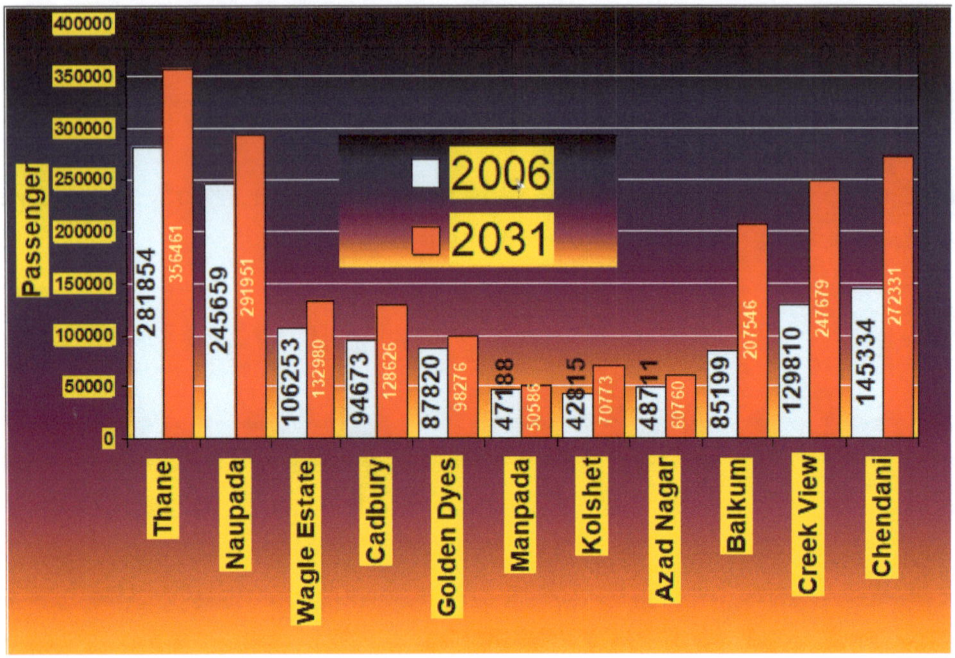

Fig.8.20: Travel Demand at Station Area

Fig.8.21: Alternative View of proposed Thane Metro Station

The state-of-the-art station complex will become a landmark in the vicinity. All the required amenities have been proposed on the site along with landscaped areas to provide relief from the congested surroundings. A certain amount of openness is introduced without being lavish. The circulation pattern has been established without disrupting the existing road layout and traffic flow.

Fig.8.22: Connecting Main with Metro Station

Fig.8.23: Concourse Area Plan

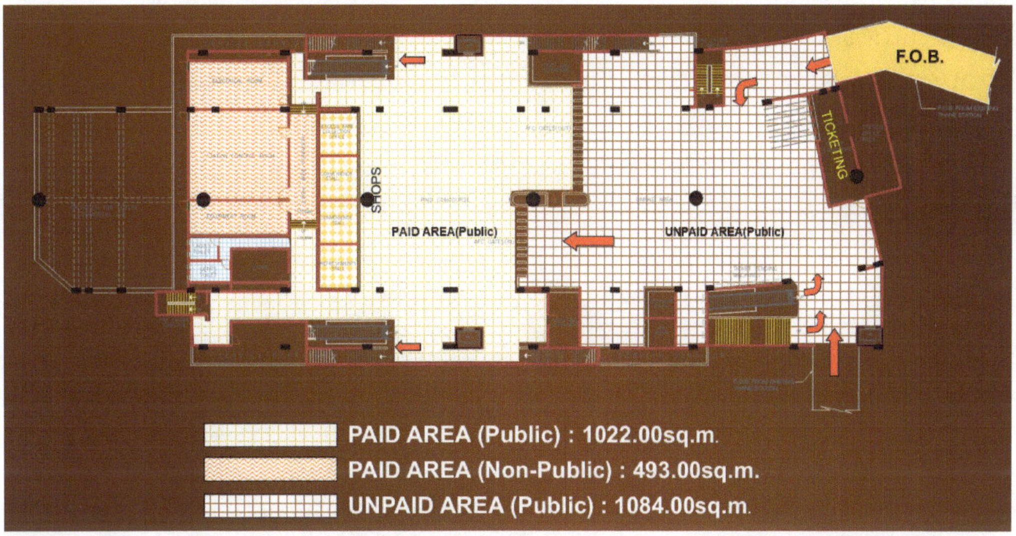

Fig.8.24: Concourse Area Plan for Proposed Thane Metro Station

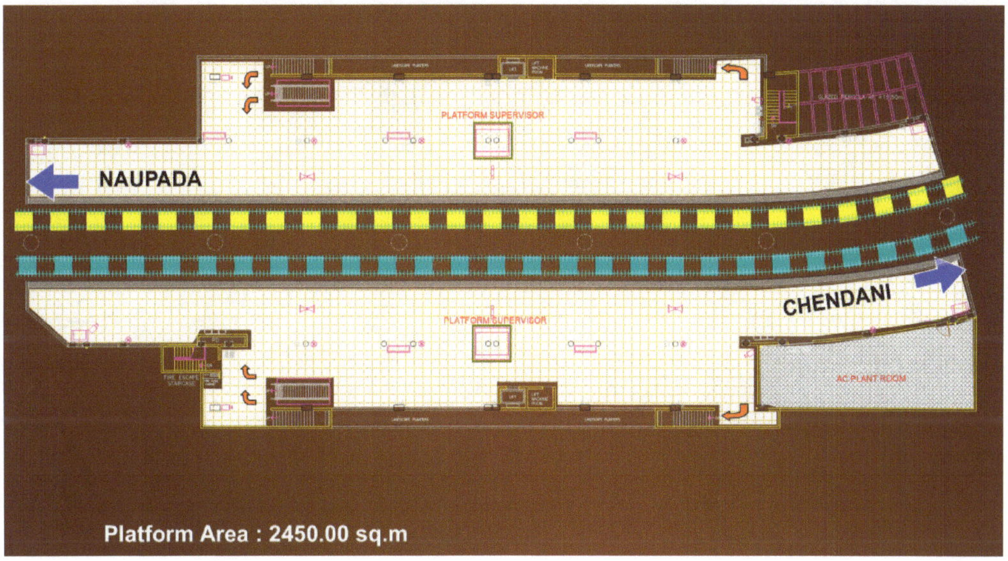

Fig.8.25: Platform Level Plan Proposed Thane Metro Station

8.12.2 Naupada Station (Interchange Station with Elevated tracks)

The proposed site for Naupada Station is on the existing Eastern Express Highway. It has residential establishments on the east and west sides. The highway runs north-south. The station area is planned on the service lane at the eastern edge and covers the green space between the service lane and the highway. The station is located towards the end of the flyover. Two pedestrians Subways, each 4.5 m wide, are proposed on either side of the station for pedestrian access across the highway.

8.12.3 Wagle Estate Station (Intermediate Station with Elevated tracks)

The proposed site for Wagle Estate Station is at the junction of the existing Eastern Express Highway and the road to Thane City. The highway has a flyover in the centre and a service road on either side. There are severe constraints on the availability of land. The ground space is required for the Station building facilities are staircases, escalators, lifts, parking, and an auto rickshaw stand, an electrical substation, a water tank, a pump room, etc. All these facilities have been provided in the middle of the road without disturbing the traffic. A subway of 8.5 m width is proposed for pedestrians' access from the northwest side, as the maximum people are expected from that side, i.e. from beyond the flyover.

Fig.8.26: Proposed 3D view of Four Stations –with Nau pada as Interchange Station

8.12.4 Cadbury Station (Intermediate Station with Elevated tracks)

The proposed site for Cadbury Station is South of the existing Eastern Express opposite the Cadbury Factory. Station facilities at ground level have been provided south of 75th Service Lane. The Cadbury factory is across the highway, and many people will access the station from the factory. A subway 8.5 m wide is proposed to connect the MRTS station side to the other side of the highway. The width is designed to accommodate 90% of the peak loading. The station area on the ground has been carefully designed. An effective circulation pattern has been devised using the station as a divider between the highway and the service lane.

8.12.5 Golden Dyes Station (Intermediate Station with Elevated tracks)

The proposed site for Golden Dyes Station is located at the junction where the road to Ghodbunder bifurcates from the Eastern Highway. At this junction, there is a flyover between the Eastern Express Highway and a large traffic island diverting traffic from the Expressway to the Ghodbunder Road. Parallel to it is another divider that connects Ghodbunder Road to the Expressway, and a service lane along the northwestern edge of the expressway is separated by a divider. The divider between the service lane and the Expressway on the northwestern side is to accommodate the columns supporting the tracks above. The columns for the station building are on either side. A large, paved area is provided at the road edges and the station entry on the north-west divider. The auto rickshaw/ taxi/bus stand has been located along this road and in the paved area near the station entry.

8.12.6 Manpada Station (Intermediate Station with Elevated tracks)

The proposed site for Manpada Station is located on the existing Ghodbunder Road, which runs north-south. To the west of the road is an open area, beyond which lies a residential area. There is an Industrial area on the east side. Columns for the elevated tracks are in the road divider. Due to the large span resulting from the road layout and the fact that commuters approach the station from both sides, entry to the station is provided from both sides.

8.12.7 Kolshet Station (Interchange Station with Elevated tracks)

Kolshet, a future interchange station, will be served by both green and red corridor lines. The proposed site is planned along the existing Ghodbunder Road, which runs north-south. The site is north of Ghodbunder and Kolshet road junction. The area served by this station is residential. The proposed station is planned to serve both the red and green corridors. Station entry is provided beyond the red corridor on the west side of the main road. The divider on the road houses the columns for the green corridor. Columns for the red corridor are located at the western edge of the road and at the station entry, with a staircase, escalator, and lift provided at either end of the corridor.

The open area is available on both sides of the road. The west side is developed for service buildings, auto-rickshaw stands, taxi/bus stands, among other facilities. Stairs, an escalator, and a lift are proposed at either end.

8.12.8 Azad Nagar Station (Intermediate Station with Elevated tracks)

The proposed site is to the southwest of Kolshet Road. It has a swampy area on the north-west side and factories on all the other sides. The swampy area has dense mangroves. Beyond the swamp is the residential colony of Clariant Factory.

The identified site is an open space, and Azad Nagar Station will be constructed on virgin land. To provide access to the station, a loop road has been proposed off Kolshet Road, which will include a bus stand and parking for cars, taxis, and autorickshaws.

The tracks are elevated. The platform is 13.05 m above the ground, and the concourse floor is proposed at 6.55 m. The ground level comprises the station entry, a staircase, an escalator, a lift to the upper floor, autorickshaw and taxi stands, car parking, two-wheeler parking, fire escape staircases, and more. Electrical substations, water tanks, pump houses, lift machine rooms, and service ducts are also located at ground level. All the facilities are located on the ground level.

8.12.9 Balkum Station (Terminal Station with tracks at ground level)

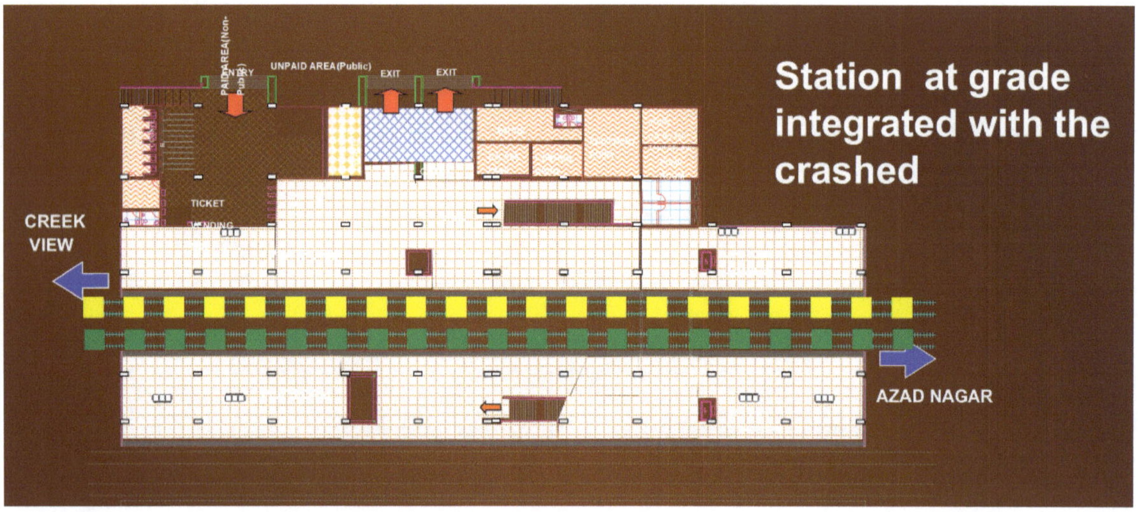

Fig.8.27: Ground Level Plan of Balkum Terminal Station

Balkum is the terminal station at ground level for the green line, including tracks. The car shed is near this station. Balkum Station will also act as an interchange station. The station is proposed on the east side. Two major water pipelines run parallel. On the western side, an overhead electric line runs in a southwest-northeast direction. A housing complex is planned for the north-western part of the site. There will be five tracks through the station—two for the green line, two for the Bhiwandi line, and one loop line. A total of five platforms is proposed. All the facilities, including ticketing, are provided along the approach road. From here, the platforms are approached by a pedestrian subway. These are linked with the station and the platform by lift and stairs. A feeder road of 8.0 m width is proposed from the northern end of the 20.0 m wide DP road to access the station building. A separate 7.0 m lane, with two bus stops, is proposed to segregate buses from other modes of transport.

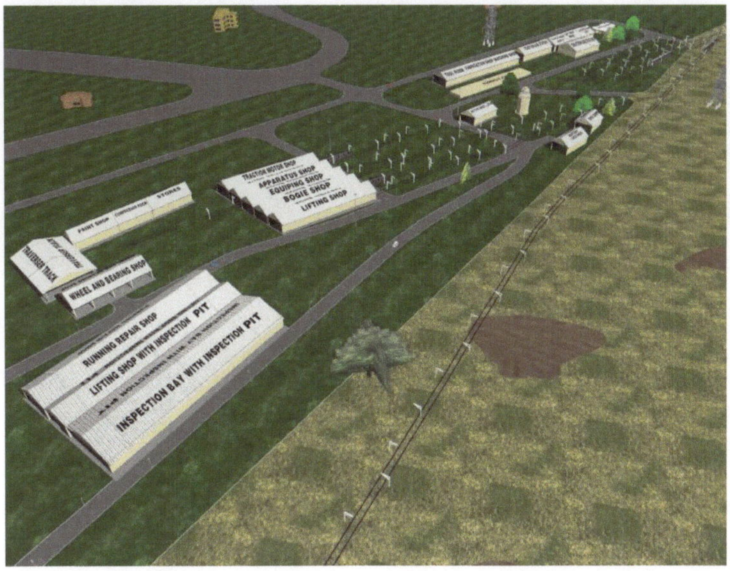

Fig.8.28: Balkum – Car Shed

8.12.10 Creek View Station (Intermediate Station with Elevated tracks)

Creek View Station is off the proposed DP road running parallel to the Thane-Balkum Road on the northwestern side. A connection is proposed between these roads, on the northeast side of the proposed site, from the station. DP road runs northeast southwest. The elevated track is southwest of the DP road. For vehicular access to the Station, a loop road is taken from the DP road to the connecting road. Parking for cars, taxis, and 2-wheelers is proposed under the station building.

8.12.11 Chendani Station (Intermediate Station with Elevated tracks)

Chendani Station is located on the east side of the road running north-south, approximately 300 meters to the south of the CIDCO Bus Stand. The station is north of a 'T' junction and opposite Dadaji Kondev Stadium. Slums currently occupy the densely populated area required for the station and must be acquired for its construction. For commuters coming from the west side of the road, a foot-over bridge is planned, along with a link to the stadium for spectators.

To provide vehicular access, a loop road with parking facilities is planned. A traction substation is provided on the east side of the station building. Autorickshaws, taxis, and bus stands are provided along the main road.

8.12.12 Golden Dyes Station (Intermediate Station with Elevated tracks)

The proposed site for Golden Dyes Station is located at the junction where the road to Ghodbunder bifurcates from the Eastern Highway. At this junction, there is a flyover between the Eastern Express Highway and a large traffic island diverting traffic from the Expressway to the Ghodbunder Road. Parallel to this is another divider that connects Ghodbunder Road to the Expressway. The divider between the service lane and the Expressway on the northwestern side used to accommodate the columns supporting the tracks above. The columns for the station building are on either side. The columns are provided at the end of the islands/ dividers below, after proposing a slight modification to their existing geometry.

8.12.13 Manpada Station (Intermediate Station with Elevated tracks)

The proposed site for Manpada Station is located on the existing Ghodbunder Road, which runs north-south. To the west of the road is an open area, beyond which lies a residential area. There is an Industrial area on the east side.

Columns for the elevated tracks are in the road divider. Due to the large span resulting from the road layout and the fact that commuters approach the station from both sides, entry is provided from both sides, duplicating the staircase, escalator, and lift for the concourse above. This eliminates the provision of either the subway or the foot-over bridge. Structures housing the services are located on the west side.

The rest of the planning concept is the same as that of the Golden Dyes station.

8.13 Engineering Design Parameters

The design is generally based on the Limit State Theory as outlined in the IRS Concrete Bridge Code (Second Revision, 1997). The design and loading requirements of the structures shall not be less than those specified in the Indian Standards and Code of Practices, together with all applicable amendments. The design is based on the following standards: IRS: Bridge Rules, IRS: Concrete Bridge, IRC: Sections 18 &21, IS: 456, 1893, 6403, 2911, and BS: 5400, as well as the ACI Code.

Grades of **concrete** have been considered in the design of the station, for Piles -M30, Pile caps, Piers & Pier heads, Concourse, Staircases & Floors - M35, Precast Prestressed I & Box Girders – 40 and others – M25

Clear Covers range from 25 mm to 75 mm.

Untensioned reinforced

High-yield strength /TMT bars of grade Fe 415 conforming to IS:1786-1985

Prestressing Steel

Cable consisting of 12 nos 12.7mm dia 7 ply class 2 strands/ 19 nos. 12.7 mm dia 7 ply class 2 strands/ 19 nos 15.2mm dia 7 ply class 2 strands conforming to IS:6006.

Clear Cover varies between 25 mm & 75 mm.

Loads

Dead loads: Dead load of structural elements as per actual

Superimposed Dead Load: Rail & track load -16 kN/m per track

Live loads

Vertical Live load of train and its equivalent dynamic effects as per the IRS bridge Rule (70% of the modified BG loading-1987)

Live load on concourse floor, platform & stairs- 5 kN/m2

Roofs (as per IS:875)-0.75 kN/m2

Design also considers: Centrifugal force**,** Temperature Loads (IRS) (for spans equal to or greater than 30 m -5-10 deg.), BS-5400, Part II

Long welded Rail force 16kN/m per track

Racking Force Not applicable for the concrete deck, but applicable to the steel deck

Derailment Load As per Bridge Rules

Overhead Line Equipment Loading- IRS Bridge Rules

Loads due to Wind, Earthquake, Braking & traction forces, Frictional Force on bearing,

Braking & traction forces, forces due to Fatigue, Erection Forces& Effects and Others.

Live Loads, as per IS- 875, except where the loadings given below are more severe.

Platforms and Ticket Hall -	5.0 kN/m^2
Staff Rooms, Toilets, Offices -	3.0 kN/m^2
Storerooms -	5.0 kN/m^2
Plant Room -	According to the self-weight of Machines (or 5.0 Kn/ sq.m

Design Life & Serviceability Requirement

The design life of all civil engineering structures other than building structures shall be 120 years unless otherwise specified or agreed.

Building Structures: minimum of 100 years, unless otherwise specified or agreed.

Bridge Bearings and Movement Joints Bridge bearings and movement--minimum design life of 100 years. Minor components that can be replaced, shall have a service life of 20 years.

Serviceability of Civil Engineering and Building Works

Shrinkage and Creep, Differential Settlement, Noise Abatement, and

Provision of Noise Barriers 7.15

8.14 Elevated Structures - Structural System

The form, dimensions, and design require special attention to structural suitability, economic concerns, and aesthetics. It is observed that PSC I – Girders with a deck slab are the economical solution under normal circumstances.

Consideration shall be given to include in the design the following requirements:

- Long welded rails with track centres at 4.25 m and suitable for the selected rail gauge kinematic profile of 2.65/3.0 metre railway coaches.
- The section with ballasted track, the top of the deck must be provided with a wearing coat of mastic asphalt of a minimum thickness of 25 mm.

Emergency evacuation provisions shall be provided along the railway for the full length of the structure.

The design shall include a width of 2.5 metres from the platform edge, insulated from the ground earth by the insertion of PVC and/or other insulating compounds. Parapets shall be provided on both sides of viaducts for the full length and shall be designed to support the railway cabling. Parapets shall be designed to resist a horizontal and vertical force of 150 kg/m applied simultaneously to the top of the railing or parapet.

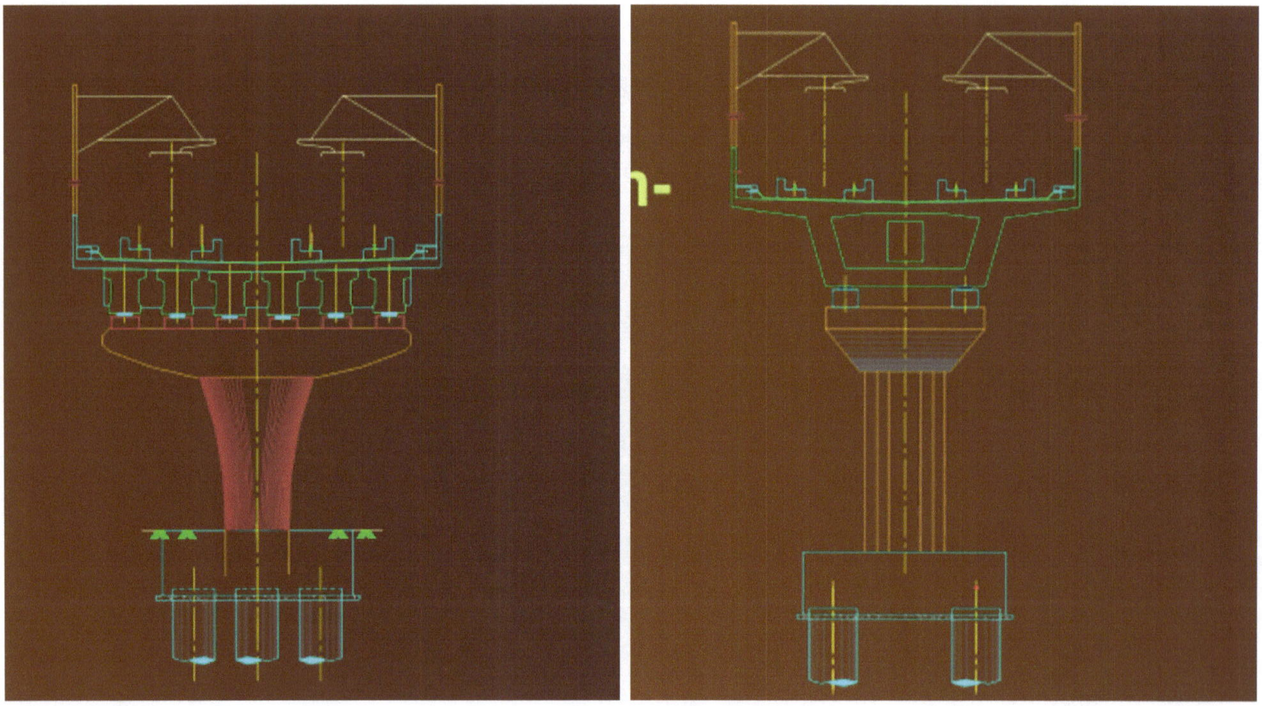

Fig.8.29: Alternative Sections of Viaduct

Provision for super-elevation shall be made as part of the track structure over the deck. Increased dead load is to be considered at such locations.

The superstructure deck, including the soffit of any overhead structure above the deck, shall be cambered to compensate for the combined effect of:

- Vertical curvature, if any.
- Dead load deflection, and

- Permissible live load plus impact deflection as accepted by the Client.

Desirable Minimum thickness of any RC member:

- Deck (above girders) 150 mm
- Web of girder 250 mm
- Web of prestressed girders 150 + d
- If there are 2 cables at any level 150+ 3d

(where d is the diameter of the cable duct)

Box Girders: minimum member thickness

- Deck slab 200 mm
- Bottom flange 300 mm
- Web 250 mm

Or as required by the IRS Concrete Bridge Code, whichever is the greater thickness. In an aggressive environment, an additional 10-20 mm of cover shall be provided for extra protection.

8.15 Design Considerations

Vibration and Deflection Limitations

The amplitude and frequency of vibrations of the viaduct and station structure shall be limited to international standards. The overall deflection of elevated structures accounts for vibration effects, along with other factors. Suitable arrangements shall be provided at the ends of beams, and jacking pads on pier caps shall be installed to allow for the replacement of bearings and any necessary repairs during service.

Reinforced and prestressed concrete elevated structures shall be designed in accordance with the provisions of the IRC Codes.

For Substructure and Foundations design, IRS bridge sub-structure and Foundation Code, and/or IRC 78-Section VI – Code of Practice, with special consideration for High-Speed Tram Loading, have been followed.

8.16 Consideration of Construction Approach

Stresses in partially complete structures shall be analysed under appropriate critical conditions at various stages of construction. Any restrictions on construction operations resulting from the design assumptions shall be specified in the contract drawings and specifications.

8.16.1 Movement/Expansion Joints

Movement/expansion joints and other necessary measures to control shrinkage and thermal effects shall be incorporated into the structural design, ensuring that the performance of architectural finishes or any services is not adversely affected.

Movement/expansion joints shall be designed for easy maintenance and replacement.

8.16.2 Design Surface Crack Width

For the serviceability limit state of cracking:

- The design surface crack width of structures shall not exceed the values given in Table 1 of BS 5400 Part 4: 1990 or 0.2 mm, whichever is less.
- Pre-stressed concrete viaduct structures shall be designed as per the provisions of the IRS and IRC Codes
- Design surface crack width of reinforced concrete station structures exposed to weather shall not exceed the values given in Table 1 of BS 5400 Part 4: 1990 or 0.2 mm, whichever is less.
- Pre-stressed concrete station structures, exposed to the weather, shall be designed as class 1 of clauses 4.3.4.3 BS 8110 Part 1. Structural elements fully protected from the weather may be designed as class 2.
- All aqueous liquid retaining structures and basements shall be designed to the requirements of IS 3370 unless otherwise varied by this specification.

8.16.3 Vibration and Deflection Limitations

The amplitude and frequency of vibrations of the viaduct and station structure shall be limited to international standards.

The overall deflection, as specified elsewhere in the contract for elevated structures, will be limited, considering vibration and other relevant factors. Suitable arrangements shall be provided at the ends of beams, and jacking pads on pier caps shall be installed to allow for the replacement of bearings and any necessary repairs during service. Provision should be made for adequate fixtures of the superstructure to the substructure, if any loading or loading combination, increased by 100% of the live load plus impact, is likely to cause uplift of any support.

8.17 Design Procedures

Reinforced and prestressed concrete members of elevated structures shall be designed in accordance with the provisions of the IRC Codes.

For Substructure and Foundations, reference was made to the IRS bridge substructure and foundation code, as well as IRC Section VI, with special consideration for High-Speed Tram Loading.

8.18 Method of Construction

Stresses in partially complete structures shall be analysed under appropriate critical conditions at various stages of construction.

Any restrictions on construction operations resulting from the design assumptions shall be specified in the contract drawings and specifications. Conversely, the designer of specified construction procedures or sequences may take advantage of affecting a more favourable distribution of loads or stresses.

8.18.1 Movement/Expansion Joints

Movement/expansion joints and other necessary measures to control shrinkage and thermal effects shall be incorporated into the structural design, ensuring that the performance of architectural finishes or any services is not adversely affected during normal working conditions.

8.18.2 Design Surface Crack Width

For the serviceability limit state of cracking:

- The design surface crack width of structures shall not exceed the values given in Table 1 of BS 5400 Part 4: 1990 or 0.2 mm, whichever is the least.
- Pre-stressed concrete viaduct structures shall be designed as per the provisions of the IRS and IRC Codes
- The design surface crack width of reinforced concrete station structures exposed to weather shall not exceed the values given in Table 1 of BS 5400 Part 4: 1990 or 0.2 mm, whichever is the least.
- Pre-stressed concrete station structures, exposed to the weather, shall be designed as class 1 of clauses 4.3.4.3 BS 8110 Part 1.
- All aqueous liquid retaining structures and basements shall be designed to the requirements of IS 3370 unless otherwise varied by this specification.

8. 18.3 Structural Members with Bearings

Consideration should be given to the easy maintenance and replacement of viaduct and station bearings.

8.19 Supporting Structures for Permanent Way

The Thane Mass Rapid Transit System is proposed to be built on elevated concrete structures, i.e., railway tracks and station buildings, except for approximately 1.4 km near Balkum station in the Green Corridor, where the track is at ground level. The remaining portion will be on the elevated concrete structures. The elevated track is a ballast-less track with UIC 60 rails, supported on the concrete bed with suitable fixtures.

8.19.1 Standard Span

It is proposed to use precast elements for the way structures. The girders are cast in the yard and placed in position by launching. The girder weight for a larger span is very high, and the launch assembly requires more space to accommodate it. It is proposed to provide a small-span superstructure with spans up to 25 m. Because girder weight is lower, girders can be erected without obstructing traffic during construction.

Preliminary design calculations for spans of 15 m, 18 m, 20 m and 25 m were made. Quantity calculations and cost estimates were prepared for these spans. Based on comparative values, it was concluded that an 18 m span is the most economical option. Thus, an 18 m span is proposed as the standard for the entire alignment, including the elevated portion. In addition, most of the alignment can be adopted.

However, obligatory spans, i.e. spans other than 18.0 m, are proposed depending upon different site conditions.

For an elevated portion of MRTS, eight span lengths have been provided: 14 m, 16 m, 17.5 m, 18 m, 20 m, 24 m, 30 m, and 50 m. Precast post-tensioned girders topped by cast-in-situ deck slabs have been adopted for spans up to 24 m, while post-tensioned box girders have been adopted for 30 m and 50 m spans. For a span of 14m cast-in-situ solid slab has been adopted.

Substructures consist of the RCC pier cap and pier. The portal-framed structure is situated along Gokhale Road, near Thane station. The height of the piers in this section of the corridor has been raised to 7.5 m to improve road illumination and provide adequate clearance between the existing roads and the crossbeam of the Portal Structure supporting the track structure. The arrangement has the advantage that the center line of the corridor does not need to coincide with the center line of the supporting system.

8.20 Foundations

A detailed soil investigation was conducted to ascertain the soil strata available for founding the foundation. As the soil strata at ground level are not suitable for supporting such heavy loads, the foundation must be taken to rock level.

8.20.1 Recommendations for Foundation System

The adopted foundation types are either open foundations, where rock is available at shallower depths, or pile foundations, where rock is available at deeper depths. Piles are cast-in-situ bored piles, with or without casing (depending on the soil strata), and range in length from 8 to 12 m. Two types of piles, viz. 1.0 m and 1.2 m dia have been provided.

8.20.2 Shallow Foundations

Rock is encountered at quite shallow depth in a few boreholes, especially those explored between Balkum Station and Golden Dyes, including Azad Nagar, Kolshet and Manpada. At these locations, the possibility of providing shallow foundations could be further explored after evaluating the space constraints and the associated costs.

The water table in all locations is observed to be near the ground level at the time of investigation. Therefore, the buoyancy effects from water and the uplift on the raft foundations must be considered when designing the shallow foundations. The minimum widths for individual column and continuous wall footings should be 1m and 0.5m, respectively, to provide a margin of safety against a local or punching shear failure of the foundation soils.

8.20.3 Deep Foundations

All the boreholes explored in the area encountered weathered rock and hard rock underlying the subsoil profile zones discussed in the above section. Structural requirements necessitated transferring heavy pier loads to the ground; therefore, most structures would require pile foundations to do so. The rock profile at each borehole location is studied to identify a suitable foundation stratum for transferring the pile loads.

The foundation recommendations are made based on subsurface conditions revealed by the Geotechnical Investigation. These recommendations are made with the assumption that the rock with similar properties continues and that a sufficient cushion of sound rock is available below the founding depth, including the locations where founding levels fall near the borehole termination depths.

8.20.4 Foundation Recommendations – Discussion

To transfer heavy structural loads, large-diameter Bored Cast-In-Situ piles are recommended for all major structures, allowing loads to be transferred through the overburden soils to the bedrock. The upper rock formation encountered at this site is jointed and even weathered in many places. Piles are therefore recommended to penetrate these zones and extend about 1.5 times their diameter.

The rock is relatively sound to support the superimposed load. The required depth of bedrock is needed to eliminate the risk of excessive settlement due to joints.

Piles with diameters of 1000 mm, 1200 mm, or 1500 mm are proposed for most major structures. For the Station Building structures, 600 mm-diameter piles are proposed.

8.20.5 Load Transfer Mechanism of a Bored Cast –In- Place Pile

The safe bearing capacity of a single pile can be estimated as the sum of the capacities in end bearing and skin friction along the periphery of the pile. The net safe bearing capacity of a single pile is the total safe bearing capacity less than the submerged weight of the pile, considering that buoyant conditions exist. Based on this, typically, the safe load bearing capacity of a 1000 mm diameter pile is estimated to be about 350 T, that for a 1200 mm diameter pile is about 475 T and the same for a 1500 mm diameter pile is about 750 T. Since the friction in the rock socket is likely to have significant contribution in the overall pile capacity, different capacities can be achieved for the same diameter of the pile by changing length of the socket into sound rock. Locations with poor core recovery should not be included when calculating the side frictional resistance.

Water tables at most explored locations are shallow and are likely to rise further during wet months; buoyancy effects due to water are considered in the design of pile foundations. Negative drag-down forces are accounted for while calculating the total safe bearing capacity at locations where soft clayey layers underlie freshly placed fill.

Design of pile foundations is carried out in general accordance with IS:2911 and IS:456. The settlement criterion is also considered while deciding the founding levels. Pile head deflections are estimated based on the anticipated loads, using field and laboratory data from the investigation.

IS: 2911 Part 1, Section 4 suggests that pile load tests should be conducted to confirm the safe bearing capacity of pile foundations. This is of utmost importance to verify the design assumptions and parameters, as well as the quality of construction.

8.21 System Selection

To carry the estimated ridership at the desired level of service, various alternative system technologies have been studied, system design parameters established, and a system selected for detailed engineering design and cost estimates.

8.22 Rolling Stock

The Rapid Transit Train was planned to utilise viaduct structures with a 750 V DC overhead traction system, which was recommended to be paired with conventional steel wheels on steel rails.

The Thane Metro train, or Rake, consists of four coaches. The train design was based on:

1. Reliability
2. Traffic needs
3. Noise reduction
4. Fire Hazard
5. Lightweight construction
6. Quick reversibility
7. Layout of equipment

The basic dimensions of the train, length, width and height, were based on a) the maximum number of passengers to be carried, for a four-coach rake was 1250 passengers of average height, b) pleasant travel without congestion or suffocation c) with longitudinal seating on two sides for the convenience of standing and accommodating more passengers d) ratio between standing vs seating accommodation 5-7:1 e) Average weight of passengers as 68 kg with 9pax/sqm in crash condition and 6 pax for normal condition f) Door opening on each side between 25-30% of coach length g) only one class and vestibuling each coach. Possible dimensions considered – length 18.2-20m, width 2.7-3.1, and height 3.6-3.8. The detailed study indicated that, at the given scheduled speed, a width of 2.76m would be most economical. The height of the empty new car is 3.55m from the rail top. The recommended dimensions of a coach were 20 m x 2.76 m x 3.6 m, and the floor height from the rail top was 1.05m, with a platform height of 1.0 m. The coach's full capacity was calculated to be 315 passengers for crash loading and 250 passengers in normal conditions.

8.23 Car (Coach) Body

Generally, car bodies are made lighter, resulting in a lightweight structure to support the system with a comparatively lighter foundation. The full approach reduces the overall cost. The manufacture of a coach has been achieved with three different materials: i) Stainless Steel, ii) Corten Steel, and iii) Aluminium. Aluminium alloys, such as those with zinc, are being used across the wall due to their lighter weight. Stainless steel is costly, so for the Thane Metro, Corten steel with approved colour painting was recommended as an alternative, given its corrosion resistance and hardness. Corten steel coaches were used in the first metro in India, the North-South Metro in Kolkata.

Doors, as per global practice, should automatically close was be controlled from the Driver's cabin for reasons of safety. For automatic running later, certain equipment mounting provisions had been recommended in the driver's cabin. Two driver's cabins were recommended —one at each end —to facilitate reverse operation. The bogie's recommendation was a 2-axle type and did not need to be sophisticated, as the maximum speed limit was 80 Km/h. However, it was recommended that the bogies' design should aim to reduce their weight.

At a crash load, the total load reached 52 tons. With a safety factor of 10-15%, the axle load calculated out to be 14 tons. It was also found that using solid wheels would be economical and improve dissipation. The use of a solid wheel was also recommended.

Concerning the monopoly of the steel-spring suspension, it is slowly giving way to interleaved rubber and pneumatic suspension for Rapid Transit Cars. The modern trend is towards increasing use of rubber and pneumatic suspensions. The advantages include reduced weight, lower maintenance costs, noise reduction, and improved riding quality.

8.24 Power Supply

The basic scheme of power supply involved more than one receiving station, where power was planned to be purchased from the supply authorities, stepped down to the transit system's HT network voltage, and taken along the MRTS alignment to traction and auxiliary substations. In traction substations, the conversion was proposed to be done to 750 V DC (nominal) for feeding the overhead traction. Five traction substations at Balkum, Chendani, Naupada, Golden Dyes, and Kolshet were required to feed the overhead system.

The power supply in the auxiliary substations was converted to a 415V 3-phase/240V single-phase AC supply for feeding various auxiliary loads. The requirement of power for the third category of load, viz. Carsheds, maintenance depots, offices, control rooms, staff quarters, etc., are designed to be arranged for convenience, either by supplying them through the transit system's auxiliary substations or by obtaining separate, isolated points of supply at different locations as needed.

A standby supply system from batteries was proposed for certain important loads. The system was equipped with its high-voltage network, ensuring reliable and flexible supply. The main factors were:

- MSEB has an adequate 11 KV cable network
- The HV 11 KV network of the rapid transit system would run along the viaduct; it would be well protected and not liable to damage, as in the normal road routes.
- The transit system's high voltage network was necessary to ensure continuity of supply in the event of failure of one source, as changing over to the other alternative routes was possible.

MSEB's supply is reliable and backed by suitable interconnection with thermal and hydro-generating stations of other authorities. They maintain a modern high-tension distribution network inside the city. Preliminary discussions with MSEB authorities indicated that they would consider favourably any proposal to supply MRTS/Thane.

The traction load takes a few years to develop to the anticipated extent; therefore, provisions were made to phase out the traction demand and the minimum guaranteed charges over a period of three years from the commencement of the supply. Whenever the supply at the receiving station fails because of the supply company's failure or a plant shut down for maintenance, the power requirement can be met by other receiving stations.

8.24.1 Scheme of Power Supply

The transit system's primary distribution network can be any one of the standard HT voltages according to the ISS, specifically 11 KV or 33 KV. The distance involved in the distribution system will not justify any voltage higher than 33 KV. Between 11 KV and 33 KV, the choice has fallen to 11 KV on the following technical considerations:

- The quantum of power to be transmitted can be effectively handled at 11 KV within the permissible voltage drop.
- As the cables will be taken along with the viaduct, a voltage higher than 11 KV is not considered appropriate.
- Considering the short length of distribution involved, there will be a marginal difference between the two systems in the total cost and given the inadvisability of adopting a higher voltage than 11 KV, this marginal benefit, if any, must be ignored.

For substations to be built overground on costly urban lands, the construction of multistory structures is proposed. The ground floor and, if necessary, the first floor can be utilised for the substation. For these purposes, the higher floors can be utilised for offices and residences. Where accommodation at the complete substation on one floor is not feasible, heavy equipment, such as transformers, was planned to be installed on the ground floor, while switchgear and control equipment would be provided either in the basement or on the first floor. For smooth operation, some of the isolators will be motorised for remote operation. The offtake from the supply authority's mains would be from three "strong" points, with two stages of transformation: from 33 KV to 11 KV, and from 11 KV to the required phased voltage for the 750 V DC conversion.

The DC output generally contains ripples to a certain extent, and such ripples would require smoothing for the proper functioning of the traction motors. Further refining/smoothing may be necessary to mitigate the impact on the transit system's communication lines, which are situated near the OHE. The specifications for the rectifier equipment must be drawn up after a due study of these two aspects.

Individual fuses are planned to protect silicon cells. The rectifier equipment should be designed to withstand the fault until it is cleared on the AC side. The rectifier must be protected against cooling air fan failure by the provision of a suitable fan-failure and temperature-rise relay. The rectifier tripping can be accomplished on the AC side or via a simple load-breaking switch on the DC side, depending on what is deemed appropriate after a thorough study.

8.25 Traction Power Supply

The traction power supply will be 750 V DC via an overhead system. As mentioned above, a 750-volt supply will be fed to the overhead traction system at five traction substations. The OHE between two traction substations will be fed from either side to ensure an uninterrupted traction power supply. The return circuit will be through the running rails.

8.25.1 Auxiliary Power Supply

The necessary auxiliary power supply will be obtained at 415 volts, 3-phase, by converting an 11 kV supply to 415 volts at auxiliary substations at all stations and traction substations. A Separate Auxiliary Substation will be established at the depot complex to meet the power requirement in the car shed.

8.25.2 Traction Sub-Station

Each of the 5(five) traction substations will have two Rectifier transformer sets of 3MW/2MW capacity to meet the traction load. In addition, two Auxiliary transformers, each with a capacity of 500/350 KVA, will be installed at each traction substation to obtain a 415-volt 3-phase power supply for various loads.

8.25.3 Auxiliary Sub-Station

At all stations without a traction substation, separate auxiliary substations will be provided, each equipped with two transformers of 500 kVA and 315 kVA, to supply various loads. As discussed in the write-up for the Power System for Thane MRTS, the system will include receiving Traction and auxiliary substations.

8.26 Signaling

The design of signaling arrangements in a rapid transit system is based on

- To maintain a minimum safe distance between two trains on the same track at every instant.
- Route trains into the correct tracks at junctions, terminals, etc., safeguarding against conflicting movements.
- To introduce means to regulate train movement to adhere to the timetable at the proper spot on the platform.

The most important characteristic of a signaling system should be its ability to work in a 'fail-safe' manner, which ensures that any failure of a vital component of the current or circuit will produce the most restrictive conditions. Human elements reduce system safety and to be kept to a bare minimum.

The signaling system is to have a very high degree of reliability. Adequate backup systems and redundancy of vital components and circuits are normally provided to ensure the objective. System design should enable replacement of faulty components in the least possible time, thereby reducing the "downtime" of the system.

8.26.1 Signaling Methods

Signal systems complying with the requirements can vary in complexity from the simplest line-side signal installation operated from a local signal cabin to a complex computer-controlled and remotely supervised system with automatic control/operation of all possible functions, including the trains themselves. The more comprehensive the signaling system, the better overall traffic regulation will be, including the operation of individual trains. On the other hand, the more complicated the system, the greater the probability of failure, thereby necessitating standby or duplicated equipment. These features introduce costly designs and require highly reliable components. The systems, therefore, become very costly, and their use must be properly evaluated against the needs of the proposed system.

The choice of control system and the degree of automation are key factors in assessing the project's economic viability and financial feasibility. This is particularly evident in the Indian context, where capital costs are comparable to international standards, while ticket prices are only a small fraction of international prices.

8.27 Telecommunication

The operation of Thane MRTS requires a reliable and quick communication system. It is necessary for efficient operation and to enable passengers and staff at all levels to be instantly informed of situations that affect the smooth running of the trains. The different communication systems proposed are as follows:

i. Train control telephone system – for communication between station masters and the control center.
ii. Train Radio System – A VHF communication system between the control center and the individual cabs of trains.
iii. Train the PA system to announce important messages to the passengers in a running train.
iv. Station PA system and Visual Train Information System
v. A centrally controlled Clock System.
vi. Optical fibre-based communication channels
vii. A private telephone network system connecting different administrative and functional units.
viii. Closed Circuit Television System for Station Master and Central Control.

8.27.1 Train Control System

For operational purposes, telephone communication is required between the train controller in central control, train dispatchers at different stations, and the car shed. This is an omnibus circuit, where all stations are connected in parallel. A selective calling system will enable the central controller to call any station, and the equipment will consist of a microphone and a loudspeaker for transmission & reception, providing a hands-free system. needed. A handset with a press-to-talk button is provided in the station master's telephone set.

All train control activities will be managed from the control center located within the administrative building near Balkum station.

For a successful communication system, including a **Train Radio System,** (operated in the VHF (Very High Frequency) communication band), Train **Public Address System** (every coach provided with Speakers), **Station Public Address and Visual Information Board** (loudspeakers are provided to all significant public areas like platforms, ticket vending area in the concourse, entry to the station building, etc.)

Centralised Clock System (master clock with 100% standby with auto changeover switching facility) **Optical Fibre Communication System**, **Telephone Network** (an intercom system of 32/64 lines are provided in the administrative building and car shed), **Closed-Circuit Television System, Control Centre** (including Train Controller, Rake Controller and Power Controller).

8.28 Traction

The economic and technical advantages of electric traction over diesel traction for high-density traffic on MRTS become more pronounced. This is due to the higher acceleration and deceleration requirements, which necessitate power distribution to all axles of the lightweight coaches to achieve the desired tractive and braking effort within the adhesion limit.

The use of diesel traction in MRTS is precluded due to the emission of exhaust gases, noise production from the diesel engine, and the fire hazard associated with diesel oil. Hence, electric traction was chosen for the MRTS Thane.

With electric traction as the preferred choice, the system of traction was worked out based on:

a. As MRTS of Thane would be a new system for the city, and there was no apprehension of obsolescence of existing assets, a proven technology and a reliable system for Thane was adopted.
b. The construction and commissioning of the MRTS was estimated to take about 3 years from the time construction is approved. Therefore, electric traction, which has been recently introduced commercially, is expected to become fully established in the next few years.
c. Indigenous availability/adaptability of the proven system and cost factor are the prime considerations in selecting the system.

Possible Systems of Traction in the development stages, there were four possible systems of traction, such as i) Conventional System ii) Use of DC Chopper iii) Pulse with Modulation, and iv) AC Thyristor Control

Conventional systems operate with DC voltages ranging from 600 to 1500, and motor control is rheostatic. The electrical braking employed in this system is also rheostatic. The considerations that lead to the adoption of the conventional system are

1. DC motors are far superior for traction duty compared to AC motors
2. As the length of the RTS is comparatively small, AC would not give any commensurate benefit.
3. The conventional method has proved reliable, and most of the rapid transit systems in the world use the conventional system.

DC Chopper: The electric traction equipment industry, in collaboration with the RTS authorities, developed a DC chopper system using thyristors, which allows for stepless variation of the voltage of the incoming supply for application to the traction motors. The system uses DC motors. The power supply is at 600V DC/1500V DC. After chopping, a low DC voltage is applied to the traction motors at start. The same chopper is used during braking to have regenerative braking down to a very low speed. The developments took place in Japan, Sweden, Switzerland, and Germany, with trials commencing in 1964. The developments were made possible by the steady decline in thyristor costs. Chopper was used in trolley buses in Switzerland, tram cars in Sweden, and on suburban DC EMU coaches in France. This system can, therefore, be considered operational for all practical purposes.

The Pulse Width Modulation system, developed by M/S Westinghouse of the USA, provides a DC input of 600V to the train. This input is then converted by semiconductor equipment into a 3-phase AC output at a variable frequency and voltage. The drive is through brushless, commutator-less, conventional squirrel cage induction motors, which are widely used in electrical drives in all industries for their cheapness and ruggedness. Regenerative braking is also possible in this system.

AC Thyristor Control: Technically possible to supply trains with AC at a voltage range of 600 to 800 volts and obtain a stepless variable DC output through the phase angle control of the thyristors used in conjunction with diodes in a bridge network. A similar traction system is already under trial with the Indian Railways, utilising high-voltage single-phase AC at 25 KV. Both overhead and third-rail current collection systems are employed in various rapid transit systems. The third-rail system is used when overhead current collection systems are restricted by clearances or other factors, especially in tunnels. As proposed, the Thane MRTS would be mostly elevated, allowing the use of an overhead current collection system, which is cheaper. It is, therefore, proposed to adopt an overhead current collection system.

8.29 System Voltage and Current Collection System

Two systems, i.e., 25 kV 1f AC and 750 V DC traction systems, were considered for Thane MRTS. 750 D.C. is in use in most MRTS systems worldwide. While 25 KV can be considered, the advantage of a 750 V D.C. system over a 25 KV A.C. 1f are as follows:

a. MRTS being in a highly built-up area 25 kV 1f system is considered more hazardous from the point of view of safety and electrical clearance.
b. The induction effect of 1f A.C. traction may affect the communication and telephone system.
c. The use of a transformer and rectifier set in rolling stock reduces passenger loading, and the coach's weight increases, resulting in more energy consumption, which is critical due to the high specific energy consumption for MRTS.

Considering the above factors, a 750 V D.C. system with overhead current collection was recommended for adoption for the Thane MRTS.

A selected **traction system** for the Thane Mass Rapid Transit system will use 750-volt DC traction with overhead current collection. As the system will be a Light Rail System (LRT)/High-Speed Tram System (HST), conventional control or DC chopper with regenerative braking may be used.

8.29.1 Alternative Traction Voltage

As an alternative to 750 V DC, 1500 V DC and 25 kV AC were also studied. 1500 V DC is being phased out and is therefore not recommended for use. Single-phase AC is the most common use of alternating current. Some systems initially used low voltages, but the most common system now distributes power at 25 KV. Overhead lines are essential at this voltage level for safety reasons. In this system, only one substation is required to convert 3-phase, 132 kV power to 1-phase, 25 kV power. As there is only one substation, an emergency supply arrangement from another source at another location in the system is necessary to avoid disruption to train services. The 132 kV power was planned to be supplied by MSEB from their substation near Kolshet. **This system can be adopted for indigenously manufactured rolling stock.**

8.30 Ticketing and Fare Collection System

In a rapid transit system, the very high passenger volume per unit time, especially during peak hours, makes fare collection from every passenger a challenging task. Various methods have been adopted worldwide, depending on the rules established in this regard. The rule of holding tangible authority during travel makes it obligatory to issue a tangible authority in the form of valid tickets to every passenger or group of passengers. Additionally, the system configuration can be adjusted to accommodate either a single fare or a multi-fare system.

The ticketing system is divided into two categories:

1. Open system
2. Closed system

In an **Open System,** a passenger may enter a tram/train without purchasing a ticket. The ticket is purchased from a ticket machine. In an open system, a conductor issues a ticket to a passenger and keeps stock of all the different types of tickets used in the system. In a mass transit system, many passengers enter a coach. His stay on the coach will be quite short. In the Thane MRTS system, the average distance a passenger travels is 4.0 km. With an average service speed of 33 km/h, his stay will be approximately 8 minutes. Within this time, it is not possible to issue tickets. Situations may be improved by positioning one conductor at every gate, which will require deploying many ticket issuers. This system is therefore not generally favoured for ticket issuance, and the MRT adopts only a closed system.

8.30.1 Closed System, Automatic, Semi-Automatic or Manual

In a closed system, a train is closed to passengers till they purchase a ticket that allows entry. In a closed system, the ticket-issuing system can be completely manual, partially manual & partially automatic, or completely automatic. The fully automatic system is costlier than a partly automatic and partly manual system. In an automatic system, all ticket vending machines must have coin-changing facilities. In present-day MRT Systems, a fully manual system is not used anywhere due to the need for many operating/checking staff. In the manual checking system, the chance of revenue leakage also cannot be ruled out. Ticket checking is done through automatic gates in the MRT System. The Calcutta Metro has done so, and the Delhi MRTS has also planned a similar system.

8.30.2 Paid and Unpaid Area

In a closed system, the main passenger concourse in any station is divided into a paid and an unpaid zone. The paid zone leads to the platform where a passenger embarks onto a train. Similarly, a passenger disembarking at a station travels through the paid area to the checking zone, where they are checked for a valid ticket. After verification of the ticket, they are allowed to enter the unpaid area and then exit the station area.

8.30.3 Ticket Checking

In a manual system, the validity of a ticket is manually checked before a passenger is allowed to transition from the paid to the unpaid area or vice versa. In an automatic system, this function is carried out by an automatic gate. In an automatic system, the same gate cannot function both as an entry gate from the unpaid to the paid area and as an exit gate from the paid.

8.30.4 Entry Gate

At an entry gate, the ticket's validity is verified. Once confirmed, the gate typically prints the station name (or code) and the date and time of entry on the ticket, then returns it to the passenger for preservation during the journey. This information is also commonly incorporated on magnetic strips on tickets, in the form of magnetic data recorded when the ticket is passed through the entry gate. All these functions are completed in approximately 2 seconds.

After successful verification, the turnstile gate (or flap gate) is released for the passenger to pass through. After the passenger passes through, the gate automatically locks. The gate reopens only when the next passenger inserts a valid ticket into the machine. These gates can handle an average of 25 passengers per minute. However, when commuters become familiar with such a gate, a maximum of 28/30 passengers may be handled by the machines per hour.

8.30.5 Exit Gate

When a commuter inserts the ticket into an exit gate to come out of the paid area into the unpaid area. When the ticket is inserted into the gate, it verifies that the passenger has travelled up to the authorised distance. If the passenger travels beyond the authorised distance, the machine rejects the ticket and emits a warning sound. He will need to contact the station staff to pay the fine/extra fare.

8.30.6 Reversible Gate

A third type of gate is also used for ticket checking. These are known as reversible gates. These gates can function as both entry and exit gates, depending on the mode they are set to. Reversible gates can be cost-effective at stations where the difference between entry and exit passenger counts is significant during peak hours. A characteristic of MRTS is that the direction of passenger movement in the morning peak hour is opposite to that of the evening peak hour. Thus, if more passengers enter a station in the morning, more passengers will exit the same station in the evening.

8.30.7 Ticket Vending Machines

Ticket vending machines may be of two types, viz.

1. Automatic Vending Machines
2. Booking office Vending Machines.

In an **Automatic Vending Machine,** a passenger normally selects their destination station by pressing an appropriate button. Immediately, the fare for travel to that station is displayed in a small display window. One must insert coins (one or more) into the coin slot on the machine, whose total value will have to be equal to the displayed fare. He must press a button to print and issue the ticket. At this stage, he may also press a 'Cancel' button when the ticket issuing process is terminated, and the coins inserted will be returned to the passenger.

In the **Booking office,** an operator is needed to operate the machines. Passengers may use these counters to purchase a ticket. From the booking office, all types of tickets are available for use.

8.30.8 Number of Ticket Vending Machines

The quantity of Ticket Vending Machines depends on:

1. Number of passengers to be dealt with.
2. Types of multi-ride tickets to be issued.
3. Ticket-issuing capacity of vending machines.

While computing the vending machines, it was assumed that, because multi-ride tickets were used, all passengers would purchase a ticket. Sixty tickets would be issued for every 100 passengers, with the vending machines capable of issuing six tickets per minute.

If the system is designed for Smart Card operation, a separate machine for Smart Card preparation and loading will be required for ticket vending. Additionally, in such cases, a special attachment will need to be provided at the entry and exit gates to read and activate the smart cards. These cards are always used for multi-ride operations.

8.30.9 Portable Ticket Decoder

These are small, hand-held battery-operated machines that can check the data on a ticket and display it on a small window. By reading these data, the validity of the ticket may be verified. Additionally, in the event of a defective ticket, the defect can be ascertained using this decoding machine.

8.31 Signage

Mass Transit Stations need effective sign displays to serve passengers pressed for time. The signage can be a display of signs and graphics that communicate efficiently while enhancing the architectural environment. Signage systems can communicate necessary information without clutter. The signage system quickly plays a crucial role in accelerating lives. The primary function of a signage system is to guide visitors to their desired destination, providing them with proper information at the right places.

Types of Signs can be classified according to the function they serve.

1. Orientational
2. Directional
3. Informational
4. Identification
5. Regulatory

Orientation Signs locate users in an environment. Such signs include maps and exploded views at the entry and at various decision-making points. Maps generally represent the geography of the location in a Graphic form. They have a geographic view of the place and direct the viewer to various destinations. These types of signs are provided where the user is in a static position. **Direction Signs** indicate the direction of a particular area or information. Generally, an arrow indicates direction. **Information Signs** convey a specific message to the viewer. They essentially do not direct. They help reduce confusion for those around them. These signs usually display textual or pictorial information on a particular subject. These signs can be divided into two types:

Restrictive signs: These are signs with a warning and prohibited usage of the system/particular area.

> **Traffic signs are intended solely** for road users.
>
> **Identification Signs** are easy labelling devices, confirming destinations or establishing recognition of a particular location. The sign may be labelled with a structure, a building or a group of buildings.
>
> **Statutory or Regulatory Signs** of order, as conduct or prohibited activity, are prescribed by local regulations. They also safeguard and protect people against danger. They inform people about the procedure to follow in an emergency, such as a fire.

Signages can be **Top Hung, Free Standing, Surface Mounted / Flat Fixed, Cantilevered or Side Fixed** and **Illuminated** or **Non-Illuminated Signs**

Consistency in the signage design is achieved by incorporating elements such as layout, letter size/ style, arrow style, illumination, and other visual components. An attempt is made to convey the message in several ways —text, symbols, and direction —so that it gets across even if one mode fails. The amount of information in each type of sign is kept to a minimum. Destinations are grouped under general headings. Abbreviations are avoided to reduce confusion.

The typeface is to be selected to enhance readability. Very few typefaces have been specifically produced to meet the special requirements of signage. The ratio of a letter's width to its height is typically 1:8 to 1:10, and the text is broken into groups to facilitate better message recall.

8.32 Operation

Thane Mass Rapid Transit System's (MRTS) alignment, known as the green corridor, will be 6.050 km long, mostly elevated, with a short distance of track converging at grade in the Balkum Station area. There are 11 proposed stations along the circular alignment route, with a major interchange with the mainline of Indian Railways at Thane Station. The system is conceived as a self-contained railway with no interconnections to other rail networks.

The level of service provided has been estimated based on the predicted passenger demand. The Thane MRTS would initially operate 10 four-car light rail vehicles, each with a passenger-carrying capacity of approximately 1,250. All trains would be stabled at Balkum depot, and the train crews would pick up their trains at the start of service. Train operators will operate the trains according to the published schedule/ timetable. The initial train service will require 10 trains in each direction during peak hours. The trains will run at a maximum line speed of 60 km/h. All trains will be stabled at Balkum Depot overnight. Train operators will undertake a routine safety inspection and test the train's essential features before putting the train into service. All train servicing, repair, overhaul, and cleaning will be conducted at Balkum Depot, with trains rotated to facilitate maintenance access. Train crews would have the opportunity for relief and crew changeover at the terminus station on the route, i.e. at Balkum. Crews would work on a shift basis, and spare/relief train operators would be available at Balkum if required to place an additional train in service for regulation, changeover, or to replace a defective train.

The train service was planned to commence in the early morning and will run continuously throughout the day until midnight. The traction current on and off times would be adjusted as required during the day's operational hours. The remaining period traction current was planned to be discharged to conserve energy and allow access to traction equipment as required for maintenance. Traction within the depot generally remains "ON" all 24 hours to facilitate train movements within the yard.

Normally, in MRTS, the loading in trains changes distinctly in four distinct periods:

i. Lean Period
ii. Off Peak Period
iii. Peak Period
iv. Peak of peak period

The peak-hour peak-direction volume refers to the traffic volume during the peak period and determines the system's requirements. Passenger demand and system usage fluctuate throughout the day and on weekends. Therefore, the train service was planned to be tailored to demand, resulting in intensive peak services. The time between these periods is known as an off-peak period, during which the train service can be reduced appropriately.

The parameters are crucial for **headway**, which is the minimum time interval between two successive trains that can run in the system when both trains operate at the maximum sectional speed.

8.32.1 Maximum Permissible Speed

The rolling stock was designed to reach a capacity of 80 km/h. The permanent way would also be suitable for speeds of up to 80 km/h. The maximum sectional speed was proposed to be kept at 60 kmph to reduce power consumption by rolling stock, which is the main source of expenditure on power. The table illustrates the increase in power requirements with a reduction in travel time.

Table 8.8: Power Requirement

Speed kmph	Running Time	Traction Power Consumption
60	100%	100%
65	96.70%	117%
70	95.50%	137%

According to the table above, if the trains run at 70 km/h, the reduced time would be 4.5%, accompanied by a 37% increase in power consumption. Considering a station stop of 30 seconds and a maximum sectional speed of 60 kmph, the commercial speed achievable is 34.3 kmph.

The **system of working** is a significant determinant for headway. In block working, only one train is allowed in any section at a time, and the signal operator operates the signals at each station. The headway is dependent on the inter-station distance and the intersection. The Thane MRTS initially required a headway of 6 minutes in 2006, and by 2031, this had been reduced to 4.5 minutes. As a headway that cannot be achieved by block signaling but can be achieved by line-side automatic signalling. This is a cost-effective system that requires less maintenance.

The costly CATP/CATC systems have not been considered essential for the proposed MRTS system. It was planned to run 4-car trains approximately 80 metres long, and the signalling, including overlaps, was designed to accommodate trains of similar length.

8.32.2 Station Stop Time

In the MRTS system, where stations are closely spaced at distances of 1-1.2 km and the rolling stock has a high rate of acceleration and retardation, the station stop time is a very significant factor in total running time. It constitutes between 20% & 25% of the time if a 30-second stop time is allowed. The total running time of the Thane MRTS, with 11 stations, has been estimated at 1 hour, 24 minutes, and 4 seconds. This figure is encouraging because the average inter-station distance is 1.46 km, resulting in a better run-to-stop time ratio. An average of 30 seconds was used as the station stop time. This would need to be suitably modified after gathering train detention data following the system's commissioning. A variation of stop time that does not significantly impact headway and punctuality.

8.32.3 Terminal Detention

In a close headway system, the terminal detention of a train for changing the line and direction of train movement, the typical time for the above in a terminal station consists of:

i. Stoppage for discharging passengers
ii. Time to travel to the reversing siding
iii. Time to change control from one cab to another
iv. Time to travel to the opposite platform
v. Stoppage for the entraining of passengers

For Thane MRTS, the line is in the form of a closed loop, and a rake does not need to come from one line to the other; the movement of rakes on a particular line is always unidirectional. Terminal detention is not required. Being a starting station, Balkum Station stoppage time might be between 90 & 150 seconds.

8.32.4 Station Operation

There are 11 stations planned for the Thane MRTS, including Thane, which will be the interchange station with main line services and is predicted to be the busiest. Passengers would Use the interchange at Thane at the concourse level, without transfer to the street level. The stations were designed to facilitate smooth access and egress between the street and platform levels. The stations are also provided with a few features to assist the efficient running of the station, which are:

- Automatic ticketing machines
- Ticket barrier/ gates
- Public address system
- Passenger information signage
- Train information on the platforms
- Escalators and lifts as required

Each station will have a Station Master, who will be responsible for day-to-day management on a shift basis. Station Assistants will be present at each station to undertake several defined operating duties. These include staffing the ticket office, providing passenger assistance at the ticket gates, monitoring the platform train interface, overseeing the access and activities of other staff working on the station, and managing crowd and evacuation control.

8.32.5 Train Control

An automatic signalling system was planned to control train movements and authorise trains to proceed, or not, using colour signals. A signal control panel is proposed for installation within the central control room to remotely control signals. This type of intervention will only be required during an emergent situation. A panel was also planned at Balkum station to control the relay interlocking for the points and crossovers in this vicinity. Ordinarily, the section would be fully automatic for service trains continuing around the Thane MRTS loop. The control room is to provide a dynamic display of train positions from a mimic panel.

The frequency demand for trains on Thane MRTS would be met with the provision of Automatic Signalling with 3 aspects. Nobody is required to control the display of signals. But it is essential to enable operating staff to stop a train at a station in case of an emergency. Thus, a control was recommended for the Station Master and the central controller.

It is proposed to run 4-car trains in the Thane MRTS. In a lean period, the loading will be less. The alternative proposal was to run short trains of two cars at a frequency that would ensure each train was adequately loaded. This is possible by using separate 2-car and 4-car train sets. If 2-car trains are operated during the peak period, a frequency of approximately 135 seconds will be required to carry the projected traffic. A headway of 135 seconds can be achieved only with the costly CATP system, and all cars must be equipped with a driving cab, resulting in the following disadvantages.

1. Due to the presence of driving cabs in all coaches, the carrying capacity of the rake will marginally reduce.
2. Each cab will have to be equipped with costly CATP equipment.
3. It will require double the number of drivers to work the train, thereby increasing the operational cost.
4. If one car gets defective, the rake will have only 50% power to propel. The steep gradients (4%) may be too steep for the defective rake to negotiate, resulting in serious service disruptions.

Given the above drawbacks, two-car operation is not recommended for Thane MRTS.

8.32.6 Balkum Terminal

Balkum station, proposed to be built at ground level, would be the terminal station. This interchange station would involve rakes operating from the proposed car shed near Balkum station. The entry point to the car depot would be at a short distance from the centre of Balkum station. To avoid any detention of running trains, a loop line had been provided.

8.32.7 Revenue Collection

The primary revenue source would be passenger fares. The fare structure for Thane MRTS was derived through studies and discussions with potential users, based on affordability and people's ability to pay. The fare structure proposed Rs 4.00 for a journey up to 3 km, and Rs 6.00 for trips beyond this distance.

Revenue would be collected using:

- Automatic fare machines
- Booking/ ticket office
- Mobile revenue inspectors/ conductors

Cash and other transactions should be stored securely. A security firm would be hired to collect revenue at designated locations and times, then deposit it into the bank.

The gross margin is a measure of profitability, calculated as revenue minus operating and maintenance costs. Most MRTS Projects around the world require some level of subsidy to offset any cost differentials between costs and revenue.

Secondary revenue will contribute to overall revenue, enhancing gross margin and the bottom line. This can come from.

i. On-station advertising
ii. In train advertising
iii. Train exterior advertising
iv. Letting of station retail space
v. Thane MRTS product/ gift range
vi. Advertising on tickets/ maps
vii. Station theme sponsorship
viii. Renting space (viaduct) to utility/ IT companies

8.32.8 Passenger Interface

In the Mumbai Suburban System, there exist two classes of travel. There are two half-car compartments in a 9-car train for this purpose. The Thane MRTS was planned to operate with 4-car trains. A structured class travel system would complicate ticketing arrangements, possibly extend platform dwell times, and would be difficult to enforce due to the lack of ticket checking. As mentioned above, and in line with the trend of the RTS system, only one class of travel is recommended. Due to heavy rush, it is normally difficult for ladies to embark or alight from the train. Hence separate ladies' compartment was proposed to be earmarked.

The **incident management system** is a structured approach to incident management and escalation. If an incident occurs on a train, the train operator must manage. Similarly, if an incident occurs at a station, the Station master is to manage it locally. Depending on the severity of a given incident, a Duty Manager will assume responsibility for incident management. The Line Controller is to act as the focal point for communication regarding and during the incident, and to ensure attendance of emergency services, where possible, to maintain the train service in areas where safe to do. He must coordinate with the Traction Power Controller and ensure that power is switched off as required.

8.32.9 Passenger Evacuation System

In case of any holdup of a train between two stations due to rake failure, accident, derailment, power failure, fire or any other reason, it may not always be possible to bring the affected train to the nearest station for evacuation of passengers on the platform. In such a case, it may be necessary to detrain the passengers from the coaches and guide them to the nearest station platform.

The MRTS trains are in a 4-car formation, having a driving cab at either end. The coaches are vestibuled, and the driving cabs have a door connected to the passenger compartment. The cab is provided with doors on either side with steps for the entrainment/detrainment of the motorman. In the event of an emergency, passengers can be evacuated from either side of both cabs after a suitable announcement. Additionally, one ladder is provided inside each of the two central coaches, which can be hung in one of the doorways for passenger disembarkation after the door is opened.

Additionally, one ladder is provided inside each of the two central coaches, which can be hung in one of the doorways for passenger disembarkation after the door is opened.

In the event of a fire on the train during a run, the train should normally be taken to the station ahead and passengers evacuated to the station platform. If the train is unable to move, the evacuation of passengers can be facilitated between stations using the methods explained above. The passengers may take the pathway on the elevated structure to reach the station.

8.33 Standards and Procedures

The Thane MRTS will have to develop its own operating rules and procedures, which may be based on established Mass Transit operating practices and principles. The rules will draw upon best practices from other similar operators, while also incorporating the specific requirements and constraints of the Thane MRTS. Particular attention must be given to safety-critical operations such as train movements, track access and electrification.

8.34 Early Termination of Green Corridor

The traffic study of the Green Corridor indicates that the traffic demand is concentrated in the area between the proposed Balkum and Golden Dyes stations. The originating traffic of Man Pada, Kolshet and Azad Nagar is quite small. It was suggested that the lines be terminated at Golden Dyes station for the present. As the city expands and ridership increases, the system may be extended beyond Golden Dyes to terminate at Balkum, thereby completing the total loop.

In case of early termination, the additional facilities needed at Golden Dyes station are:

- The station should have a rake turning facility, requiring two crossovers.
- To operate the crossovers and arrange shunting of rakes from one line to another, a signal operating panel and the necessary panel operators will be required.
- Shunting lengths of approximately 300 m are required on both lines to facilitate shunting operations at both stations, namely Golden Dyes and Balkum.
- Shed for night stabling of 2 rakes minimum at Golden Dyes Station to enable morning start of the services from both ends at the same time.

To facilitate the **rake reversal**, a local signal panel needs to be provided. In addition, signal panel operators must also be employed at the Golden Dyes station. As the rake length will be approximately 80 meters, and the terminal detention can be contained between 2.2 and 2.5 meters. Thus, terminal detention will not affect the headway required by the system.

8.34.1 Night stabling of rakes

It is customary to start the morning service at the same time from both ends. To achieve this, a minimum of two rakes must be stabled at Golden Dyes Station, although three stable cars will be more desirable. For this, a shed of 100 meters in length covering both lines may be constructed at the end of the shunting neck so that the rakes remain protected from the vagaries of nature at night.

8.35 Engineering Safety in MRTS

As MRTS carries high-density intra-city passengers, the safe operation of the system is of paramount importance. Carrying passengers safely to their destination in the quickest possible time, without interruption, is the prime objective of any MRTS, and Thane will not be an exception. Apart from the above, the safety of staff working on the MRTS and nearby structures is also an important factor to consider in the design, construction, and operation of any MRTS.

8.35.1 Safety in Civil Engineering

i. Adequate safety precautions are important in structural design, structures carrying the track of MRTS. The factor of safety, considering dynamic loading, will ensure safety. Proper supervision and quality control during construction will also be needed to ensure structural safety.
ii. Safe track structures to avoid derailment of any type - Proper laying of ballastless track with almost zero tolerance will ensure. Special attention needs to be given to track laying on curves and grades. The use of prefabricated templates for track laying is recommended.
iii. Proper design of the station building, including passenger dispersal arrangement to ensure safe dispersal of passengers in emergencies like fire, stampede, derailment, etc. For this purpose,

specified parameters are to be followed during design. A certain amount of redundancy in design, depending upon their space availability, will ensure this

iv. Proper vertical and horizontal alignment to ensure safe negotiability of rolling stock at the specified speed with adequate tolerance. A speed limit of 30/35 kmph for a 55 m radius curve with transition, and 20 kmph without transition, is recommended to ensure safe running. Special care and strict supervision should be ensured during track laying on curves.

v. Proper design of auxiliary structures, pumping installations, water supply for firefighting, etc. An adequate margin will be kept in capacity to ensure safety.

vi. As the passenger travel time and waiting time are very small, no W.C or urinal needs to be provided to ensure cleanliness and avoid a slippery surface due to water spillage, etc. However, at Thane Station, where a large volume of transit traffic will be, a few urinals and WCs are recommended.

8.35.2 Safety in Electrical Engineering

i. Provision of proper clearance of overhead equipment from nearby structures – in heavily built-up areas, additional clearance over and above The specified criteria are recommended wherever possible to ensure safety.

ii. Proper installation coordination of electrical installations – to ensure failure of less vital and costly equipment, and isolation of electrical circuits at the earliest to ensure safety.

iii. Provision of proper fault isolation devices like HSCB and proper relays in the HT, LT and traction power system. Periodical testing of relays is recommended.

iv. Earthing and bonding of electrical equipment and structures likely to be affected by electrical failure – such bonding and earthing are essential to ensure the safety of passengers as well as staff. Two parallel earthing systems are recommended for redundancy. Double earthing will need to be maintained to ensure safety.

v. Ensuring a proper path for the DC return circuit to ensure proper flow of return current to the substation. This is ensured by providing longitudinal and transverse bonding as required. Additionally, a neon earthing device with a relay is to be installed at the rectifier substation.

vi. Provision of fire alarm and fire protection system in important and vulnerable locations – At all the traction substations, fire detection and fire alarm systems may be provided. Vital installations, such as the central control room, the remote-control centre, various equipment rooms, and station masters' rooms, may be equipped with automatic protection alarm systems as required.

vii. Provision of proper insulation to the return circuit to avoid/minimise leakage current – As MRTS, Thane will be mostly an elevated structure, the chance of drainage leakage current will be less provided proper maintenance of insulation of the return circuit, i.e. running rail is ensured.

viii. Provision of cathodic protection against electrolytic corrosion for equipment and structure

ix. It is recommended that a special study be conducted for leakage current in areas where pipelines are run parallel near the MRTS alignment, and cathodic protection to be provided if required, only after the system is commissioned and the leakage current determined.
x. Use of fireproof fire-retardant materials in the equipment for their safe functioning.
xi. Provision of emergency lighting systems in the station for safe dispersal of passengers in case of power failure – Provision of 20 to 25% of the lights connected directly to battery-backed UPS will ensure uninterrupted lighting during power failure. No auto changeover is recommended for this purpose.
xii. Proper and adequate lighting at stations and workplaces for ensuring safe dispersal of passengers and safe working conditions for staff.

8.35.3 Safety in Rolling Stock

i. Safety in the structural design of bogies and car bodies to carry dense crush load. Apart from this, the car body should be constructed with anti-telescopic features and capable of withstanding a buffing load of over 100 tonnes, ensuring passenger safety in the event of an accident.
ii. Proper design of bogie and suspension to enable the rake formation to negotiate the horizontal and vertical curves at the specified speed with a safety margin – proven design of bogie working in a similar alignment is recommended.
iii. Use of fireproof, fire-retardant and non-inflammable material inside the coach to prevent/ spreading of fire – In no case, bitumen or wood should be used for flooring. Materials like fibreglass or similar materials should be used. Material with very low smoke emission will be a desirable feature.
iv. Protection against short circuit in electrical circuits by provision of suitable circuit breakers/ live contactors and relays should be ensured – As far as possible, simple circuits should be used with a minimum no. of equipment with adequate capacity and of proven design.
v. Vestibule arrangement will ensure passenger detrainment in emergency – This provision will also ensure dispersal of passengers to adjacent coaches from a very crowded coach to avoid a stampede and ensure safety.
vi. Provision of passenger alarm system on both sides inside the coach – This will attract the attention of the motorman, who will respond and stop the train at the nearest station and ensure the safety of passengers
vii. Provision of proper ventilation in each coach – This is considered essential not only from the point of passenger comfort but also from the point of biological needs of passengers, especially in the case of running MRTS coaches in a door-closed condition.
viii. Provision of emergency lights in coaches is essential to ensure passenger convenience in case of power failure.

8.35.4 Safety in Signalling

Safety has been defined as a situation with a risk level below a given threshold. Hence, the risk limit established by the rules becomes very important. For this purpose, the Signalling and Telecommunication codes need to be issued before the system is commissioned for passenger traffic.

In a rail-based system, the driver cannot choose the route they must travel on. Hence, when a car moves on the track, it should be ensured that its path is safe and that it can move without encountering any obstructions that could lead to an unsafe situation. In cases where the railroad condition (such as an improperly set point machine or broken track) prevents the car from moving over the section, there should be a device to stop the car's movement. The function of signalling is just that. When the signal is in the "ON "condition, it proves that the track ahead is unsafe. Two functions control the safety of movement.

1. Track function
2. Signalling function

When the track is safe to run over, the signalling function may be unsafe at that time (the Off aspect of a signal is an unsafe signalling situation). When the track function is unsafe, the signalling function must be put into a safe condition (On aspect). When a signal is taken off for the passage of a tram/ train, it must also ensure that this safe condition is maintained until the entire train has passed over the section. This is achieved using an interlocking system. This system ensures that when the signal is taken off -

i. The track section over which the movement is authorised is clear of all obstructions.
ii. No conflicting move can take place over the section till the authorised movement has been completed
iii. The authorised route cannot be altered till the train movement is completed.

8.35.5 Electromagnetic Signalling

Present-day signalling is primarily based on electromagnetic principles and is known as electromagnetic signalling, which has nearly reached the stage of absolute safety. The electromagnetic signalling system is based on the principle of fail-safe design, where any failure should not create an unsafe condition for train movement.

Electromagnetic relays, which form the backbone of the signalling system, are so designed that the chance of contact welding is absent. This is achieved using carbon-carbon or carbon-metal contact combinations. The design of vital circuits is also based on the principle of fail-safe operation. The track circuit, which detects the presence of a train on the track-circuited section, is designed so that any system failure will cause the track relay to drop, thereby inhibiting the signals from being taken off.

In electromagnetic signaling, final safety depends on the driver, who controls the train.

Electronics-Based Signaling

Due to the demand for trains travelling at very high speeds and in close intervals, constant monitoring of train movement is necessary to assess the performance of the train and its driver. The system automatically detects any unsafe situation and takes corrective action. For this purpose, the use of electronics is inevitable. Also, to monitor various parameters of train performance, the use of a microprocessor has been introduced in the signaling system.

In the present-day electronic equipment, the design is based on

i. Single channel structure
ii. Two / multi-channel structure

A single-channel structure comprises a single functional unit that performs a specific function. To prevent dangerous failure in the event of a fault, the correct functioning of such a system is tested periodically or continuously monitored. Another method for a single-channel structure is using diversified software. The two versions of the programme run concurrently and determine the outputs relevant to safety, which are then compared. A command output is sent out only when the two outputs are compared favourably.

A two- or multi-channel structure comprises two independent functional units for performing a specific function. The functional structured units can be identical, or they may be built up in different ways using the principle of diversification (hardware diversification, software diversification or time diversification).

The introduction of the above method has significantly increased the safety of this equipment and is increasingly used in advanced signalling systems to handle high-speed trains or to achieve extremely high train frequencies.

MRTS Thane will have an initial frequency of 6 mts, which will gradually increase to 4.5 mts. Such a frequency does not warrant an electronic signalling system.

8.36 Balkum Depot

Balkum Depot will support the maintenance of trains for the Thane MTRS system. It is located to the northeast of the alignment, near Balkum Station. Several proposed extensions to the MRTS system include a link to Bhiwandi, which will originate at Balkum. Therefore, the Depot is strategically located for both the planned current and future train services.

The main features of the depot include stabling for up to 18 x 4-car formations, inspection and cleaning facilities, repair and lifting shops within the main shed, and a site to the west of the shed that supports all activities required for a planned overhaul.

8.36.1 Train Maintenance

Metros are composed of complex mechanical and electrical systems, with hundreds of moving parts. If the metro service is to be reliable, the equipment must be kept in good working order, and regular maintenance is the essential ingredient to achieving this.

Rolling stock requires the most intensive maintenance in the metro system and is the most vulnerable if maintenance is neglected. Reliability is the key to successful railway operation and maintenance.

8.36.2 Maintenance Facilities

Trains require special facilities for storage and maintenance. The layout of a maintenance facility or depot typically consists of a storage yard, a car-cleaning area, an inspection and light-maintenance shed, a heavy-maintenance shop, and, possibly, a separate locomotive shed for housing special rolling stock used for shunting and other purposes.

8.36.3 Access

An essential feature of any depot is good access, both road and rail. Good rail access means trains can enter and exit the depot without delaying trains on the main line or disrupting operations within the depot. Usually, a long access track into (and out of) a depot is required if space is available. If the system is equipped with ATP (Automatic Train Protection), the changeover between ATP and manual operation will likely need to occur on this track. This must be carefully incorporated into the depot track design.

Road access is equally important. Large equipment (transformers, pre-assembled traction units) may need to be delivered to the depot, and space must be provided to allow heavy trucks to enter, turn, unload, and exit the depot. In the case of Thane MRTS, it is necessary to provide car delivery access by road. Hard-standing floor areas and unloading facilities, such as cranes or gantries, have been considered in the design of this depot. The hard floor needs to be located over or near a suitable track so that cars being delivered can be craned off the road vehicle and placed on a running track.

8.36.4 Cleaning and Stabling

Trains are stored in depots or sidings when not in use and require cleaning and servicing. Cleaning involves a regular exterior wash, interior sweeping and dusting, or vacuuming. At longer intervals, seating upholsteries and carpets may be shampooed. Exterior washing typically involves a drive-through washing machine, which cleans the sides and, in some casesss, the roof.

Suitable facilities must be provided in the stable areas where trains are stored.

Water and equipment must be provided in these areas. Access to trains must be designed so that cleaning staff can reach them safely whilst carrying their equipment. This usually means walkways at floor height alongside trains.

Trainable areas are traditionally outdoors, largely because of the expense of building large sheds. However, covering the stable areas with some sort of weather-proof structure is always preferable. It protects trains and staff working on or around them and reduces contamination from pollutants and wind damage. A covered area will also provide some benefit in hot conditions and could help to reduce the air temperature inside coaches.

8.36.5 Train Washing Machines

Washing machines require that the track on either side is straight for at least one car's length. There will be a need for proper drainage facilities, including wastewater management and waste retrieval, for the chemical wash using a clarifier or separator. Currently, recirculating systems are used, with a preference for reusing water from the final rinse or even the 'ready mixed' water.

Typical drive through train washing plant. Note special track drainage.

Fig.8.30: Automatic Car Wash

Chemical washes are used for heavy cleaning, and the chemicals will often require the train to stand for some time while they react with the dirt on the car body. The stage must be protected against drips, and the waste collected. To reduce costs, it is recommended that a facility use a single washing machine for both chemical and water washes.

8.36.6 Wheel Lathe

Most modern depots are equipped with a wheel-profiling facility, commonly referred to as a wheel lathe. These are normally designed so that the wheels can be reprofiled. Removing the wheels requires lifting the train, which is both costly and time-consuming. The underfloor wheel lathe, also known as a "ground" wheel lathe, was developed.

Wheels can be removed from a train by a "wheel drop", where the wheelset is lowered underneath the train into a basement below the depot floor. Sometimes entire toolrooms are provided in such areas, but ground conditions can make these places difficult to keep dry and challenging to meet modern evacuation requirements.

Modern wheel lathes can also reprofile a wheelset that has been removed from the train. Otherwise, a separate wheel-turning facility must be provided in the workshop. Wheels on a bogie or on a single vehicle must be reprofiled within limits that are comparable to one another. Wheels on the same coach must not vary by more than 10 mm between bogies. The most modern vehicles might require a tolerance as low as 3 mm.

8.36.7 Inspection Sheds

Special facilities are required to carry out rolling stock inspections. A properly constructed shed, capable of accommodating a whole train, is to be provided. Access to the underside of the train is essential, and this must be designed to provide reasonable working conditions and safety.

The most commonly used method involves a pit between the rails of the maintenance tracks, with additional pits sometimes located on either side of the track to facilitate access to the sides of the underframe equipment.

Fig.8.31: Rake inspection Bays

Shore Supplies

Inside train sheds and shops, it is necessary to provide shore supplies for trains and power for tools and maintenance equipment. Where overhead electric Traction is used, the overhead wires are usually installed inside inspection sheds, but not in shops where vehicles are lifted. If it is necessary to access the roofs of trains, the overhead current must be switched off and the switch secured by a lock.

8.36.8 Lifting

The traditional method for accessing bogies was to lift the car bodies off the bogies by using an overhead crane or cranes. Each vehicle to be lifted must be separated from its fellows in the train and dealt with separately. If one car in a set is defective, it must be uncoupled and pushed into the shop for lifting. To access the bogies, the overhead crane lifts the bogey, and the body is lowered onto two more stands. For overhauls, the bogey may be moved to a special area where it is placed on stands for stripping and refitting.

A quicker lifting method is to use two cranes that lift both ends of the car body together and free both bogies at the same time. The body can then be removed to another part of the workshop for maintenance. Jack is the usual method of lifting nowadays. Vehicles can be lifted, of a Coach individually or, if fixed training is used for normal service, more recent practice has been to lift and are controlled by a single person from a control desk. The major advantage of this system is that it doesn't require disassembly.

8.36.9 Maintenance Workshops

It is still common to see workshops for metros equipped with tooling and equipment, enabling a full range of engineering tasks. This will include milling, boring, grinding, planning, and cutting machines, as well as cleaning facilities (including bogie washing and car underframe cleaning, also known as "blow-out"), plus electronic and pneumatic testing shops. Good storage and materials management facilities are also needed. Computerised systems are now widely available.

London Underground train (1995 Tube Stock) raised on lifting jacks in Golders Green Depot. The pit below (an old design dating from 1906) is not deep enough to allow a comfortable working position. The jacks raise the train as a set (a 3-car unit) to give more clearance. Four jacks are required for each car and they are synchronised to operate from one control position.

Fig.8. 32: Lifting of Coach for inspection, in addition to the inspection Pit

Maintenance is also important for trackwork, traction power equipment, signaling, communications equipment, fare collection systems, and all types of electronics, as well as building maintenance. The main depot must be equipped to handle the necessary tasks. Storage for hazardous materials and lubricants must be in a secure area with adequate fire protection facilities. Waste disposal must also be properly managed.

8.36.10 Signal Workshop

The overhaul and major repairs of the RTS signaling equipment will be carried out in this Shop, located in the car shed at Balkum. With suitable expansion, this shop will be capable of carrying out similar work on future RTS lines as well. A distilled water plant may be installed to supply distilled water to the storage batteries used for the signaling and communication system.

8.36.11 Electronic Laboratory

To support periodic testing and overhauling of specialised electronic equipment in the Signal & Telecommunication branch, a laboratory will be established at Balkum for repair, maintenance, testing, and other purposes. The actual facilities to be provided will depend on the equipment manufacturers' recommendations.

8.36.12 Laboratory and Test Room

The main works that are proposed to be done in the laboratory and test room are (1) periodical testing and calibration of all electrical meters, relays and delicate instruments of the RTS, (2) overhauling and repairs of all electrical meters, relays and delicate instruments of the RTS, and (3) day to day maintenance and overhauling of all equipment at the Remote Control Centre. This laboratory will be in the administrative building, near Remote Control.

Fig.8.33: Planned OCC for Ptopiosed Thane MRTS

8.37 Operation Control Center (OCC)

An Operation Control Centre, also known as a Command-and-Control Centre, is a centralised facility that serves as the operational hub of an organisation. It acts as the central nervous system, collecting and processing information from diverse sources such as sensors, cameras, databases, and human operators. OCCs utilise sophisticated software systems, visualisation tools, and communication networks to provide real-time insights and decision-making capabilities for complex operations.

8.37.1 Key Functions and Responsibilities

1. Monitoring and Surveillance
2. Incident Management:
3. Performance Optimisation:
4. Communication and Coordination:

Operation Control Centers (OCCs) have emerged as critical components in various industries, ensuring efficient and safe operations. By centralising monitoring, surveillance, incident management, and coordination, OCCs enable organisations to respond swiftly to evolving situations, optimise performance, and enhance safety.

8.38 Cost Estimate

The cost estimate for the Thane Mass Rapid Transit System (MRTS) as developed, has the following major divisions.

i. Civil
ii. Mechanical and Electrical
iii. Signalling & Telecommunication System
iv. Ticketing System
v. R & R and Environmental Mitigation

CHAPTER 9

Delhi Metro

9.1 Introduction

The plan for the Mass Rapid Transit System (MRTS) first emerged in the late '60s based on traffic and travel characteristics. Since then, the concerned authorities have studied and discussed issues related to alignment and technology. In 1984, the Urban Art Commission proposed a multimodal transport system that would include MRTS route corridors, an existing transport network, and its proposed development. During the studies, the population and traffic more than doubled, resulting in overcrowding and unreliable service. As such, Delhi roads were chaotic.

To address the chaotic and accident-prone transport infrastructure, the Government of India and the Government of Delhi established the Delhi Metro Rail Corporation (DMRC) in May 1995. The scope of the DMRC was to build an MRTS network integrated with the existing road network, which was also undergoing upgrades. The Union Cabinet approved it in 1996, with three corridors, as part of Phase 1. Later, the metro network extended considerably, and construction was divided into four phases. Currently, three stages have been completed **(Fig. 9.1),** and phase four is under implementation. Each phase comprises a different route; some are new, while others are extended corridors. Each route was provided with a colour code. The route description is added in Chapter 2, in Part 1.

Fig.9.1: Detail Metro

Delhi boasts the largest metro network in India, spanning over 350 km with 256 stations, and has plans for an additional 206.5 km in Phase V, and 128 planned stations. While passenger coaches for the main lines of one of the world's longest rail networks are manufactured in India and operated, others are imported for standard-gauge lines.

The Delhi Metro **(Fig. 9.2)** has received awards for environmentally friendly practices from organisations including the United Nations, RINA, and the International Organisation for Standardisation; it is the second metro in the world, after the New York City Subway, to be ISO 14001 certified for environmentally friendly construction. By March 2023, 64 metro stations, four sections on the central verge between piers, and 12 other Phase I and II locations on the network had implemented rainwater harvesting for environmental protection. Additionally, all 27 Phase IV elevated stations were planned to incorporate rainwater, and 52 recharge pits were being constructed for this purpose. It is the world's first railway project to earn carbon credits after being registered with the United Nations under the UN's Clean Development Mechanism. It has earned 400,000 carbon credits through the use of regenerative braking systems on its trains. The DMRC installed the metro's first rooftop solar power plant at the Dwarka Sector 21 station in 2014. The network received 35 per cent of its energy from renewable sources **(Fig.9.3)** by April 2023, which it intends to increase to 50 per cent by 2031.

The Delhi Metro initially used broad gauge (approximately 148 km) before transitioning to standard gauge. The Delhi Metro procures coaches from various international manufacturers, including Hyundai Rotem, Mitsubishi Corporation & MELCO (Phase I), Bombardier Transportation (Phase II), and CAF Besain of Spain (Airport Express). Later, Bharat Earth Movers Limited (BEML) began manufacturing coaches in technical collaboration with Hyundai Rotem.

Fig.9.2: Aerial View of Delhi Metro During Night

Fig.9.3: Rooftop Solar-Power System at the Anand Vihar Metro Station

The Delhi Metro was implemented in four phases, and the fifth phase is under planning.

i. **Phase I**

A 64.75-km network of 59 stations was constructed in Delhi, encompassing the initial sections of the Red, Yellow, and Blue Lines. The stations were opened to the public between December 2002 and November 2006.

ii. **Phase II**

A total of 123.3 km long network of 86 stations and 10 routes, and extensions were built. Seven routes were extensions of the Phase I network, three were new colour-coded lines, and three routes connected to other cities (the Yellow Line to Gurgaon and the Blue Line to Noida and Ghaziabad) of the national capital region. At the end of Phases 1 & 2, the network of 188.05 km and 145 stations became operational between June 2008 and August 2011.

iii. **Phase III**

Phase I (Red, Yellow and Blue Lines) and Phase II (Green, Violet, and Airport Express Lines) focused on adding radial lines to expand the network. To further reduce congestion and improve connectivity, Phase III included eight extensions to existing lines, two ring lines (the Pink and Magenta Lines) and the Grey Line. It has 28 underground stations, 81 elevated stations, three new lines and seven route extensions, totalling 162.08 km. The three new Phase III lines are the Pink Line on Inner Ring Road (Line 7), the Magenta Line on Outer Ring Road (Line 8) and the Grey Line connecting Dwarka and Najafgarh (Line 9).

iv. **Phase IV**

Phase IV, spanning 141.21 km with 115 stations and six lines, was finalised by the Delhi Government in December 2018. The Government of India approved three corridors in March 2019. Construction of the 65.1 km began at the end of December 2019. The metro's total length to **exceed 389 kilometres at the end of Phase IV, excluding the operated systems in the Noida-Greater Noida Metro** and the Rapid Metro Gurgaon.

v. **Gurugram (Gurgaon) Metro Line**

Rapid Metro Gurgaon is a light metro system, 12.85 Km in length with 11 Stations, serving the city of Gurgaon. Rapid Metro connects the commercial areas and serves as a feeder link to the Delhi Metro, with an interchange on its Yellow Line at Sikanderpur metro station. Built by Rapid Metro Gurgaon Limited (RMGL), the system was the world's first fully privately financed modern light metro system, developed by IL&FS.

vi. **Delhi Metro Phase V**

The Delhi Metro Rail Corporation (DMRC) has finalised Phase 5 to provide faster, more efficient, and sustainable transit options across the city. The important link is the 21 km corridor from Mayur Vihar Phase 3 to Loni Border in East Delhi. Phase 5 is planned to roll out in two sub-phases:

Phase 5(A): Focuses on completing 15.8 km of Phase 4's remaining corridors, including Indraprastha–Central Vista, Aero City–Airport Terminal 1, and Tughlakabad–Kalindi Kunj.

Phase 5(B) includes 15 new corridors, including the expanded Mayur Vihar Phase 3–Loni Border line, which was previously proposed as a light metro and is now being developed into a full-scale rapid transit corridor.

With a length of 74.6 Km and 128 new stations (79 elevated and 49 underground) planned, is designed to enhance last-mile connectivity, cut commute times, and strengthen intercity metro links. The Phase V network includes six corridors.

9.1.1 New Corridors

Out of the proposed 18 corridors in Phase 5, four will be entirely new. These include the Mayapuri to Kashmiri Gate, CGO Complex to Surya Vihar (Faridabad), Mayur Vihar Phase-3 to Loni Border, and the Najafgarh-Nangloi Metro Corridor. The remaining 14 corridors will be extensions of existing metro lines. Some of these extensions have already submitted DPRs, while new DPRs are being prepared to further extend the lines. These extensions are designed to ease congestion on current metro lines.

9.2 Metro Corridors

Delhi Metro currently comprises a network of 389km with 285 stations. It stretches to Noida and Ghaziabad in Uttar Pradesh, as well as Gurgaon, Faridabad, Bahadurgarh, and Ballabhgarh in

Haryana. Construction works to build the rail mass transit system began in October 1998, following more than 40 years of studies. Phases one and two of the Delhi Metro were fully operational by January 2013. A Japanese loan and an equity infusion from Indian public funds provided the principal funding. The Delhi Metro was designed to be integrated with other forms of public transportation. DMRC signed an agreement with the bus operator, Delhi Transport Corporation (DTC), to incorporate management and through-ticketing. In March 2010, DMRC partnered with Google India (through Google Transit) to provide train schedules and route information to mobile devices with Google Maps. The free service allows commuters to get the latest service information and plan their journeys.

Delhi enjoys a good infrastructure level, but the rapid rise in population and the extent of the urban area is putting tremendous pressure. Transportation, considered a linchpin in the city's development program, has not kept pace with its growth. The clear indication is that the number of vehicles plying in Delhi outnumbers that in Mumbai, Chennai, and Kolkata. Delhi Metro is India's largest and busiest rapid transit system, connecting the country's capital region with satellite cities.

The metro system is operated by Delhi Metro Rail Corporation (DMRC), a public sector company established by the Government of India and the Government of Delhi in March 1995. The project was developed in multiple phases. Phase one (65.11km) and phase two (124.9km) are completed, while phase three is underway. Phase four is also planned, with work scheduled to begin in 2020. The Delhi Metro project became the first railway project in the world to be certified for carbon credits by the United Nations for its reduction of greenhouse gas emissions in 2011. DMRC saved 112.5GW of power by using regenerative brakes in the trains and reduced carbon emissions by 630,000 MT annually. Delhi Metro was conceived to alleviate traffic congestion in the city, providing commuters with a faster and eco-friendly transportation alternative. 12.2 Features of elevated stations & station buildings. While the write-up would discuss four stations along two distinct lines of the Delhi Metro, it is appropriate to mention here that considerations of the economy have played a part in opting for a station design system that accommodates a generic design, which is nonetheless adapted to suit the unique nature of each site, but without Features of elevated stations & station buildings. While the write-up would discuss four stations along two distinct lines of the Delhi Metro, it is appropriate to mention here that considerations of the economy have played a part in opting for a station design system.

Nonetheless, it is adapted to suit the unique nature of each site, but without losing the recognisable image of the station. This is especially so as most commuters would have a definite station at each end of their commutes, and the individuality of a station enhances the comfort factor. Out of the two considerations of a station regarding platforms for an elevated line, there are islands and sides. A straight alignment of the track is preferred for a metro facility operating in a congested environment with frequent stations. For this reason and considering space availability, side platforms have been chosen, especially for on-road stations.

9.3 Metro Routes
(Reproduced for ease in reading)

Line 1 (Red Line) runs between Dilshad Garden and Shaheed Sthal (New Bus Adda). The section from Shahdara to Tis-Hazari was opened in December 2002, becoming the first stretch to be commissioned in Delhi. The line was extended to Shaheed Sthal (New Bus Adda) in multiple stages, with the latest section from Dilshad Garden made operational in March 2019. Line 1 (Red Line) runs 34.7km, serving 29 elevated stations.

Line 2 (Yellow Line) runs between Samaypur Badli and the HUDA City Centre. It stretches 48.8km and serves 37 stations. The first section, between Vishwa-Vidyalaya and Kashmere Gate, opened in December 2004. Extensions to Samaypur Badli were opened in six stages over three phases.

Line 3 (Blue Line) runs between Dwarka Sector 9 and Noida City Centre. It covers 49.93km, serves 44 stations, and has a rolling stock of 70 trains. The first section of the line was opened in December 2005 and ran from Barakhamba to Dwarka. Subsequent sections were opened between Dwarka and Dwarka Subcity in March 2006, between Barakhamba and Indraprastha in November 2006, between Indraprastha and Yamuna Bank in May 2009, and between Yamuna Bank and Noida City Centre in November 2009. The 2.76km section from Dwarka Sector 9 to 21 was opened in October 2010. Operations on the 2.5km extension from Anand Vihar ISBT to Vaishali and Ghaziabad began in July 2011. The section between Noida City Centre and Noida Electronic City became operational in March 2019. Lines 2 and 3 pass through the city centre and business district at Connaught Place, served by Rajiv Chowk station.

Line 4 (Blue Line), opened in January 2010, runs between Yamuna Bank and Anand Vihar. It covers 8.74km and includes eight stations, serving as a branch line of Line 3. Line 5 (Green Line) initially ran between Inderlok and Mundka. An 11.18km section from Mundka

Line 5 (Green Line) initially ran between Inderlok and Mundka. An 11.18km section from Mundka to City Park (Bahadurgarh) was opened for passenger services in June 2018. The Green Line currently stretches 29.64km with 23 stations.

Line 6 (Violet Line) is 47 km long and runs between Kashmere Gate and Raja Nahar Singh. The first section between Central Secretariat and Sarita Vihar was opened in October 2010. The latest extension on the line, from Escorts Mujesar to Raja Nahar Singh (Ballabhgarh), began commercial operations in 2018.

Line 7 (Pink Line) runs between Majlis Park and Mayur Vihar Pocket I. It was opened in four stages between March 2018 and December 2018. It now includes 38 stations, of which 26 are elevated.

Line 8 (Magenta Line) runs between the Botanical Garden to Janakpuri West, covering 38km and serving 25 stations. The line was opened. The Botanical Garden- Kalkaji Mandir section began operations in December 2017, followed by the Kalkaji Mandir-Janakpuri West section in May 2018. Opened in October 2019,

Line 9 (Grey Line) connects Dwarka and Najafgarh. The 22.7 km-long Airport Express Line (Orange Line) was opened in February 2011. It runs between New Delhi railway station and Dwarka Sector 21, linking Indira Gandhi International Airport with six stations.

Airport Express Line (Orange) operates between New Delhi and Yashobhoomi, Dwarka Sector -25. The total length is 22.7 Km and includes seven stations, including Indira Gandhi Airport.

The author's team was involved in the detailed design of several stations, the elevated corridor. DD1B (DD for Pul Bangash with Property Development (PD), Pratap Nagar, Shastri Nagar & Indro Lok.), DD1J (DD for Subhas Place station including PD and Rithala stations). Extension of Line 3 Detailed Design of 3km Elevated Section and Stations for Pragati Maidan and Indraprastha, Fob., Detailed Design of Ghitorni Depot at Sultanpur, Detailed Design of BC 20, 8 K Viaduct and 5 stations and others. Tender design, BOQ & Cost Estimate for Design-Built Tender submitted by Consortium led by Dickerhoff-Wideman, for MC 1 A and MC 1 B. Later, two consortia joined together and included a design consultant from the other group.

9.4 Brief on MC1A and MC1 B

Phase I involved the construction of an 11.1 km underground section, first for the Delhi Metro. The contract was awarded in two sections: MC1A (4.5 km long, 4 stations) and MC1B (6.6 km long, 6 stations) **(Fig.9.4).** Key aspects of MC1A specifically referred to the underground portion of the Yellow Line, connecting Vishwavidyalaya and Kashmere Gate. The challenges were to execute through highly populated areas without affecting traffic.

MC1B from the central secretariat to Kashmiri Gate, a length of 6.6 km. Six stations is all underground and constructed using bottom-up construction. The corridor was built using two hard-rock TBMs with an outer diameter of 5700mm. Several segments were 5+1 with a length of 1400mm and a thickness of 275 mm.

Solutions were implemented, including utility and traffic diversions, utilising advanced technologies such as TBM and top-down construction methods. Design and build contract was awarded to Kumagai-Skanska-HCC-Itochu• Architect: Maunsell Australia and LPT• USP.

Encardio was awarded the contract for the comprehensive monitoring work for MC1A and MC1B, encompassing the supply, installation, monitoring, and data analysis of geotechnical instruments. A team of experienced engineers and technicians were stationed at the site for monitoring work. The instruments used included:

- **Inclinometer system** for diaphragm walls and subsurface monitoring around stations and tunnels.
- Magnetic **extensometer** for subsurface settlement monitoring around stations.
- **Piezometers** (vibrating wire, Casagrande and standpipe) for monitoring groundwater level and pore pressures.

- Compression **load cells** for struts, compression load cells for pile testing, center hole load cells for anchors.
- **Strain gages** (spot weldable) for strut monitoring, strain gages (arc weldable) for pile testing and diaphragm walls.
- Borehole **extensometer** for subsurface settlement and ground settlement points for surface settlement monitoring.
- **Crack meters**, tilt meters and building settlement points for structure monitoring.
- Vibration monitoring.
- Automatic data acquisition system as well as portable readout units.

Phase II consisted of an underground corridor approximately 36 km long, with 20 underground stations. The instruments used were the same as those of Phase I.

Fig.9.4: Station Internal Views -MC 1B (Shimizu Corporation)

9.5 Airport Express Line

The **Airport Express Line (Orange Line), of** Delhi Metro starts from New Delhi to Yashobhoomi Dwarka Sector - 25, linking Indira Gandhi International Airport. The total length is 22.7 km, of which 15.7 km is underground and 7.0 km from Buddha Jayanti Park to Mahipalpur, elevated. In January 2008, the DMRC awarded a 30-year build-operate-transfer PPP contract to the Reliance Energy-CAF consortium. However, DMRC paid half of the construction cost and executed tunnelling and civil works. In June 2013, Reliance Infrastructure Ltd. informed DMRC of its inability to operate the line beyond June 2013. Following this, DMRC took over operations from July.

Fig.9.5: Station Entry for Aerocity

The express line operates at 120 km/h, and takes 15 minutes to IGI Airport from New Delhi Station. The elevated section of the Airport Express Line was built over eight busy roads, the Ring Road, National Highway 48, Ridge Road and Sardar Patel Marg. The Airport Express line began operation in February 2011, following the 2010 Commonwealth Games.

Services were suspended in July 2012 due to several technical problems, including cracks in the girders —the iron beams that support the tracks, dislocated bearings that support the train, defective tracks, and water seepage in tunnels. A joint inspection by the Delhi Metro and Reliance Infrastructure team revealed that 540 bearings need rectification, and a few cracked girders must also be replaced. The line reopened in January 2013, operating at a reduced speed of 50 km/h (31 mph). At the reduced speed, the time travel from the New Delhi Metro Station to the Airport was around 40–50 minutes. In 2022, the DMRC decided to increase thIn September 2023, the Dwarka Sector 25 metro station opened to the public, extending the Airport Express Line two kilometres into the India International Convention and Expo Centre (also known as Yashobhoomi). The station is located underground within the new convention centre, with entrances providing a quick connection to both current and future exhibition hallse operating speed from 100km/h, reducing travel time from the current 20 minutes to less than 15 minutes. Views of Airport Express Metro line stations are included in **Figs.9.5, 9.6 & 9.7,** Internal view of Coach **in Fig.9.8** and Elevated Corridor **in Fig.9.9.**

Fig.9.6: Airport Terminal (Metro Rail Guy)

Fig.9.7: Entry to Terminal (IGI Airport)- Wikipedia

Fig.9.8: Airport Express Line- Internal View of Coach

Fig.9.9: Airport Express Line- Near New Delhi Station

The seven metro stations of the Express line are called City Airport Terminals. Special features of the stations include.

- Explosive detectors
- Large X-ray baggage scanners
- Under-vehicle scanners
- Quick reaction teams
- Dog squads to ensure the safety of commuters and their luggage
- Full-height screen doors to prevent incidents
- To save air-conditioning loads
- CCTV cameras to monitor and send images to the Station control, Security control, and operation control rooms.
- The Central Industrial Security Force to manage security
- Stations are accessible to disabled persons, with ramps, elevators and elevators to accommodate stretchers
- The Braille letters on elevator buttons for visually impaired persons

Additionally, for hassle-free travel, two stations, New Delhi Station and Shivaji Stadium Station, have been equipped with baggage check-in facilities. Additionally, Cloakroom Facilities have been provided at New Delhi Station.

Figure 9.10 includes launching of coach and **Fig.9.11** includes transportation of first coach.

Fig.9.10: Launching of Coach (Financial Express)

Fig.9.11: Transportation of First Metro Coach (DMRC)

9.6 Design Features

This section includes the general design aspects of the four stations built during the initial stage of the Delhi Metro. Structures are supported on central piers and side columns, mainly with pile foundations. In Pulbangash, the foundations are partially exposed due to the presence of rocky strata. Precast prestressed concrete girders are erected on cast-in-situ cross girders, which are supported by central piers and side columns. Precast girders with projecting shear reinforcement are placed contiguously, and the cast-in-situ deck slab provided on top acts as the flooring of the station concourse. The side platforms are supported on precast prestressed concrete girders placed on the pier arms with a cast-in-situ deck slab on top of the pier cap. The platform roof is provided on steel members erected on concrete girders on either side.

For faster implementation, the standardisation of structural elements of the Indraprastha station was considered. The sleekness of form was a significant consideration in the design of the roof systems. Stations, including those at Pratap Nagar, Pul Bangash, Pragati Maidan, and Indraprastha, were involved in studying several alternatives. The detailed design of the stations included the viaducts within the station limits. The platform length is 185 m, which would accommodate an 8-car consist, and the station building is 67 m long. Two alternatives for Off-Road and On-Road stations were developed. The essential difference lies in the vertical and horizontal clearance required for on-road stations to accommodate road vehicles below.

9.7 Elevated Stations – Phase 1

The Delhi Metro network, which serves Delhi and its satellite cities in the National Capital Region (NCR), serves 257 stations, including 24 interchange stations. Out of 257 stations, 179 are elevated,

71 are underground, including 7 transfer stations and 7 at-grade stations. In addition, 44 stations are under construction and 39 are in the planning Stage. This section includes the stations at Pratap Nagar, Pulbangash, Pragati Maidan and Indraprastha, designed by CES, led by the Author. Pratap Nagar Station on Line 1 and Pragati Maidan Station on Line 3 are on-road stations, while Pulbangash Station on Line 1 and Indraprastha Station on Line 3 are off-road stations. Commercial development is an essential feature at Pragati Maidan Station. **Figure 9.12** shows the Elevated Station with side platforms.

Fig.9.12: Elevated Station with Side Platform

The facilities on the metro system were designed to be modern, aesthetically pleasing, and easily accessible to everyone, including individuals with disabilities. There are escalators and accessible elevators at all stations.

Discussing four stations along two distinct lines of the Delhi Metro, the Red Line and Line 1 would be meaningful, as stations were the first in the series of Delhi Metro Stations. It is appropriate to mention here that considerations of the economy have played a part in opting for a station design system that accommodates a generic design adapted to suit the unique nature of each site, but without losing the recognisable image of the station. This is especially so as most commuters would have a definite station at each end of their commutes, and the individuality of a station enhances the comfort factor. The Delhi Metro system features a combination of underground, at-grade, and elevated stations. All stations have escalators, elevators, and tactile tiles to guide the visually impaired from station entrances to trains. Many stations are equipped for rainwater collection as part of their environmental policy. The DMRC has also deployed rooftop solar power plants at many of its stations. In Phase 1, 58 stations across three lines were built, and an additional 85 stations were constructed under Phase 2. Most of Line 2 runs underground. And stations were built approximately 13m below ground using cut-and-cover methods; Chawri Bazar, which lies roughly 20m below, required tunnelling. The Mandi House station on Line 3 is situated beneath a busy thoroughfare, with much of the station constructed top-down. The diaphragm wall panels are built from ground level to form the permanent walls of the station. Subway facilities were provided in all the underground metro stations of phase two. Bicycle rental is also available at Vishwa-Vidyalaya, Pragati Maidan, Patel Chowk, and Indraprastha Station. **Figure 9.13** includes viaduct approach to elevated station.

9.7.1 Features of elevated stations & station buildings

While the section includes only four stations along two distinct lines of the Delhi Metro, it is appropriate to mention that considerations of the economy have played a part in opting for a station design system that accommodates a generic design, which was nonetheless adapted to suit the unique nature of each site, but without losing the recognisable image of the station. This was especially so as most commuters would have a definite station at each end of their commutes, and the individuality of a station enhances the comfort factor.

As mentioned earlier for an elevated line, two considerations for a station regarding platforms are an island platform and a side platform. A straight alignment of the track is preferred for a metro facility running through a congested environment punctuated by frequent stations. For this reason and considering space availability, side platforms were chosen, especially for on-road stations.

Fig.9.13: Approach to Elevated Station

9.7.2 General design details

Line 1 (Red Line) runs between Dilshad Garden and Shaheed Sthal (New Bus Adda). The section from Shahdara to Tis-Hazari opened in December 2002, becoming the first stretch to be commissioned in Delhi. The Structures of the four stations are supported on central piers and side columns from the concourse, mainly with pile foundations. In Pul Bangash, the foundations are partly exposed due to the presence of rocky strata. Precast prestressed concrete girders are erected on cast-in-situ cross girders, which are supported by central piers and side columns. Precast girders with projecting shear reinforcement are placed contiguously, and the cast-in-situ deck slab provided on top acts as the flooring of the station concourse. The side platforms are supported on precast prestressed concrete girders placed on the pier arms with a cast-in-situ deck slab on top of the pier cap. The platform roof is provided with a steel structure **(Fig.9.14) erected** on concrete girders on either side. For faster implementation, the standardisation of structural steel sections has been considered. The sleekness of form was a significant consideration in the design of the roof systems.

Fig.9.14: Steel roof system

Stations, including buildings at Pratap Nagar, Pul Bangash, Pragati Maidan and Indraprastha, were involved in studying several alternatives. The detailed design of the stations included the viaducts within the station limits. The platform length is 185 m, which would accommodate an 8-car consist, and the station building is 67 m long. Two alternatives for Off-Road and On-Road stations were developed. The essential differences are the vertical and horizontal clearances. Entry for on-road stations to allow road vehicles to pass through.

9.7.3 Design of elevated station structures, Pratap Nagar Station

Pratap Nagar Metro Station is located on the Red Line of the Delhi Metro. It is within walking distance of Sabzi Mandi Railway Station and Kishan Ganj Railway Station, both operated by Indian Railways. It is a common metro station for residents of Pratap Nagar and Gulabi Bagh. An elevated station opened in 2003. It has side platforms, Platform 1 for Rithala and Platform 2 for Shaheed Sthal.

The structural supporting system provides space for the free flow of traffic below the concourse floor. In the concourse area, the whole floor, track and platform were supported on one central pier, on the road median and two side columns, whereas beyond the concourse, the track and the platform were supported on a single central pier.

A structural system for the different components was provided as follows:

Precast pre-tensioned I-girders with a cast-in-situ top slab supported on an RCC box (trough beam) at the concourse floor were provided. The cast-in-situ PSC box girder supported on central piers was provided for the railway track. All the box girders were simply supported on bearings. An expansion gap of 25mm was provided between the cantilever portions of the deck slab of two box girders.

Watertight compression seal expansion joints were used at each pier location. **Figure 9.15** includes roadside view of Pratap Nagar Station and **Fig.9.16** includes Metro at Station during operation.

Fig.9.15: Roadside view of Pratap Nagar Station

Fig.9.16: Metro car at Platform

Precast post-tensioned I-girders with cast-in-situ deck slabs were supported on the pier arm at the platform level. The expansion gap of 25mm, like that of the concourse level, was filled with watertight sealant between cross girders at every pier location.

9.7.4 Pul Bangash Station

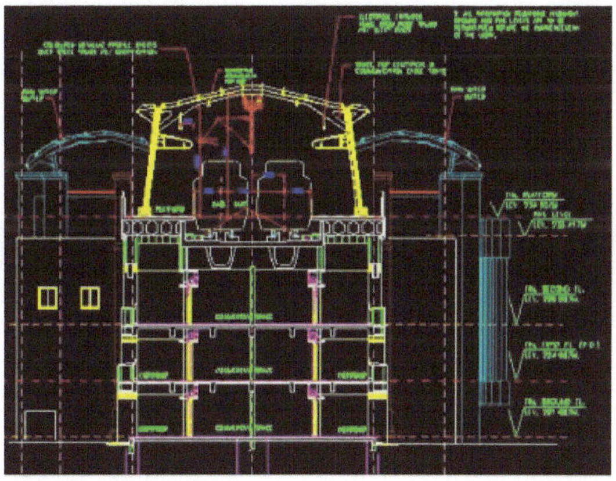

Fig 9.17: Pulbangash Station- Designed Section

Pul Bangash Station **(Fig.9.17)** is located at Aram Ganj, Sabzi Mandi, Old Delhi. An elevated station in Line 1 with side platforms.

The structural arrangement adopted in the concourse, track, and platform areas was like that of Pratap Nagar station, except as noted below.

Precast pre-tensioned twin U-girders with cast-in-situ deck slabs were provided for the railway track, and the girders at the platform level were Precast pre-tensioned I-girders.

The space on the ground and first floor for commercialising as an additional revenue stream was an essential feature of this station.

Figure 9.18 shows the Entry Structure for Pulbangash Station and a photo **(Fig.9.19)** of the station during operation. Entry Structures are important features of station design.

Fig.9.18: Entry Structure for Pul Bangash Station

Fig.9.19: Pul Bangash - Metro Car at Platform

5. 9.7.5 Traffic circulation & Architectural layout

A suitable circulation plan was adopted for the stations to minimise conflicts with the surrounding network, resulting from the dispersal of metro traffic, as well as conflicts between vehicles and pedestrians within the station area. It provided organised parking space in the station area, smooth entry/exit points from/to the Station area and access to the fire tenders. The concourse spanned across the lower-level roads.

The architectural designs were prepared, considering the stipulated criteria, including the site's characteristics, surroundings, and the stipulations of local authorities, as per the design criteria by DMRC. The consideration of design was to achieve a contemporary station character with a distinctive identity and sophisticated grandeur in tune with the image of the capital city.

9.7.7 Pragati Maidan Station

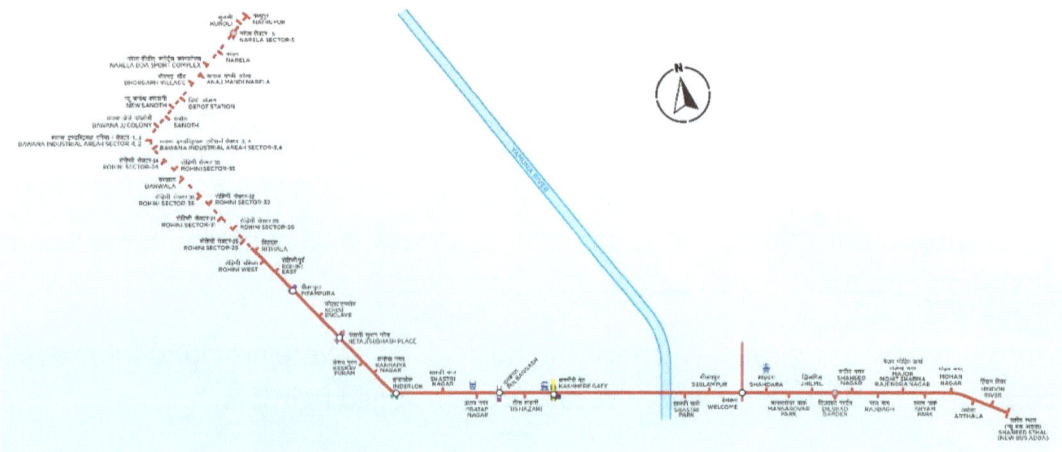

Fig.9.20: Alignment of Metro Red Corridor

The extension of Line 3 beyond Barakhmaba Road Station emerges from underground at Mandi House Station and goes elevated to Pragati Maidan Station. The station abuts an amusement park, Appu Ghar, and the Pragati Maidan, a significant exhibition centre. During the Trade Fair and the Book Fair, the region experiences a high floating population, and the metro line has helped to relieve traffic congestion in the latest edition of the Trade Fair. The anticipated footfall in the station and the potential for commercial exploitation have led to the construction of a sizable, multilevel station that accommodates a commercial complex. The Pragati Maidan station is the largest in the section, spanning over 6,800 sq metres. It had direct access via a 300-metre-long walkway to Pragati Maidan, where the India International Trade Fair is held. The roofing for the platform was made of structural steel, covering an area of 6,800 square metres. It will have direct access via a 300-metre-long walkway to Pragati Maidan, where the India International Trade Fair is held. The roofing for the platform is of structural steel.

Fig.9.21: Views of Pragati Maidan Station

Parking to meet the demand of the proposed commercial development was worked out based on the Delhi building bylaws. Parking requirements were estimated and prepared considering the broad modal split observed in the surrounding area. Shares of two-wheeler and car parking areas out of the total parking area are 28% and 72%, respectively.

9.7.8 Architectural aspects for both Pragati Maidan & Indraprastha Stations

Fig.9.22: Pragati Maidan Station – Commercial building

The elevation of the building has been conceived by creating a façade of rectangular solids and voids at different elevation levels. Structure glazing is introduced to break the monotony of the whole scheme. The external surface of the central lift is displaying the DMRC monogram.

The platform's roof has been proposed with profiled coloured metallic sheets supported on steel trusses, imparting lightness to the whole mass. For Indraprastha Station, the vertical distribution of space is done by having a platform level at the top. Durability, ease of maintenance and cost aspects have been considered for finishes. All the elements/ finishes of the station building have also been provided with the appropriate protection against vandalism

9.7.9 Traffic management for both Pragati Maidan & Indraprastha Stations

Major components addressed while preparing traffic circulation plans were lane configuration for the road network in the influence area, vehicular traffic and pedestrian movements, parking provision and circulation plans, location of bus bays and intersection control.

9.7.10 Lane configuration

The Indraprastha station area is an off-road station, and no intersection is there at the entry. The Ring Road will finally be developed into an eight-lane divided facility with service roads on both sides. The road within the station complex has a 6m carriageway to provide access for the fire tenders.

The Ring Road, through which the station is accessed, is designed for two-way traffic, allowing vehicular movement in both directions.

9.8 Indraprastha Station

Indraprastha Metro Station is located on the Blue Line. From the station, one can change the metro without changing platforms for Noida Electronic City or Vaishali.

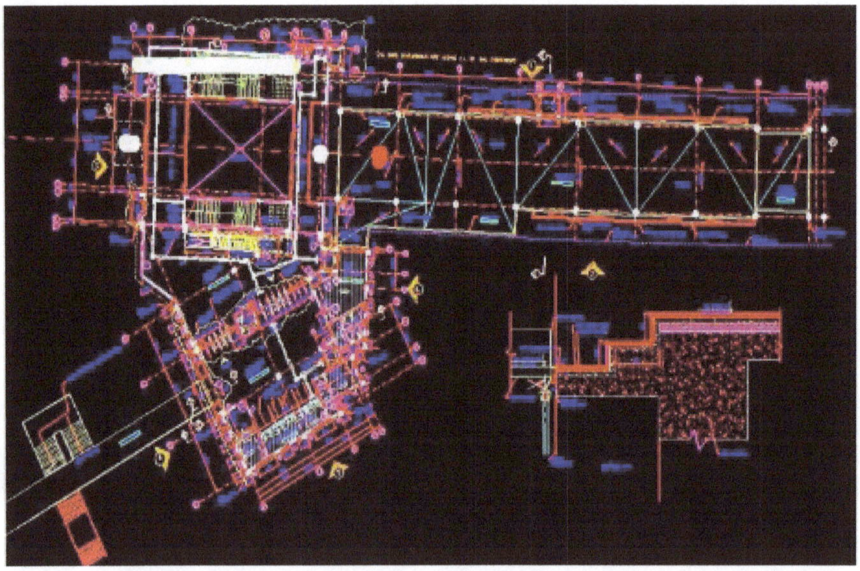

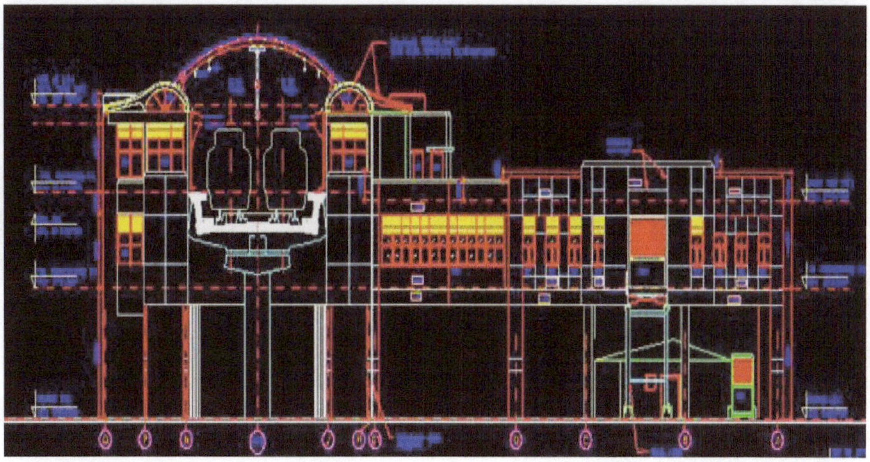

Fig.9.23: Design Drawings of Indraprastha Station

A large, open stormwater drain was located near the alignment in the Indraprastha Station area and extended westward up to the west bank of the Yamuna River. Another open drain, coming from the east, was located along this drain. In addition, water pipelines and electrical lines crossed the site.

The available space had a severe constraint on locating station facilities at the Indraprastha site.

One of the drains had been channelised through a rectangular box section to facilitate the circulation area around the station building. The distinct feature of this station was that the structure was structurally independent of the track and platform. A unique feature of the station was a foot over-bridge connecting to the station across Ring Road, the city's main arterial. **Figure 9.23** includes the design drawings of the station. The platform, including the deck, was fully covered. The roof was with pre-coated profiled galvalume sheets in combination with coloured polycarbonate sheet bands to enhance the appearance and natural lighting of the platform area. The roof is imparting lightness to the whole mass. The sheeting was supported on steel trusses, imparting lightness to the entire mass. Station operation areas were housed in a single storey building under the railway track, away from the passenger area. **Figures 9.24, 9.25, 9.26 & 9.27** include different view and a perspective view of the Indraprastha Station.

Fig.9.24: Aerial View of Indraprastha Station Area

Fig.9.25: Inner View of Indraprastha Station

Fig.9.26: Different Views of Indraprastha Station and Station Area

Fig.9.27: Perspective of Indraprastha Envisaged During Design Stage

Rithala is the elevated western terminal metro station of the Red Line of the Delhi Metro. It is in Bhagwan Mahavir Marg, Varun Kunj, Rithala, Sector 5. The station was inaugurated in March 2004. The station includes two tracks with a side platform. Platform 1 is for trains terminating at Rithala, and Platform 2 is for trains to Shaheed Sthal.It is within walking distance of Metro Walk and Adventure Island.

Fig.9.28: Rithala Station Entry and Parking Area

Fig.9.29: Rithala Metro Station

Entry to station and parking area included in **Fig.9.28** and a station view during operation in **Fig.9.29.**

The **Netaji Subhash Place** metro station **(Fig.9.30)** is an interchange station on the Red Line and Pink Line, catering to Netaji Subhash Place. The interchange, which started operating to the public in March 2018, serves as an alternative route for passengers travelling on the Blue Line, thereby reducing the crowding on the heavily congested Line. It's an elevated station with four tracks and side platforms for both the red and pink lines. Platforms 1 and 2 are used for Rithala and Shaheed Sthal, the terminal station at Ghaziabad. Platforms 3 and 4 are used for Shiv Vihar and Majlis Park in the Pink Line.

Fig.9.30: Subhas Place Station

9.9 Corridor BC 20

Contract no. BC 20, the yellow line, is an 8 km long elevated corridor with five stations, Chhaturpur, Arjangarh, Ghitorni, Sarvodya School and Sultanpur Station. This elevated corridor, featuring elevated stations, was designed by the author for CES and constructed by Simplex Infrastructure. Viaduct spans vary in length, with the longest span measuring 37 m. The viaduct was designed with segmental box sections and post-tensioned I girders. **Figures 9.31 to 9.43** include views of elevated corridors, Sultanpur Metro station, piers, pier caps completed & during construction, as built plans, sections, reinforcement details etc. and design drawings of BC 20.

Fig.9.31: Completed View of Viaduct with Pier & Pier Cap

Fig.9.32: View During Construction with Precast Box Segment Eccentrically Placed

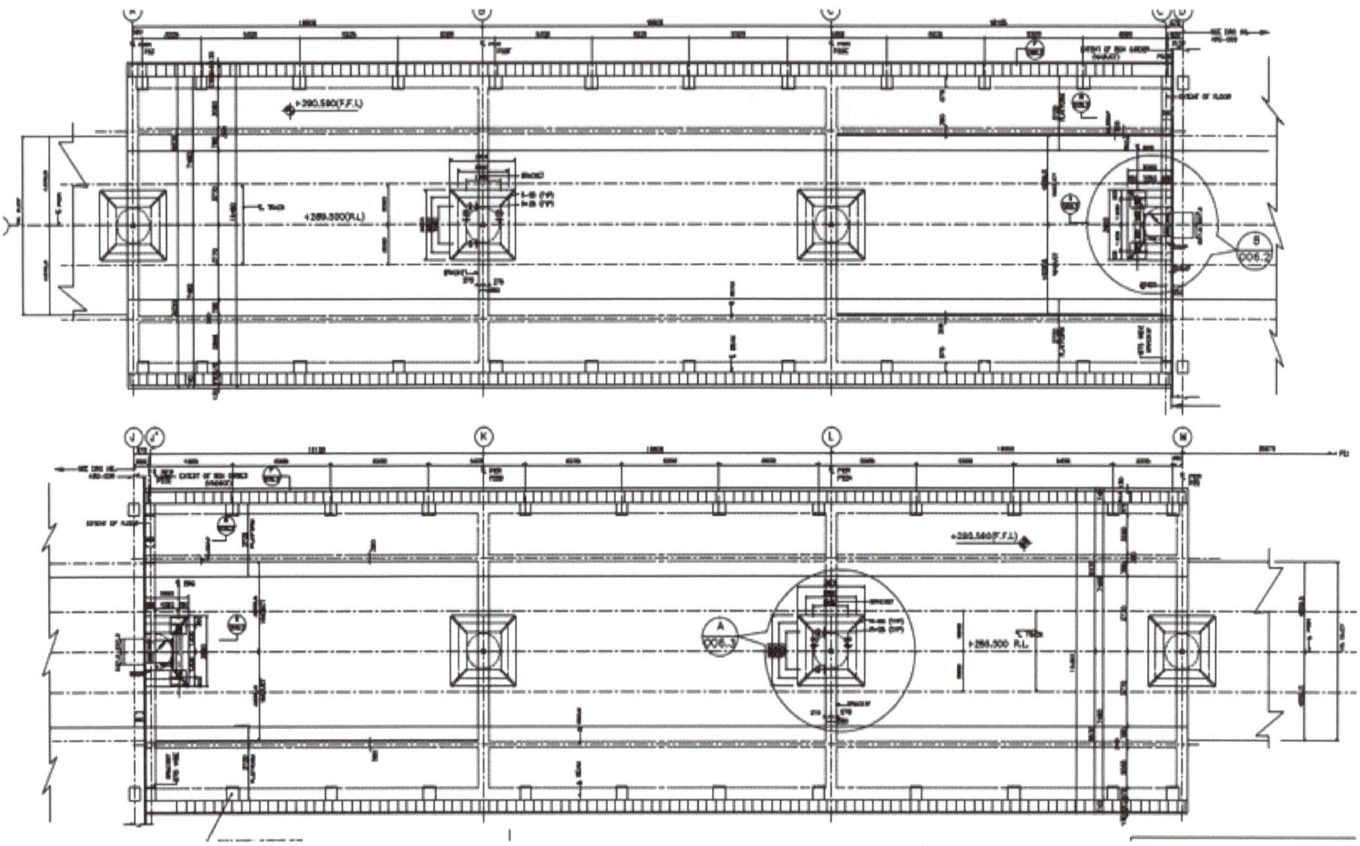

Fig.9.33: Plan of a Station Structure Viewed from Concourse Level

Fig.9.34: View of Sultanpur Metro Elevated Station at Platform Level

Fig.9.35: Completed View of Viaduct

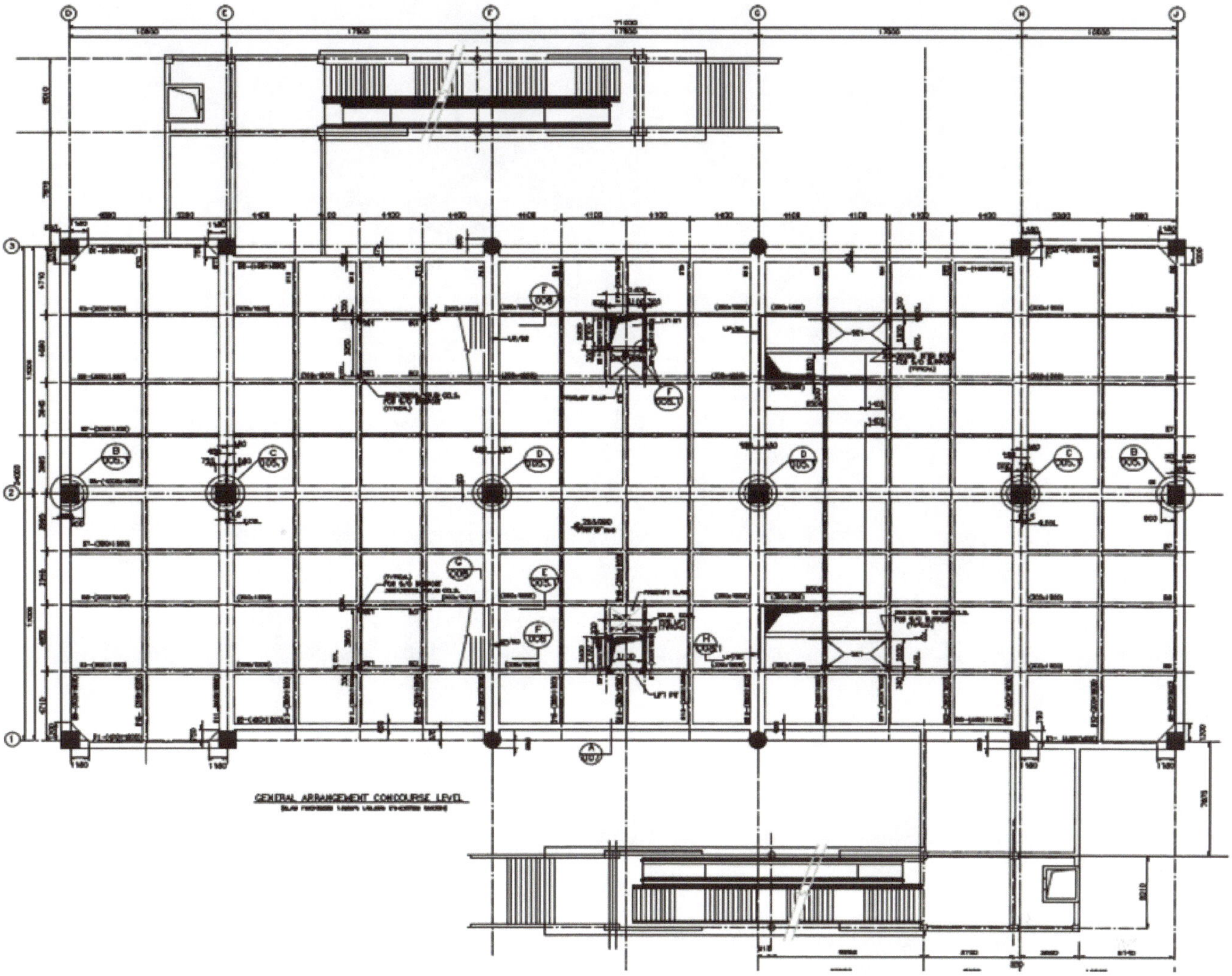

Fig.9.36: Structural Arrangement at Concourse Level (as Built)

Fig.9.37: Viaduct Constructed on Single and Portal Pier

Fig.9.38: Photo Showing Pier Segments Supported on Pier Cap

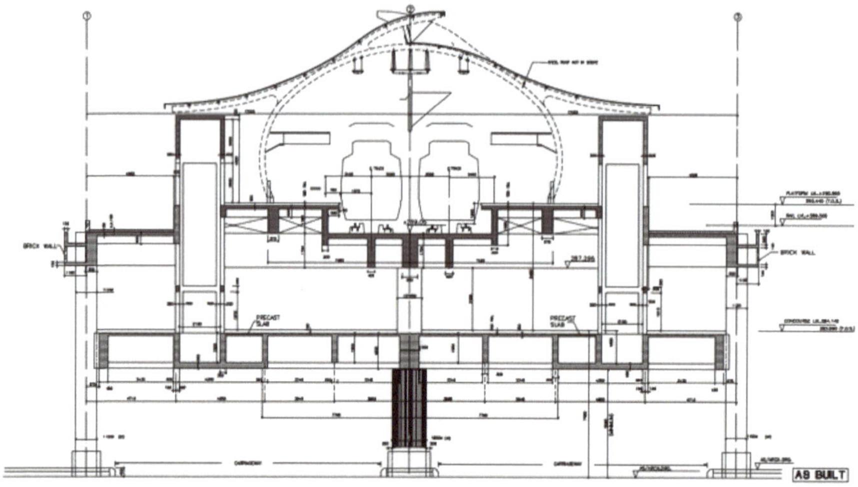

Fig.9.39: Section through Concourse, Platform and Roof (as Built)

Fig.9.40: Piers with Serrated Design to Protect from

Fig.9.41: Viaduct supported by Eccentric Pier From Postering

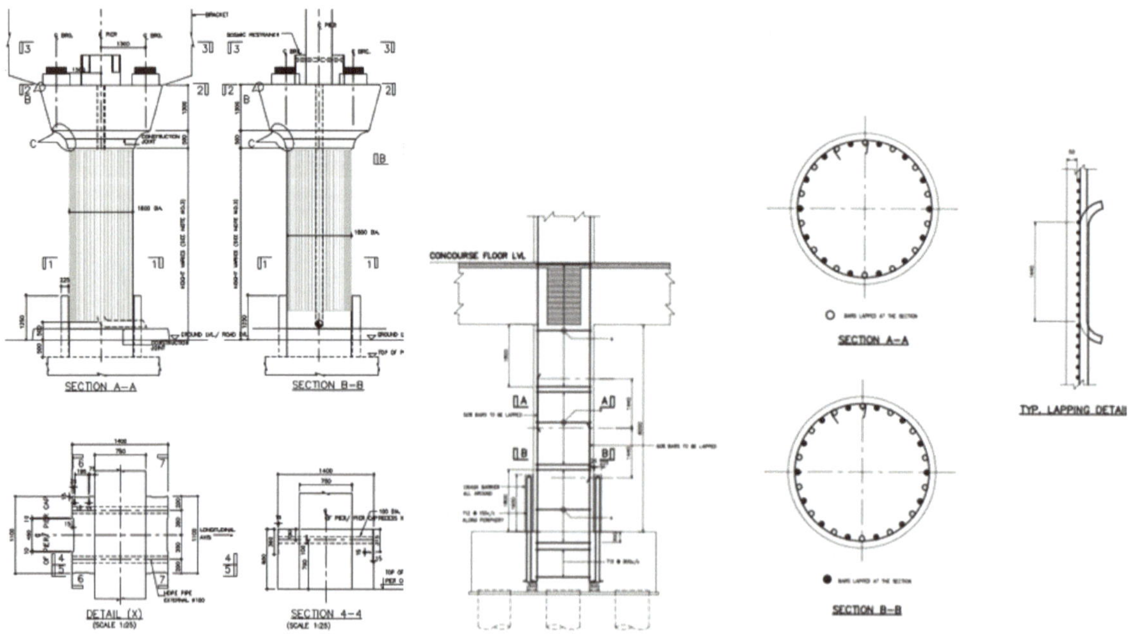

Fig.9.42: Elevation of Piers and Sections (as Built)

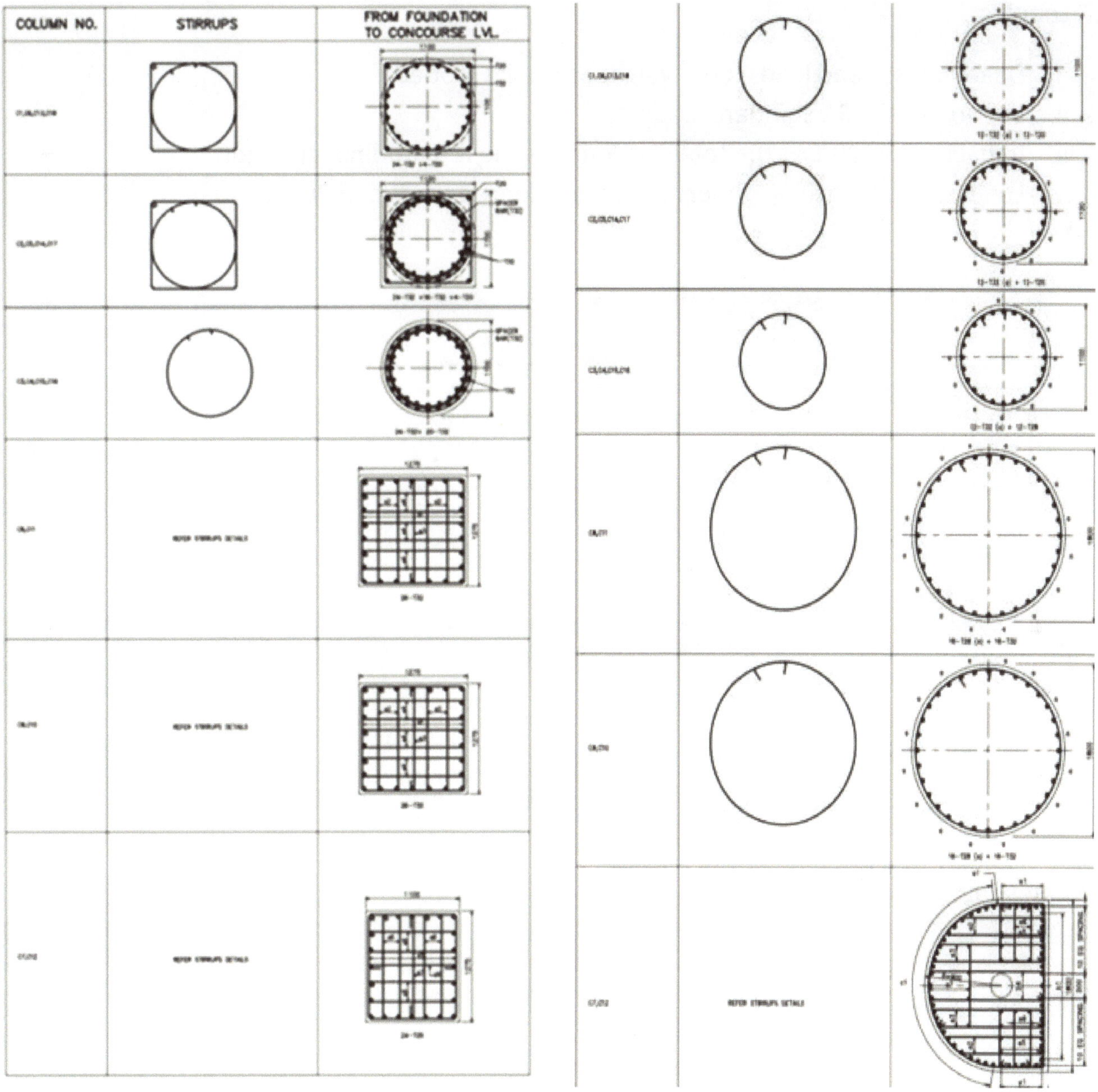

Fig.9.43: Piers and Pier Reinforcement

As built drawings of beams are included in **Figs.9.44 & 9.45** and of foundations in **Figs.9.46, 9.47 & 9.48.**

9.10 Design basis using Precast Prestressed Box Segments

The precast prestressed box segments were used for spans of 35.658 m (center-to-center of piers) and 33.958 m (center-to-center of bearings). A simply supported, single-cell, post-tensioned box girder was designed for the spans, and 19 K15 and 6 K15 cables were used for prestressing. Basic design data used were:

a. Span length (simply supported): 35.658 m c/c of piers
b. Overall width of box girder: 9.059 m

c. Loading:

   i. Dead loads: Dead loads, unit weight as per IRC code.
   ii. Live loads: DMRC standard loadings
   iii. Impact force: 20% of the total live load for the longitudinal direction.
   iv. Load due to launching girder: values considered as UDL of 30t/m.

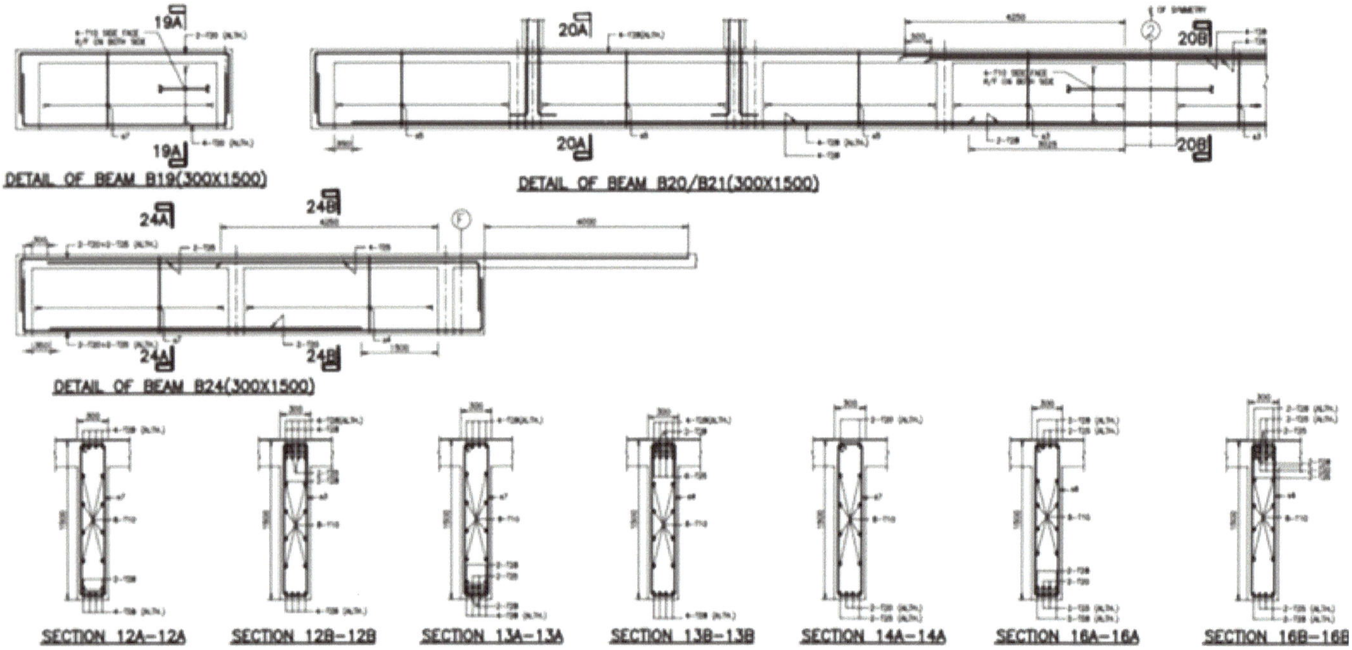

Fig.9.44: Beam Details (As Built)

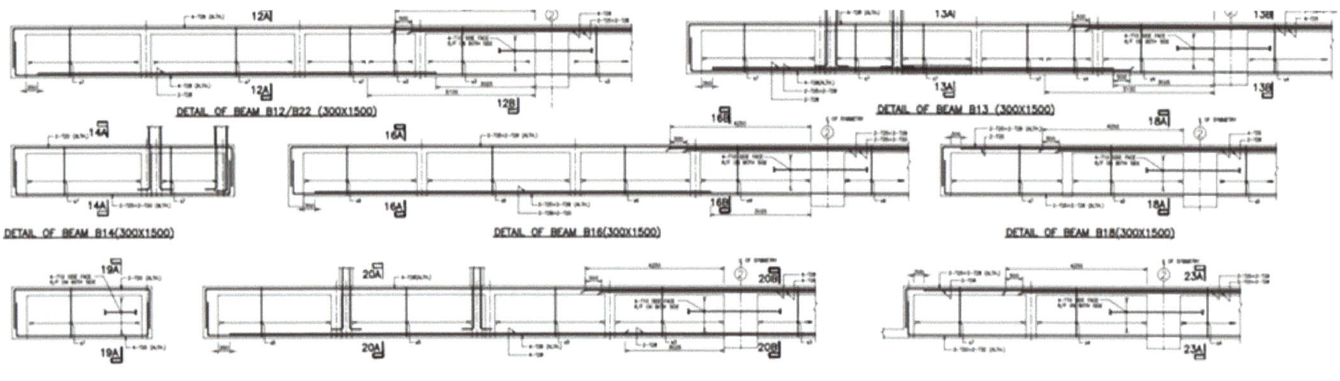

Fig.9.45: Longitudinal Sections of Beams (AS Built)

9.11 Design Basis

BM and SF, due to dead and live loads, worked out based on simple beam theory, considering the c/c of bearing as span length.

Design was based on 19 K 15 & 6 K15 cables with galvanised/metal sheathing for prestressing.

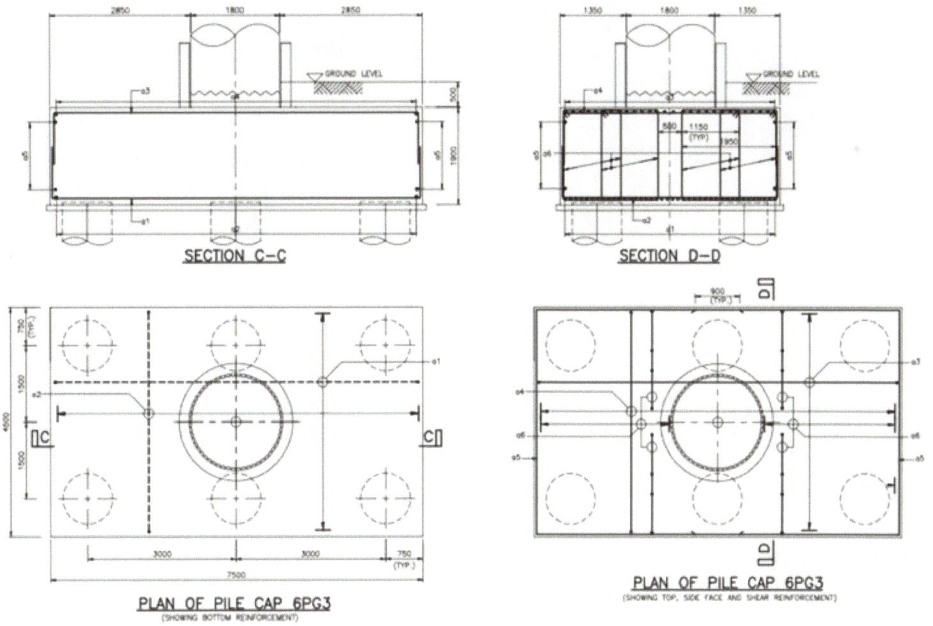

Fig.9.46: Details of Pile Foundations

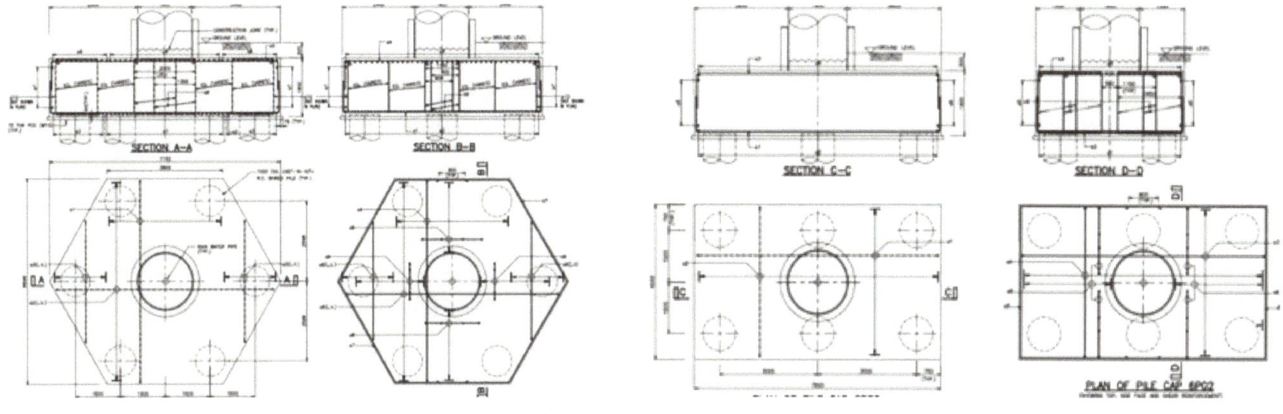

Fig.9.47: Details of Six & More Pile Group Foundations (As Built)

Values of 'friction m and 'wobble coefficients k' for prestressing strand, considered:

m = 0.20; k = 0.00328/m

Relaxation loss computed considering low relaxation steel according to 'Table 4A IRC:18-2000'.

Concrete grade box girder; M-45

Modulus of elasticity of concrete = $5700\sqrt{45}$ MPa = 38236.76 MPa

Compressive stress in concrete during service ($0.33f_{ck}$) = 0.33×45 MPa = 15 MPa

Tensile stress in concrete during service = Nil

Modulus of elasticity of steel = 1.95×10^{5} MPa

Anchorage slip = 6 mm

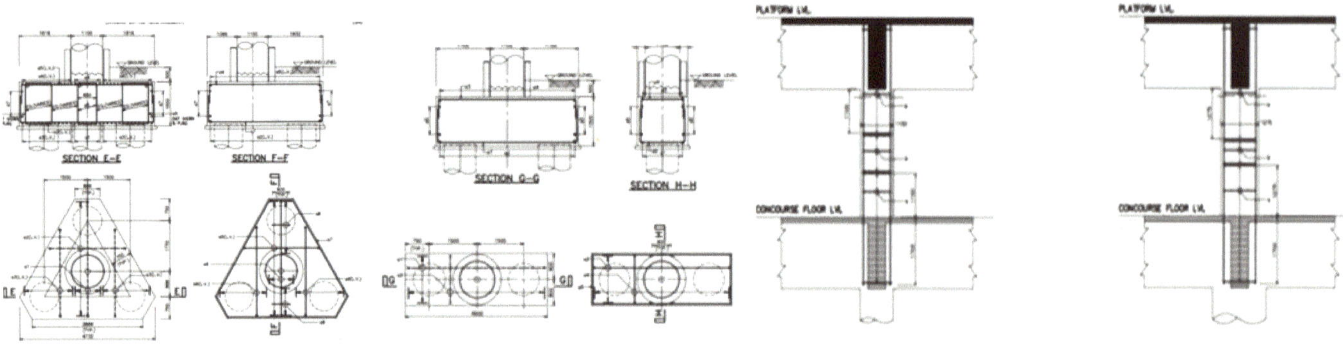

Fig.9.48: Details of Other Pile Groups and Pier & Beam Connections (As Built)

Ultimate breaking load for one cable for 19 K 15 = 506.83 t

Permissible load at jack end for 19 K 15 = 0.765 U.T.S. = 379 t

Ultimate breaking load for 6 K 15 = 160 t

Permissible load at jack end for one cable for 6 K 15 = 0.765 U.T.S. = 120t

Area of one strand = 140 mm^2

Area of one cable 19 strands = 2666 mm^2 / one cable 6 strands = 842 mm^2

Piers were generally at the central verge, shapes are Circular and Elliptical, supporting Pier Caps.

Rectangular for concentric superstructure with a segmental box girder. In cases where Superstructures are eccentric pier caps with one-sided cantilever. In some cases, two piers with rectangular pier caps are also used to reduce eccentricity.

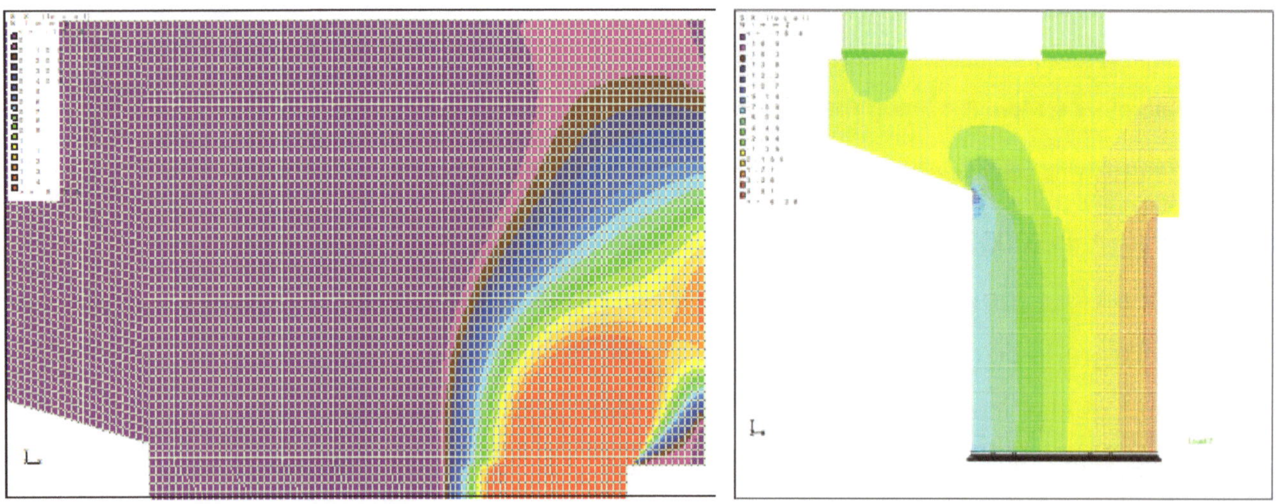

Fig.9.49: Views of Stress Distribution at Pier & Pier Cap Junction

FEM analysis **(Fig.9.49)** was carried out for the pier and pierhead in a few cases to determine the proper anchorages for the pierhead top reinforcements.

9.12 Typical Sections, Photographs to Describe Sequence of Viaduct Construction

The Delhi Metro is not only the longest metro in India, but it also utilises various types of Superstructures, including both in-situ and precast construction. Additionally, various types of Construction Equipment and approaches are employed. This section features various forms of structures and the sequence of their erection, presented through site photographs. Detailed erection system using overhead girders and erection using the underslung approach discussed in Volume 1 of 2.

Most of the Photographs are from the presentation made by Rajan Kataria, former C.E. of the **Delhi Metro.**

9.12.1 Segmental Box Casting and Erection

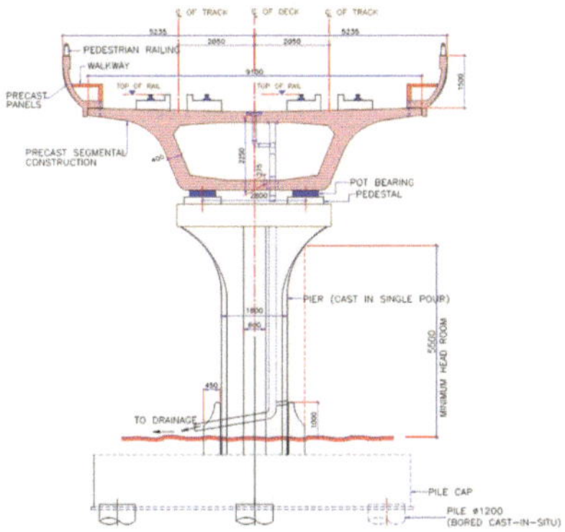

Fig.9.50: Typical Sections of Elevated Corridors

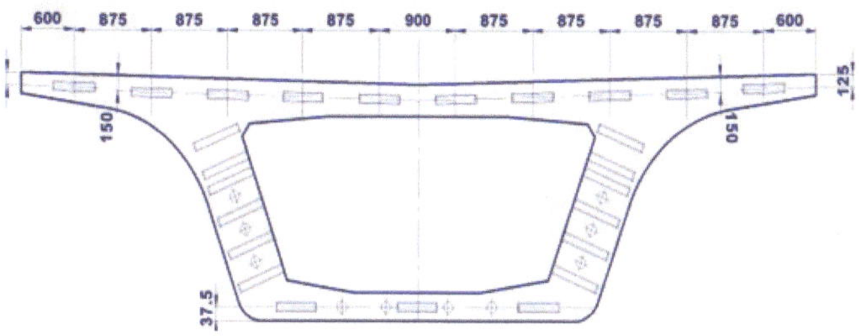

Fig.9.51: Multiple Shear Key Arrangement in Box

Fig.9.52: External Shutter and Bottom Bed

Fig.9.53: Prefabricated Reinf. Cage-lifted using Gantry Crane

Fig.9.54: Match Casting of Segments – Short Line Method

Fig.9.55: Curing of Segments using Sprinkler System

Fig.9.56: Erection of Precast Segments

Fig.9.57: Launching Girder in Position Over Temporary Structure

Fig.9.58: Temporary Prestressing inside the Box Girder

Fig.9.59: Temporary Prestressing at the Deck of the Box Girder

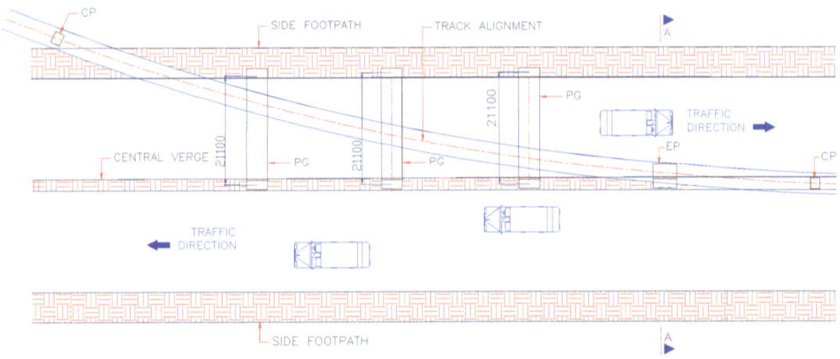

Fig.9.60: Alignment in Curve Supported on Pier Portal

Fig.9.61: View During Erection of Elements on Portal Pier

Fig.9.62: Completed Section – with Different Types of Pier Caps

9.13 Casting and Erection of U-Girder

Fig.9.63: Arrangement & Casting of U-Girder

Fig.9.64: Reinforcement of U-Girder During Fabrication

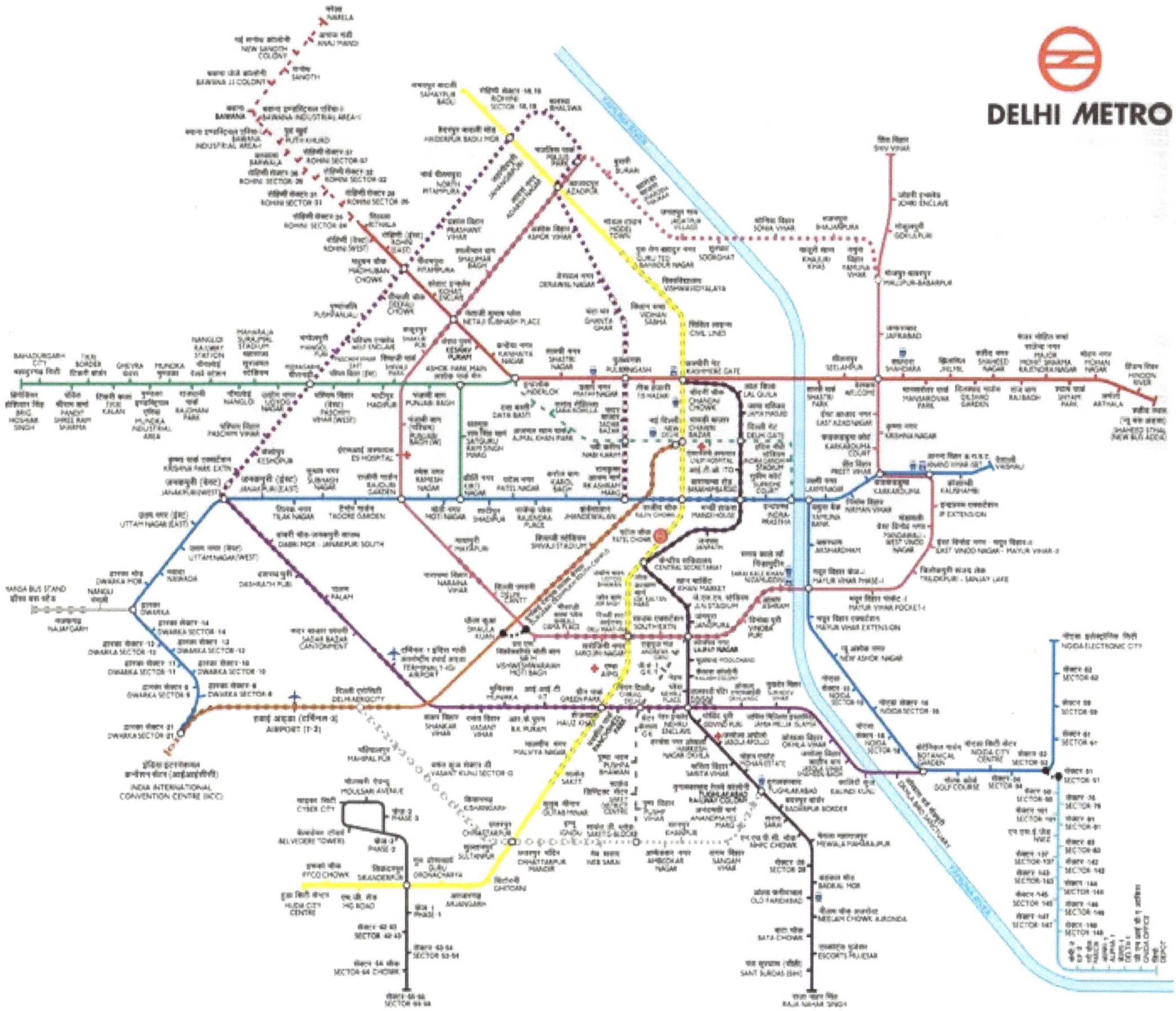

Fig.9.65: Delhi Metro Network-2025

Fig.9.66: Girder Showing Provisions for Shear Transfer & Prestressing

Fig.9.67: Stacking U-Girders at Casting Yard

Fig.9.68: U Girder-Pier Segment During Transportation

Fig.9.69: Transportation of U Girder

Fig.9.70: Erection of U-Girder in Segments

Fig.9.71: Erection Full span U Girder Prestressed at Casting Yard

Fig.9.72: Pre-Tensioned Full Span- U Girder

Fig.9.73: Precast Pier Caps for Future Use
(In use in Dubai, Taiwan and Other Places)

9.14 Other Types of Structure Used in Delhi Metro

Fig.9.74: Pretressed Box Girder- In situ Construction

Fig.9.75: Shows in-situ Construction in busy City Road

Fig.9.76: Superstructure using Cantilever Construction

Fig.9.77: Formwork for Elevated In-situ Box Girder

Fig.9.78: Shows Pier Cap During Cantilever Construction

Fig.9.79: Elevated Metro Structure over Underpass During Construction

Fig.9.80: Full Span U Girder

Fig.9.81: Transportation of Full Span U -Girder

9.15 Magenta Line (Line 8) - Engineering

Fig.9.82: Photos of Stations with Train Line 08

Ayesa, a Spanish engineering firm, was contracted to design and construct Delhi Metro Line 8 (the Magenta Line). Ayesa's involvement in designing stations included architectural, structural, and facility elements, ventilation system for underground sections. The inauguration of Line 8 in 2017 significantly reduced traffic congestion and improved connectivity to previously underserved areas.

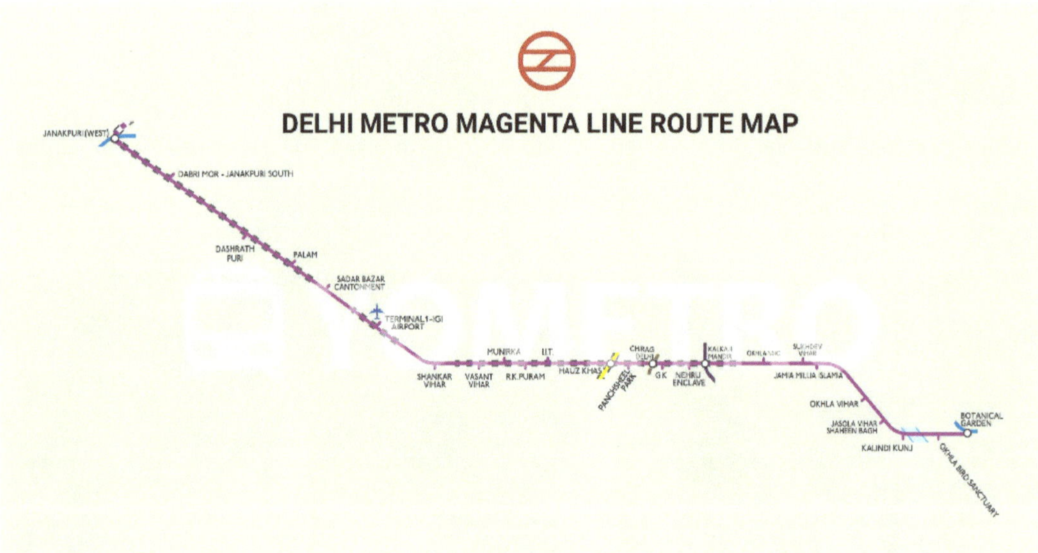

Fig.9.83: Magenta Line Route Map (YOMETRO)

The **Magenta Line (Fig.9.83),** is the first driverless metro in India. The line, which runs across South Delhi and is primarily underground, spans 40.26 km and consists of 26 metro stations: 10 are elevated, and the rest are underground. Unlike the Airport Metro Express, this line serves Terminal 1 of the Indira Gandhi International Airport.

The Janakpuri West metro station on the line contains India's tallest escalator with a vertical height of 15.6 meters. The Magenta Line has interchanges with the Yellow Line at Hauz Khas, Blue Line at Janakpuri West and Botanical Garden and the Violet Line at Kalkaji Mandir of the Delhi Metro network.

Shankar Vihar metro station is a station where civilian movement is restricted by the military. Located within the Delhi Cantonment, it primarily serves defence personnel due to its position in the heart of the defence zone. The Hauz Khas metro station on this line is the deepest metro station in the network at a depth of 29 meters. For the first time in India, the construction of two tunnels was completed concurrently at the Dabri Mor–Janakpuri South station.

Ayesa's involvement in designing stations included architectural, structural, and facility elements, ventilation system for underground sections The Janakpuri West station on the line holds India's tallest escalator. Ayesa's work involved planning, architectural design, E&M (Electrical and Mechanical), ECS (Environmental Control System), PHE (Public Health Engineering), Firefighting, BMS (Building Information System), Tunnel Ventilation System including detailed structural design of elevated station (built-only construction contract) and proof checking of underground station (design-built construction contract) – both geotechnical and structural elements.

Fig.9.84: Station Platform with Screen Door

Fig.9.85: Tunnel Section

Fig.9.86: Okhla Bird Sanctuary Station - External View

Fig.9.87: Janakpuri West Station

9.16 Kalindi Kunj – Botanical Garden Corridor -Line 8-ph 3

This Corridor is aligned across River Yamuna is a part of the Janakpuri West-Botanical Garden Corridor and runs between Janakpuri (West) to Botanical Garden, across the Yamuna River near Kalindi Kunj

(New Delhi) and Botanical Garden (Noida) as an elevated viaduct, with two stations. The length of the elevated viaduct is 2.977 km & the length of each station is 144m. **M/s S. P. Singla Constructions Pvt. Ltd., Panchkula, was the Contractor, and M/s B&S Engineering Consultants Pvt. Ltd., Noida** (BSEC) was the detailed designs and working drawings.

The alignment of the viaduct had curves of radius varying from 250m to 1500m. A vertical clearance of 6.0m from the road surface was ensured at the outer face of the protective crash barrier. The geometric design of the alignment adopted for the viaduct was supplied by DMRC. The viaduct had a total deck width of 8.844m. with standard spans of 31m & 28m and shorter spans as necessary. The standard span superstructure proposed consists of simply supported, pre-tensioned girders (3 nos..) with cast-in-situ RCC deck slab & diaphragms/cross girders supported over elastomeric bearings. A pair of RCC shear keys protruding from the pier cap transferred the lateral loads. For obligatory spans of 56m, a steel-concrete composite superstructure consisting of prefabricated steel T girders with cast-in-situ RCC deck slab was provided. As per the geotechnical investigation along the corridor, rock was not encountered down to the intended depth of the foundation. Considering the nature of the loads, a bored cast-in-situ pile foundation was adopted.

The standard spans for the superstructure comprise three pre-cast, pre-tensioned, simply supported PSC T girders with a cast-in-situ RCC deck slab. Two intermediate cross girders and two end diaphragms were introduced. The total width of the deck is 8,844 mm, which accommodates Standard Gauge tracks, along with brackets for supporting various cables, a precast parapet, and a walkway & railing on top. The precast girders were supported by permanent elastomeric bearings. The superstructure for 56m steel-concrete composite bridges at obligatory locations was designed with prefabricated steel T girders, with a cast-in-situ RCC deck slab on top, maximum compressive stress was restricted to 0.4fck and zero tension to satisfy SLS conditions. Finally, based on the prestressing provided, ultimate stage checks were carried out for flexure, shear & torsion. The reinforcement required for shear and torsion was worked out.

For transverse analysis of the RCC deck slab, a unit slice of the slab was idealised in STAAD Pro as a two-span continuous beam with overhangs resting on unyielding supports. The effective width method of IRS Bridge Rules (and IRC:112) was adopted for working out the effect of a concentrated live load on the deck. Various loads (including temperature load) were considered on the idealised frame, and analysis results obtained required reinforcement at sections of the deck slab.

9.17 Model for Substructure & Foundation – Line 8

a) Substructure

Solid circular RCC piers with long cantilever rectangular pier caps were provided. Depending on height, span length and horizontal curvature, the size of piers was 1.8m to 2.2m. For standard spans, the pier cap width and length were sufficient for placing girders, bearing pedestal, shear key, etc.

b) Foundation

Piles used were of 1.0m dia. bored cast-in-situ piles for all the spans, with a minimum of four piles under each pier. As per the geotechnical report, the subsoil strata were. predominantly sandy soil with small percentages of silt. up to 30m and some clayey layers with silt at mid & lower depths. The rocky stratum is not encountered during subsoil investigations. The top of the pile cap was kept at a 500 mm minimum below the existing bed level. To keep the seismic forces low, piers were designed as flexible. For requirements, the theoretical pile capacity was worked out based on the IRC: 78-2000 (Appendix 5).

The model used a stick model (line members) in STAAD Pro for each pier as a vertical cantilever and included pier cap, pier and pile foundation. To account for soil-pile interaction, Winkler's springs were applied at every one-meter spacing (stiffness calculated according to Annexure-C to IS:2911 Part 1/Sec 2-2010). The upper strata of soil had liquefaction potential, and modifications in the soil springs were introduced in the idealisation under seismic conditions. All primary loads, such as dead load, superimposed dead load, live load, RSI force, seismic load, etc., were applied on the model, and results obtained from the analysis were used for checking of various critical sections of the substructure and foundation. The reinforcement in the pile was designed as per the bending moment along the depth of the pile obtained.

Since the water table was assumed at ground level, the submerged weight of the pile cap was considered for load evaluation purposes. and the vertical load on a pile worked out by rivet theory (rigid body rotation). The resultant lateral force was divided equally among all the piles. Under fixed conditions, the lateral capacity of the pile was determined by the spring method and as per Appendix C of IS: 2911 (Part 2), allowing a deflection of the pile head (at the cutoff level/scour level) up to 1% of the pile diameter. The least of the two values was considered as the lateral capacity of the pile. The vertical capacity, as per the Geotechnical Consultant's recommendation and capacity under seismic conditions, was increased by 33%.

Fig.9.88: Elevated Corridor for Line 8

Initial load tests, 2.5 times of load based on a static formula, were conducted for the determination of safe vertical and lateral loads and the safe capacity was considered lower between the test result and theoretical calculation. The safe load shall be taken as the least of (a) the load arrived at from the initial load test and (b) the calculated safe load based on a static formula. The initial test shall be conducted for a load of 2.5 times the safe vertical load based on a static formula. The working load on the pile for vertical and horizontal loads was checked by conducting routine load tests during construction. Based on the load evaluated at the bottom of the pile cap, the pile cap was designed by truss analogy and by bending theory (as per IRS: CBC) for ULS condition. The top of the pile cap was kept (500mm minimum) below the existing ground level, and the weight of the earth cover shall be applied on top of the pile cap when unfavourable. Elevated Corridor of Line 8 is included in **Fig.9.88.**

c) Bearings

The superstructure was supported by elastomeric bearings under each girder for carrying vertical loads. A pair of RCC shear keys protruding from the pier cap were used to transfer the lateral loads from the superstructure due to braking/traction, centrifugal, wind, seismic, etc. The design of the elastomeric bearings (meant for vertical loads only) was carried out as per UIC-772-2R. The arrangement of the bearings over the pier cap is shown in Fig.7.4. The bearings for steel-concrete composite bridges were POT-PTFE type. The bearing design was done as per IRC:83 (Part-III) in the absence of specific provisions in the IRS Code.

Fig.9.89: View During Erection

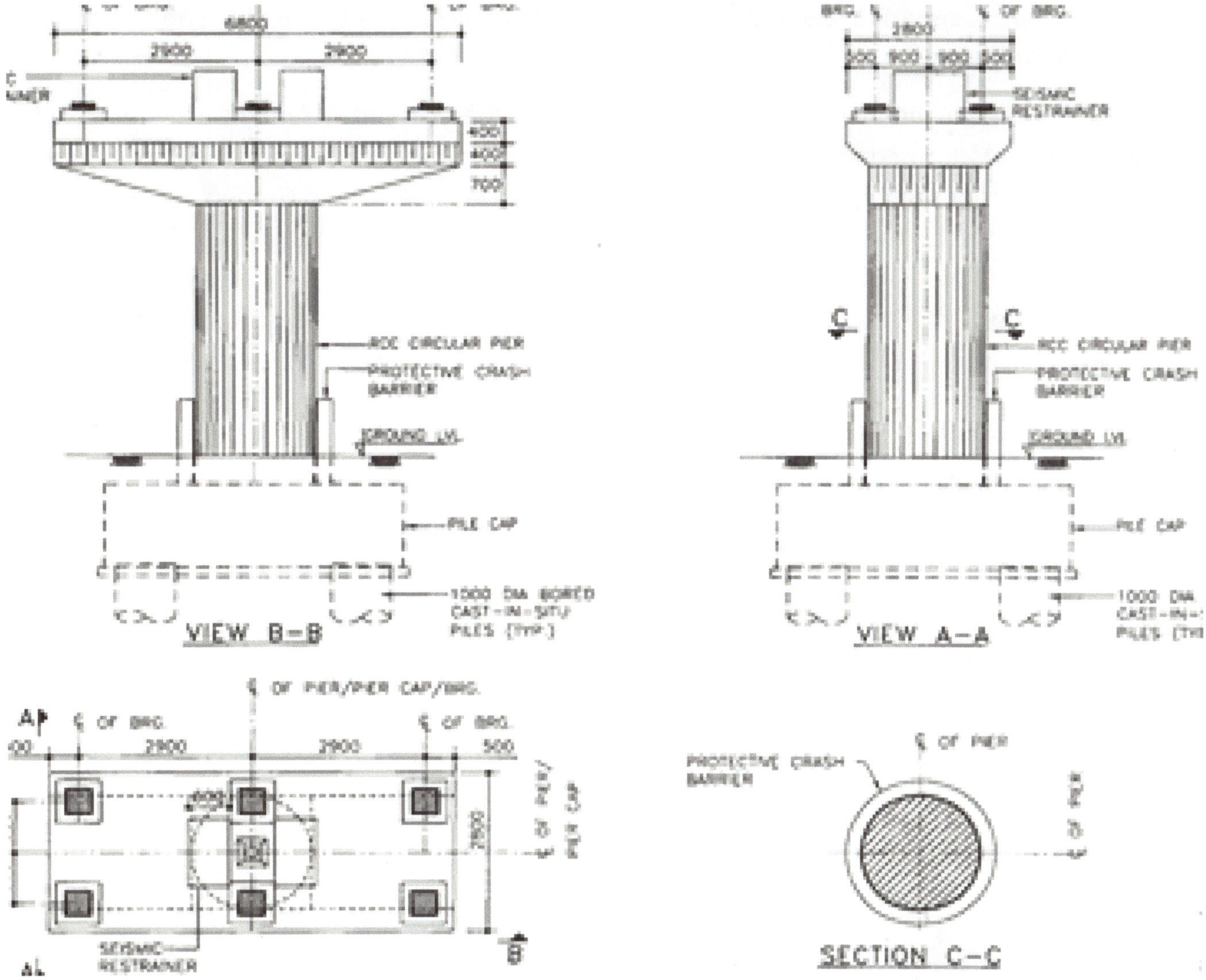

Fig.9.90: Pier, Pier Cap Plan & Elevation including Bearing Locations

The grade of Concrete used ranged from 35 MPa to 45 MPa, with cover thicknesses between 35 mm and 75 mm, except for elements in contact with nallah water, which had a cover thickness of 90 mm. Plan, Cross Section of Pier & Pier Cap including bearing locations are included in **Fig.9.90.**

One of the most challenging projects of DMRC was a part of the Mukundpur–Yamuna Vihar Corridor of Line 7 of Phase III. The corridor lies in the Eastern part of Delhi and passes through one of the busiest corridors on Road nos. 56 & 57 of the trans-Yamuna. The alignment crosses the Shahdara drain & Northern Railway main line (connecting Ghaziabad with Delhi Main station) and Metro line 1 (Dilshad Garden to Rithala). The total length of the viaduct structure is 9 km. The superstructure comprises spans ranging from 19.0m to 37.0m precast, single-cell, segmental, post-tensioned box girder in general and special spans at obligatory locations as per site requirements. The alignment of the viaduct has many curves with the sharpest radius of 200m. Solid circular RCC piers with rectangular pier caps were provided on top. Bored cast-in-situ piles have been provided.

9.18 Viaduct for Automatic Walkway / Travellators at Dhaula Kuan, Punjabi Bagh and Rajouri Garden Corridor Phase-3

Delhi Metro decided to interconnect the Metro Stations of Phases II and III. And the corridor included a skywalk **(Fig.9.91)** for pedestrian traffic with travelators, connecting the existing Phase II metro station with the proposed Phase III metro station at Dhaula Kuan & Rajouri Garden, for commuters. BSEC was assigned to do the detailed design. The total length of the skywalk was 1200m at Dhaula Kuan and 250m at Rajouri Garden. Span length varied from 8.126m to 29.414m, and a total width of 6.9m, 2 nos. parallel, Precast, RCC, U-Trough superstructure housing travellator units with cast-in-situ slab in between. The maximum height required was 18 m. The superstructure is supported by an RCC pier cap on the square-shaped pier. The structure was supported on 0.8m dia. bored cast-in-situ piles at Rajouri Garden & an open foundation at Dhaula Kuan and was covered by a PEB roof structure. The maximum weight of a precast unit was 70t.

Fig.9.91: Photos of Skywalk

The corridor is part of the Metro connection between Ghaziabad and Delhi, under Phase II. The station is an elevated off-road station, 185m long, and operates on a broad-gauge metro. The trains consist of eight coaches. The station included a concourse level where all the amenities required for operation were provided. The total width of the station was designed for 30m. The other side of the road was connected to the station through a steel foot overbridge. The height of the platform level was 13.0m above road level. The station design was divided into three segments, e.g. a central building with a concourse floor with extended platforms at either end. The entire central portion of the building was made of a cast-in-situ RCC framed structure (including track bed) supported on a 0.9 m-dia pile foundation. Provisions were made for staircases, lift shafts, escalators, underground water tanks, pump houses, etc. The extended portions of the platform covers were built with precast PSC girders with a cast-in-situ deck slab for the track bed and platform having an 18.5m span. Circular square columns were provided as per aesthetic requirements.

9.19 Karkarduma- Gokulpuri Corridor- Line 7- Ph-3

The Delhi Metro, Karkarduma - Gokulpuri corridor, part of Line 7 Mukundpur - Yamuna Vihar, spans a total length of 55 km. Karkarduma **(Fig.9.92) –** Gokulpuri **(Fig.9.93)** is an elevated corridor with six elevated stations: Krishnanagar, East Azad Nagar, Welcome, Zafarabad, Maujpur, and Gokulpuri. DMRC entrusted the construction work to **M/s JMC-CHEC (JV)** of Mumbai, with **M/s BSEC as the** design consultant. The basic plans of the stations were done by **Systra.**

Fig.9.92: Karkarduma Station.

Fig.9.93: Gokulpuri Station

The Geotechnical investigation was carried out by a separate organisation appointed by the JV. Bored cast-in-situ bored pile foundations were adopted for stations, as rock was not available at a reasonable depth. The main buildings were proposed with one-meter-diameter pile foundations. The building was designed using a grillage of cast-in-situ RCC beams supporting an RCC slab at levels as per approved plans. The foundation for columns of entry structures and other ancillary structures was with open foundations at a depth of 1.5 m of ground or finished grade level, whichever was lower. The portion of extended platforms on both sides of the central area was designed with precast, PSC, simply supported segmental box girders, like viaduct structures, for the track bed. For the platform, Precast RCC girders, monolithic with RCC in situ cross beams supporting RCC slabs, were used. Expansion joints were introduced at the extended portion. All designs were based on the approved Design Basis Report. The structure was analysed using Staad-Pro software. Different views during construction are included in **Fig.9.94.**

Fig.9.94: Different Stages of Construction of Karkarduma-Gokulpuri Corridor

9.20 Mukundpur-Yamuna Vihar Corridor – Line 7-Phase 3

The viaduct corridor, designed by **BSEC and constructed by SP Singla Constructions P. Ltd.,** includes a major portion of Delhi over Ring Road, spanning more than 55 km. The elevated corridor interconnects various running lines of the Delhi Metro and starts between the Ring Road and the Yamuna River (where the corridor changes from 'underground' to 'Elevated') and terminates before Link Road near Mayur Vihar-I. The section of the contract did not have any stations. The corridor crosses the Yamuna River (the bridge over the river is not included in this package) and runs through the flood zone of the river. The Eastern side of the river is mainly agricultural land. The corridor does not cross any major road or overhead utility.

The deck width is 8.844 m. The viaduct was designed with standard spans of 31.0m, except where a 31m span did not fit, shorter spans were adopted. The standard spans were proposed as simply supported, precast, pre-tensioned girders (3 nos..) with cast-in-situ RCC deck slab & diaphragms/cross girders supported over elastomeric bearings, and a tie bar-shear key arrangement was adopted for lateral forces. The alignment consisted of curves of radius 450m, 500m & 650m. Vertical clearance of 5.5m below was not mandatory as the corridor was on agricultural land.

The viaduct was a single arm, supporting two tracks with center-to-center distance between tracks of 4.1m. The rail level of the Standard Gauge track was fixed as per the functional requirement of DMRC. The standard span length was 31m, with nonstandard spans adopted as per site constraints. The ramp at H. Nizamuddin end started where the rail level was 1.0 m (approx.) below ground level and terminated where the rail level was 5.25m (approx.) above ground level. Based on the known history of the area, it was envisaged that rock was not likely to be encountered down to the intended depth of the foundation. Considering the nature of the loads, bored cast-in-situ pile foundations were adopted.

a) Structural Scheme-Superstructure

The superstructure was designed considering three precast, pre-tensioned, simply supported, PSC **'I'** girders with RCC cast-in-situ deck slab and two intermediate cross girders and two end diaphragms. The total width of the deck was 8.84m to support two tracks along with hangers for supporting various cables and a precast parapet (with a walkway & railing on top). The precast girders were supported by elastomeric bearings.

b) Bearings

The superstructures were supported by elastomeric bearings, and a pair of RCC shear keys protruding from the pier cap shall be used to transfer the lateral loads from the superstructure due to braking/ traction, centrifugal forces, wind, seismic forces, etc. One end of the PSC girder was provided with high tensile bars, which connected the superstructure with the shear key to provide a fixed-free arrangement of bearings.

c) Substructure

Solid circular RCC piers with long cantilever rectangular pier caps were provided. Based on height, span length and horizontal curvature, 1.8m to 2.2m dia. piers were designed. For standard spans, the pier cap width and length provided sufficient space for placing girders, bearing pedestals, shear key, etc. The Minimum seating width of the girder was to conform to clause 16.3 of IITK-RDSO Guidelines.

d) Foundation

1.0m Dia bored and a minimum of four cast-in-situ piles were provided for all the spans. The top of the pile cap was fixed at 500mm (minimum) below the existing bed level, and to keep the seismic forces low, the piers were flexible, and the pile cap was lowered suitably below ground level (wherever required). As per Clause 2.1.2 Note (1) of Employer's Requirements - Functional (Volume 3) page no. 14R (issued vide Addendum No. 1), the top 4 m layer of soil was considered as `scourable' and without soil resistance was considered. The depth of pile caps was based on the results of the subsoil investigation. The coefficient of earth pressure, K, was considered as 1.0; besides,

i. The density of earth fill corresponded to full saturation, considering the water table at ground level.
ii. The overburden shall be restricted to 15 times the diameter of the pile (for both end bearing and side friction).

In cases where rock was encountered in any borehole along the corridor, the methodology for pile capacity calculations in rock was addressed.

e) Ramp

The ramp, where necessary, was designed as an RCC box-type structure resting on an open foundation.

Fig.9.95: Elevated Station (Kaushambi) on Extension Project up to Vaishali (Ghaziabad)

Fig.9.96: Bank-Anand Vihar Corridor of Delhi MRTS

9.21 Elevated Corridor between Indraprastha and New Ashok Nagar

Fig. 9.97: Viaducts and Ramp on H. Nizamuddin – Mayur Vihar Corridor

The Elevated corridor, 5.5 km long, included 167 Spans with span lengths varying between 10 m and 34 m. The standard spans were 31m, simply supported, with PSC I Girders and an in-situ deck supporting both tracks, resting on pile foundations that included piers and pier caps.

Fig.9.98: View of Station During Construction

Fig.9.99: Corridor During Construction

9.22 Chawri Bazar Station – FEM Analysis and Design

Chawri Bazar **(Fig.9.100)** is the second-deepest metro station, after Hauz Khas, on the Delhi Metro. It is located approximately 25 metres below ground level. It is located close to the monuments of Jama Masjid, the largest mosque in India and the Red Fort, a UNESCO World Heritage Site. Major stations, including Old Delhi and New Delhi railway stations, are located near Chawri Bazar. The station is 25m deep, includes two tracks on an island platform **(Fig.9.101)** with 25 kV 50 Hz AC through overhead catenary and opened in July 2005.

Due to the challenging geotechnical features, the Design and construction process was quite involved. Analysis of structure was carried out using the Finite Element Approach.

Fig.9.100: View showing Metro Train at Chawri Bazar Station Platform

Underground metro stations are subjected to earth pressure, hydrostatic pressure, bedding spring stiffness and backfill soil cover. Constructability and sustainability are essential and can be achieved through critical analysis and design. The effect of variations in design parameters, such as bedding spring stiffness and earth pressure coefficient, can be determined through a detailed analysis. and lower bound parameters for finite element analysis. This section includes practices using finite element modelling, including mesh configuration, loading, geometry, support arrangement, analysis that requires parameters, solvers, and design idealisation of critical forces, bending moments, shear forces, and axial forces at wall slab and slab column junctions.

The load combinations are analysed, considering unidirectional compression- bedding springs. For the ultimate limit state (ULS), the strength criteria are evaluated to ensure the safety of users and structures. The Serviceability Limit State (SLS) checks are carried out to assess deformations that affect appearance, vibrations causing discomfort, and cracking that impact the durability and functioning of the structures. For the metro structures, as shown in **Fig. 9.101**, the plate elements are subjected to out-of-plane loading, and a nonlinear analysis is performed to determine the load effect. Concrete slab monolithically cast with a supporting wall; the relevant moments in the slab are those at the face of the supporting wall, where a failure mechanism occurs.

The metro stations are designed to meet crack-width criteria due to exposure class, durability criteria, and a design service life of 120 years. Proper analysis and design assessment may achieve optimisation, constructability, and sustainability. The analysis and application presented here are based on Eurocodes and Indian codes. 3D Finite element models are developed and analysed using Autodesk Robot Structural Analysis software (Version 2018). The Geometrical Parameters for a cut-and-cover underground metro station consist of three levels: the entry level, the concourse level, and the platform level, as shown in **Fig.9.101**, based on functional requirements. The columns are generally placed at an 11m spacing.

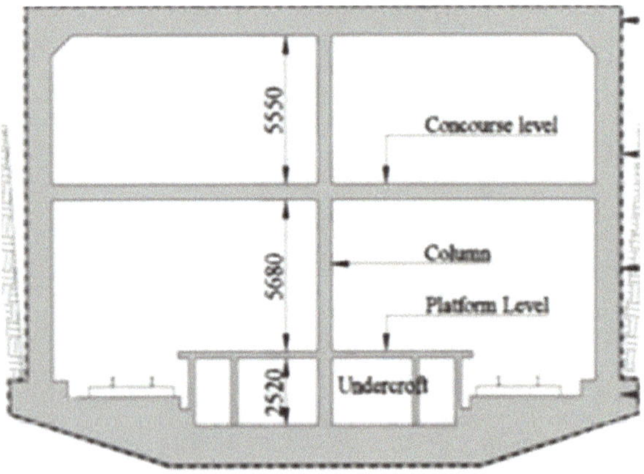

Fig.9.101: Typical cross-section of Station

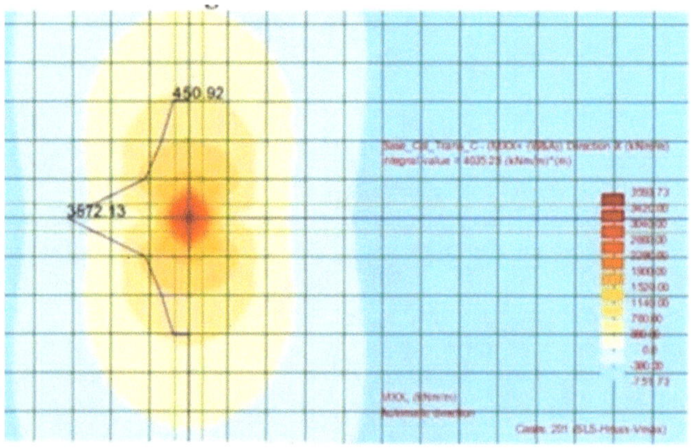

Fig.9.102: Mesh Size of 0.5m x 0.5m

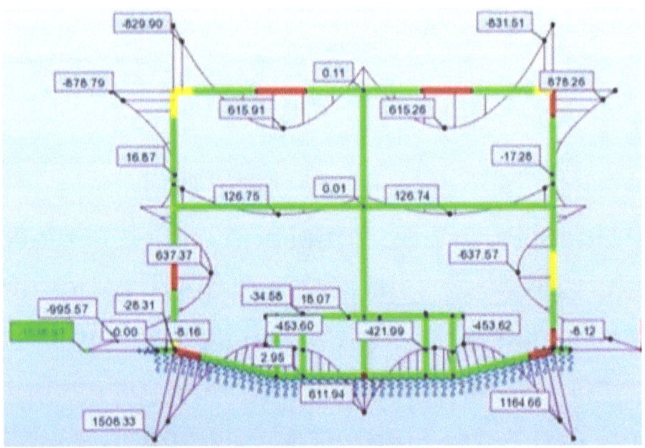

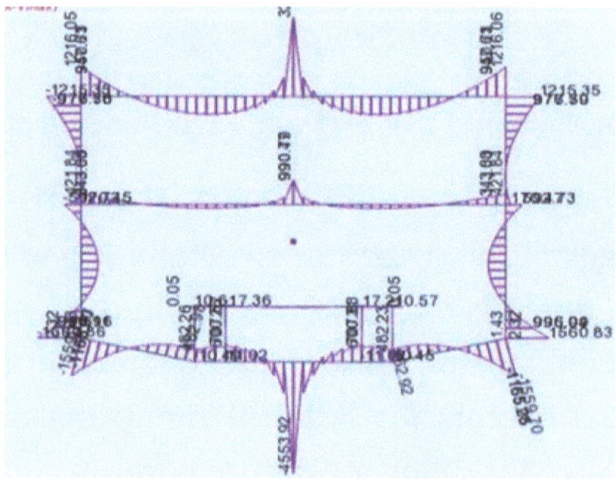

Fig.9.103: 2D Model – Bending Moment

The 3D model shows realistic behaviour based on the element and applied load, resulting in two-way behaviour along transverse and longitudinal directions. The wall and slab elements are treated as plate elements, whereas column elements are treated as bar elements. The plate elements are discretised with mesh sizes of 1.0 m × 1.0 m, 0.5m × 0.5m, and 0.25m × 0.25m. It is noted that a 0.5m x 0.5m mesh size shows acceptable results and variation **(Fig.9.102).** A Simplified 2D model is prepared for a unit length and analysed for the same loading to compare results. The bending moments at the critical location are worked out and discussed. The rotational stiffness of the wall provides a reasonably accurate support condition for the slab. Loading Parameters For underground structures, governing design loads are earth pressure (EPh), water pressure (WP), backfill (EPv) along with typical basic loads such as self-weight (SW), superimposed dead load (SIDL), live load (LL), traffic surcharge (Ts), train load (TL), building load surcharge (Bs). Backfill as per soil cover on the roof slab, Water Pressure (WP) Variation in the groundwater table, superimposed dead load (SIDL), concourse and platform Live Load (LL), above and adjacent at-grade building surcharge (Bs)

In the case of FE centerline modelling, this results in larger spans compared to the clear span, which increases the loading and consequently the bending moment and other forces.

Fig.9.104: Views, one at the entry and the other at the platform Level

The 3D FE model is analysed with a static-nonlinear analysis, considering the support nonlinearity of compression-only bedding springs. The structure includes non-linear elements such as unilateral compression-only support; calculations are performed using an incremental method in Autodesk Robot Structural Analysis Software. In the incremental approach, the structure is modified once the state of equilibrium for the previous increment is achieved. The norm of unbalanced forces is specified for each step, allowing for monitoring of the structure's force-deformation relations. Structural Model Validation is conducted through geometrical and loading validations.

The idealisation is presented as:

i. the bending moment in the face of support is either idealised as the square ratio of spans or according to the structural model. The bending moment at the face of the monolithic junction shall not be less than 0.65 of the fixed bending moment at the center of the support (EN 1992-1-1, 5.3.2.2.2).
ii. Slab and column junction – The bending moment at the center of support shows a peak. The bending moment above the column shall be averaged over the width of (b+2d), where b is the width of the column and d is the effective depth of the slab. The bending moment at the face shall be a maximum of the average bending moment at the face or (Mc-Pb/8), as shown in **Fig.9.103**.

Underground metro stations are structurally designed for a service life of 100-120 years. The reinforcement cover and grade of concrete shall be determined based on the exposure class and durability assessment. Based on Eurocodes, the structures are designed for four design situations: permanent, transient, accidental, and seismic cases. Design situations for ULS based on EN 1990-2002+A1:2005 Design Situation Verifications, Persistent Service stage ULS and SLS, Transient Temporary conditions during construction, maintenance or repair. The ultimate limit state (ULS) design includes a check for strength criteria to ensure the safety of users and structures.

For ULS, the bending and shear checks are carried out for critical load combinations. The Serviceability Limit State (SLS) checks are carried out to assess deformations affecting appearance, vibrations causing discomfort, and cracking impacting the durability and functioning of the structures. SLS load combinations are mainly classified as characteristic

The following conclusions are derived. The underground stations must be modelled as finite element 3D models for realistic behaviour compared to 2D models.

i. The walls, columns, and slabs should be included in the FE model to obtain rotational Stiffness at the joint and the corresponding moment at the joint. The joints should be modelled with rigid links.
ii. The mesh size of 0.5m is the optimum size to derive a smooth variation in results.
iii. The design bending moment at the face of the monolithic slab wall junction should not be less than 0.65 of the fixed bending moment at the center of the support.
iv. The design bending moment at the face of the slab column junction should not be less than Mc-Pb/8.
v. The extent of loads must be applied on the clear span, while the point load and corresponding moments can be applied at the support for the extent of the load over the junction thickness. This showed about a 6% reduction in bending moments for shallow underground metro stations.
vi. Euro-codes allow 10% lower load factors for the ULS condition compared with the relevant Indian. Code.

9.23 DC-09- Underground Section

It runs from Kishangarh to Saket, covering 6.51 kilometres. The package runs on 4 TBMs deployed by L&T. It was awarded to L&T to be completed within a 42-month deadline, equivalent to 3.5 years. This has four stations, which are Chhatarpur, Chhatarpur Mandir, IGNAOU and Neb Sarai. It comprises the design & construction of twin tunnels (for both up and down lines) **Fig.9.105**, utilising Shield Tunnel Boring Machines (TBMs), a Cut and Cover Tunnel box, an Underground Ramp, and the establishment of four Underground stations, namely Chhattarpur, Chhattarpur Mandir, IGNOU, and Neb Sarai. The project, implemented by L&T, also involves creating entry and exit points, as well as connecting subways linking Aerocity to the Tughlakabad Corridor, as part of Phase 4.

Fig.9.105: TBM Breakthrough at Kishangarh

Kishangarh Jat lies in the Southwest district of Delhi and, like the rest of the city, has a semi-arid climate with a significant difference between summer and winter temperatures. The soil of Kishangarh Jat consists of sandy loam to loam texture. The water level has gone down in the recent past, hovering between 45 m to 50 m. **Larsen & Toubro (L&T)** recorded a first breakthrough for its TBM at Kishangarh Station's Crossover for the 23.622 Km Silver line. The TBM was commissioned in April 2023 from Chhatarpur station's west end to build a 1274 m tunnel, in phase 4, for the down line in 8 months.

The major design parameters, in general, are adopted to plan, analyse and design the Metro Rail Underground Tunnels are:

i. Characteristics of rock/soil mass and collection of engineering geological databases from surface exposures and boreholes.
ii. Rock mass/soil classification and identification of potential failure modes.
iii. Structurally controlled and gravity-driven failures and stress-induced, gravity-assisted failures.
iv. Evaluation of static and dynamic possible failure modes.
v. Determination of in situ stress field in the surrounding rock/soil mass.
vi. Checking of shear strength to the potential failure surface.
vii. Calculation of the factor of safety or risk of potential failures.
viii. Analysis of the size of overstress zones around excavations
ix. Determination of support requirements.
x. Non-linear support-interaction analysis to design support.
xi. Design of support, considering excavation sequences, availability of materials and cost effectiveness of the design.
xii. Installation of support with strict quality control to ensure correct bolt and cable lengths, anchoring, tensioning and grouting, and effective Shotcreting as required.
xiii. Monitoring of excavation and support behaviour to validate design and permit future design modifications.

Design Methods are divided into three main categories for Metro Rail Tunnel Linings. One among them is Direct Loading Method. This method calculates the load based on the tunnel's rock/ soil depth using the finite element method. This load is independent of the rock/soil-structure proportional stiffness. In this method, the restraining effects of the rock/soil around the structure can be modelled by Winkler springs (compression only). In cut-and-cover sections, considering the total weight of rock and soil on the tunnel crest, this method is realistic for determining the internal forces on the tunnel.

9.24 Design of Underground Metro Stations

The design of the underground corridor, employing the cut-and-cover technique, is like that used in other metro systems in the country. The cut–and–cover method was used in the Kolkata North-South Metro, the first metro in India. The Delhi Metro also used the same approach and adopted both bottom–up and top-down strategies. Delhi Metro deals with the analysis and design of typical underground metro station structures in two types of analysis. One is the construction stage analysis, and the other is the permanent stage analysis. Construction stage analysis is carried out using WALLAP software, and permanent stage analysis is typically performed with STAAD Pro software. The retaining structure used is typically a Diaphragm wall. Calculate the maximum moment and shear force in the active and excavated face of the diaphragm wall in both construction and permanent analysis results. Taking the maximum moment from the two results to design the diaphragm wall. The load combinations are provided as per IS:456, the ultimate limit state and the Serviceability stage. The various components of the underground metro station box **(Fig.9.105)** are as follows. • Diaphragm wall • Roof slab • Concourse slab • Base slab. Designing the roof, concourse, and base slabs using STAAD Pro software.

Fig.9.106: UG Saket Metro Station View at Platform Level

Diaphragm walls (DW) are generally reinforced concrete walls constructed in the ground using slurry technique. These structures can be used for depths of up to 100 meters and for both temporary and permanent applications. For permanent structures, it can be used either as a semi-integrated or fully integrated system. The semi-integrated and integrated systems have been discussed in detail in volume 1 of 2 and have been used in the North-South Metro, Kolkata. The advantage of DW over other shoring structures is that it has fewer joints and can go to greater depths. It improves watertightness.

The roof slab of the Station Box is below ground, and the topmost section of the underground station box. Above the roof slab, the soil backfill is placed. The concourse slab is placed below the roof slab. The base slab is the bottom-most layer of the underground metro station box. The base level holds the tracks laid over the plinth concrete. The station box is analysed and designed, considering the entire structure. The diaphragm wall, roof slab, concourse slab, and base slab are designed based on the moment and shear force results from WALLAP and STAAD Pro, with the maximum value compared to the independent design of the diaphragm wall.

Systems

9.25 Permanent Way

9.25.1 Choice of Gauge

Standard Gauge (1435mm) is invariably used for metro railways worldwide due to its inherent advantages. During the last decade, 20 new metros have been constructed in various cities around the world. All these metros have gone in for Standard Gauge, even though the national gauge for mainline railways in some countries were different from Standard Gauge. In India, the national gauge is Broad Gauge (1676mm)

9.25.2 Track Structure

Track on Metro Systems is subjected to intensive usage with very little time for day-to-day maintenance. Thus, the track structure must be selected so that Metro Systems should be long-lasting and should require minimal or no Maintenance is performed to ensure the highest level of safety, reliability, and comfort, with minimal noise and vibrations. Ballastless track with continuously Welded head-hardened rails have been proposed as the mainline track in elevated and underground stretches. However, for the at-grade section and at the depot, the track structure shall be ballasted.

Technical details and design parameters of the Delhi Metro track structure are briefly given below. Delhi Metro uses both Broad Gauge (BG) and Standard Gauge (SG). Accordingly, the metro system uses a track structure compatible with both gauges.

Broad Gauge Track, tracks spaced 4.10m apart, lower compared to what the Indian Railway uses. This is primarily due to the Metro system operating in a closed-door system. The steepest grade permissible is 3 %. Curves of radius used 450m

Mostly, but not lower than 300m. Gradients at stations are not steeper than 0 1% (i.e. 1 in 1000). The vertical curve at the change in gradient has a radius of 2500 m. Track structure uses 60 kg.

UIC long welded rails on the level track and supported on concrete sleepers with elastic fastening. Points and Crossing are in 1 in 12 and 1in 8.5. To enhance the service life of rails, on sharp curves and in steep gradients 60 kg head hardened rails have been used. To reduce track maintenance, ballastless tracks have been used in elevated and underground structures, whereas tracks at grade and in Depot ballasted tracks have been used.

9.26 Rolling Stock

The Metro uses rolling stock of two different gauges. Phase I lines use 1,676 mm (5.499 ft) broad gauge rolling stock, while three Phase II lines use 1,435 mm (4.708 ft) standard gauge rolling stock. Trains are maintained at seven depots at Khyber Pass and Sultanpur for the Yellow e, Mundka for the Green Line, Najafgarh and Yamuna Bank for the Blue Line, Shastri Park for the Red Line, and Sarita Vihar for the Violet Line.

Rolling stock is the main component of a Metro System. Metro coaches are lightweight and air-conditioned for comfort. Coaches are made of Stainless Steel, equipped with three-phase AC motors. A microprocessor-based Integrated management system monitors the coach's system. Coaches are 22 m in length, 3.2 m in width, and 3.2 m in height. In the Delhi Metro, rolling stocks were different for broad gauge and standard gauge. Coaches are equipped with an automatic door closing system, an emergency brake, an emergency evacuation facility and an emergency power system.

Rolling stock supplied by two Manufacturers. For Phase 1, rolling stock was supplied by Hyundai Rotem, Mitsubishi Corporation, and the MELCO Consortium. The coaches were manufactured in Hyundai Rotem and later by BEML through technology transfer. The trains consist of lightweight, stainless-steel coaches with vestibules. The vehicle features microprocessor-controlled brakes and secondary air suspension and has an average speed of 20 mph (32 km/h).

The train is extensible with eight coaches.

For Phase II, rolling stocks supplied by Bombardier Transportation and trains were manufactured in Gorlitz, Germany & Sweden. The trains were of six coaches with a capacity of 1178 passengers. The coaches were provided with CCTV cameras with 8 hours of backup.

DMRC metro coach cost Average per coach cost at mid value of the contract (August 2009) of RS 2 stock (3.2 m BG stock) is Rs. 7.88Cr (Year 2007), and that for RS 3 stock (2.9 m wide SG stock), it is Rs. 9.996Cr with taxes (Year 2007). The cost of Phase III rolling stock, including taxes & duties, for which the orders were placed in April 2013, is Rs 9.86 Cr. The prices per coach of Rs 10, with deemed export benefits of 13%, amount to Rs 8.58 Cr. Notably, in April, the cost per coach of rolling stock for all DMRC phases includes spares. DMRC stock of Phase III is like BMRCL in performance requirement and will provide acceleration of 1 m/sec2 with 67% powered axle, but its coach width is 3.2 m., as opposed to 2.88 m of BMRCL

9.27 Traction Power

The Delhi Metro Rail Corporation (DMRC) rolling stock utilises 25 kV AC traction with a propulsion system capable of regenerating energy on its network. A typical rake used by DMRC consists of two units, each comprising a Driving Trailer (DT) and a Motor (M) Car.

9.27.1 Brief on Traction Power System

The selection of a proper traction system has a significant impact on capital costs, operational costs, traffic growth, operational flexibility, and the system's future expandability. It is also linked to the ultimate capacity being planned and the technology available at the time of planning. The appropriate selection of the traction system at the design stage is crucial for achieving the optimal performance of a metro system.

There are 184 Metro systems worldwide. However, traction systems in use on rail-based transit systems in India are as follows.

i. 7,50 V DC third rail – Metros at Kolkata, Bangalore
ii. 1500 V DC third rail
iii. 1500 V DC with OCS
iv. 25 kV ac with 50/60 Hz – Delhi, Chennai, Jaipur, Hyderabad

The most widely used traction system at the Global level is 750 V DC third rail. In the DC third rail traction system, the following three types of current collection systems are prevalent worldwide:

i. Top current collection system
ii. Side current collection system
iii. Bottom current collection system

In India, the Kolkata Metro, which commenced services in 1984, has adopted a 750 V DC traction system with a top-contact third-rail design. In contrast, the Bangalore Metro, which commenced services in 2011, has adopted a 750 V DC system with a bottom contact design for its third rail.

To cater to higher traffic requirements, the trend shifted towards a 1500V DC catenary system, especially after the 1970s. The third rail, with 1500V DC, was not considered safe earlier, and experiences in a few metros were not satisfactory. Therefore, the overhead catenary was preferred with 1500V DC traction voltage. A 1500V DC catenary system requires a larger tunnel diameter as compared to a 750V DC third rail.

25 kV AC with OCS (rigid/flexible) is being used in the Delhi Metro (Flexible OCS). The decision was made at the time of selecting the traction system for DMRC Phase I. The Finding was "The capital cost for the traction system and the rolling stock for 1500 V dc and 25 kV ac has been worked out

for the Shahdra-Barwala section. It is noted that the initial capital cost of a 1500 V DC traction installation would be 112 Crores higher than that of a 25 kV AC traction system. In 2005, 132 coaches are required for this corridor to meet the revised traffic requirement of PHPDT 30,000 from Shahadra to Pitampura and PHPDT 15,000 from Pitampura to Barwala. The 1,500 V DC rolling stock will cost approximately Rs 53.5 Cr less than the 25 kV AC rolling stock. The overall capital investment in the 1500 V DC traction system would have been approximately Rs 58.5 Cr higher than that of the 25 kV AC traction system, including traction installation and rolling stock. As regards maintenance and operation cost of traction installations plus rolling stock, maintenance and the energy cost of a 1500 V DC traction system for this traffic level has been assessed to be higher than the 25 kV AC traction systems by about Rs 2 Cr per annum in 2005".

The requirement of a higher tunnel diameter of 6.4 m was a major cost issue for adopting 25 kV AC in the underground corridor. However, with the advancement of technology and adoption of international standard i.e. IEC and EN, when it became possible to adopt 25 kV AC traction system in a concentric tunnel diameter of 5.55m, technoeconomic benefit shifted in favour of 25 kV AC traction system as can be seen from the chronology of events and techno-economic studies carried out for Delhi Metro from 1989 to 2001.

The maximum power requirement, which contributes to the power system, is dependent on:

i. the Peak Hour Peak Direction Traffic (PHPDT),
ii. performance requirement and level of technology of rolling stock,
iii. loaded train weight,
iv. headway between the trains,
v. Rake composition of trains
vi. track geometry and
vii. efficacy of regeneration.

The design of the traction power system depends upon.

i. Maximum power to be transmitted to meet the traffic requirement
ii. The level of redundancy, e.g., the number of transformers being provided for the traction and auxiliaries, and their cabling arrangement
iii. Market for a particular type of technology, volume of requirements and orders already available with the manufacturers.
iv. Requirement of HT and LT cables and the number of gas-insulated substations (GIS)

It is considered that Technical Feasibility, economic viability, Aesthetics, Sustainability, and costs should be considered when selecting a traction system. Additional considerations include reliability, Operation, maintenance requirements, Upcoming technologies, and cost. Each system, whether 750 V DC, 1500 V DC, or 25 kV AC, has its merits and demerits, and a decision must be made regarding the adoption of a particular system over the others.

9.27.2 750 V DC

The following two metros in India are using a 750 V DC system with a third rail today:

- Kolkata Metro – revenue services introduced in 1984
- Bangalore Metro – revenue services introduced in 2011

There are major differences between the two systems as detailed hereunder:

i. Steel third rail with top contact has been used in Kolkata Metro whereas an aluminium composite third rail with bottom contact has been adopted in Bangalore Metro.
ii. Barring recently introduced new rakes, Kolkata Metro trains are not air-conditioned. Instead, the tunnel itself is air-conditioned and hence power consumption for the auxiliaries is quite different compared to recent metros.
iii. Regenerated energy in existing Kolkata Metro trains is not utilised by other trains and hence it is wasted in a resistor, which further increases the load of air-conditioning of the tunnel.

The Kolkata Metro has begun replacing its steel third rail with an aluminum third rail.

9.27.3 1500 V DC System

DMRC, at one point in time, awarded a contract for a 1500 V DC traction system for Vishwavidyalaya to the Central Secretariat Section, which was electrified on a 25 kV system. It has been observed that the cost of a 25 kV AC system was almost Rs 1 crore per km less than the rates offered for a 1500V DC system. Considering the present-day cost of a 25 kV AC system as Rs. 9.51 crore per km for the elevated corridor and Rs. 11.26 crore per km for the underground corridor, the cost of a 1500 V DC traction system works out to Rs. 10.51 crores per km for the elevated corridor and Rs. 12.26 crores per km for the underground corridor.

9.27.4 25 kV AC traction

In the case of the Delhi Metro, the cost of the traction system, including auxiliaries, is Rs. 7.3 crores per km for elevated sections and Rs. 8.7 crores per km for the underground corridor as per the last project executed for phase II. The orders for DMRC Phase II, including the elevated section, were placed in February 2007, and for the Underground Section, in September 2007, with a provision for a PVC clause on the transformer and 25/33 kV switchgear. Based on this, the current cost of the traction system, including auxiliaries, amounts to Rs. 9.51 crore for the elevated corridor and Rs. 11.26 crore for the underground section, after considering a 5% annual inflation rate.

The estimated cost of the auxiliary system is approximately 25-30% of the elevated section and 35–40% of the underground section. This is based on the Phase II contract awarded in February 2007 and September 2007, and after considering a simple inflation rate of 5% per annum. The current cost of underground rigid OCS is Rs. 1.75 Cr more per route km, after considering inflation.

9.28 Signaling and Telecommunication

The Delhi Metro uses cab signaling along with a centralised automatic train control system consisting of automatic train operation, Automatic Train Protection and automatic train signaling modules. A 380 MHz digital trunked TETRA radio communication system from Motorola is used on all 6 lines to carry both voice and data information. For Lines 3 and 4, Siemens Transportation Systems has supplied the electronic interlocking, the operation control system, and the automation control system. An integrated system comprising an optical fibre cable, on-train radio, CCTV, and a centralised clock and public address system is used for telecommunication during train operations as well as in emergencies. For Line-1 and Line-2, ALSTOM has supplied the signaling system, and for Lines 5 and 6, Bombardier has supplied signaling system.

The telecommunication systems used by Delhi Metro Rail Corporation (DMRC) include key components such as fibre optic transmission, public information displays, public address systems, master clocks, CCTV, radio systems, and telephone exchanges.

9.28.1 Signaling and Train Control

a) Continuous Automatic Train Control (CATC)

The System shall be suitable for working on Sections having 25 KV AC traction. The Continuous Automatic Train Control (CATC) Function shall be responsible for implementing Train movements in accordance with the requirements established by the ATS Function, within the constraints established by the Interlocking Function. The complete system shall operate in a Fail-Safe manner, based on the principles of Target Speed and Target Distance, with Cab Signalling. It shall continuously supervise the safe running of the train and, in case of danger, warn the driver and apply Brakes, if necessary, after a predetermined interval of Time.

The CATC System shall consist of three major Components:

i. Automatic Train Supervision (ATS)
ii. Automatic Train Operation (ATO)
iii. Automatic Train Protection (ATP)

These are the three major operations through which a train can be controlled, supervised and protected. This helps to figure out the intricacies of the train. This can be seen through the Operation Control Centre (OCC), which features an in-built screen for tracking the metro's movement.

b) Automatic Train Supervision (ATS)

The ATS system communicates with the Radio system provided by Motorola to exchange information about the Metro's whereabouts. Communication between ATS and Radio systems is available at the OCC level. The ATS system sends train location and train descriptions for the main line and depot area and receives radio parameters and train operation mode information from the Radio System, both in the main line and depot areas. The information exchanged between the Radio and ATS systems is:

- From ATS: Train Identification (Train Identity, Rake Id, Crew Id, ATP Car number) and Train Location (Train Circuit Number, Location Status).
- From Radio: Radio Parameter (Radio Identity Number), Train operating mode (Restricted Manual, Running on sight, Cut-out, ATP or ATO modes)

The Radio System provides information to ATS regarding the train equipment, including Radio Identity, Number, and ATC current mode, which are available from the ATC on-board system. The radio server sends a message to request a connection to every CATS server. Then, only the active CATS server responds to the Radio server with a Train Description message for all current active trains located either on the main line or in the depot.

Later, CATS will send the train description message for each train when there is a change of train identification and train location data. If no Train Description message is sent from CATS. The Radio server maintains communication with the CATS server through the Operational ATS session message, which is sent from the CATS if no other message is received for more than 10 seconds.

The Radio Server shall establish a new connection if it does not receive any message from the CATS server for more than 30 seconds by sending a connection request message to both the CATS server via LAN1 and LAN2. Below is the figure for information exchange between ATS and the Radio System.

c) Automatic Train Operation (ATO)

The Automatic Train Operation (ATO) System is responsible for Train Operation between the Stations. The Train Speed, acceleration, deceleration, and Braking shall be automatically controlled, without the driver's intervention, preventing unnecessary Braking, Stopping, and starting.

The functions of the ATO System are,

i. Automatic Control of Train running from Station to Station.
ii. Automatic Opening of Train Doors at the Stations. Closing of the Door to be activated by the Driver, on receipt of a Command from the ATS System to leave the Station. When the Doors are closed, the Train Driver shall start the Train. The ATO System shall then operate the Train to the next Station.
iii. Execute Programmed Stops and Control, in conjunction with the ATS / ATP equipment, of Dwell Times in accordance with the Timetable / Headway.

The ATO System shall be able to send a Control to the Train to leave the Doors closed upon arriving at the Station or to bypass/skip any Station. The System shall operate the Train within the Parameters set by the ATP System. As a Train approaches a Station, the ATO System shall reduce the Speed and control the stopping of Trains at Platforms. When the Train is properly berthed within the Stopping Window, the ATO System shall initiate a Command for opening of Doors on the appropriate Platform side (s). The ATO System shall prevent the Train from starting if the Train Doors are detected as 'not closed'.

ATO Operation Outline:

- The target speed and braking profile curve between stations are calculated based on the line speed restrictions and grade conditions indicated by the ATP.
- The braking commands are then calculated to match the target speed and braking profile, so that the train stops at the specified station-stop position.
- The target settings can be programmed to adjust automatically in response to actual conditions through communication with the line-side operating management system.
- The ATO System is also compatible with platform-gate interlocking control as well as driverless automatic starting and automatic resetting of equipment-fault indications, using a highly reliable system for the communication of safety information.

d) Automatic Train Protection (ATP)

In this Mode, the Train Driver shall operate the Train manually. Instructions shall be provided in the Cab of the Train with On-board Displays for Maximum Safe Speed (MSS), Current speed, and Target distance / Speed as deduced from the most restrictive ATP Condition, Signalling Mode, and Routes established through Interlocking. The Braking curve shall be computed continuously along the Line, to enable a minimum safety distance to be maintained. This computation is based on the Line characteristics and the parameters of the Train, with dynamic monitoring and Enforcement of changes in target distance and Speed. The ATP Equipment also indicate which side the Doors may be opened when a Train enters a Station. The ATP Equipment shall not allow the Doors to be opened on the wrong side, unless an Emergency Over-ride Control is activated.

- **Cut-Out Mode**:

The Cut-Out Mode shall be used when the On-board ATP Equipment on ATP-equipped Corridors fails, or both the On-board ATO & the On-board ATP equipment fail on corridors equipped with ATO. Automatic braking of the Train shall be initiated in the event of a fault in the On-board Equipment to an irrevocable Stop.

Under such conditions, Trains shall run in Cut-out Mode. Before entering Cut-out Mode, the Train Driver shall inform the OCC that the ATP /ATO has failed. OCC will ascertain the position of the Trains ahead and then authorise the Train Driver to enter cut-out mode. The Delhi Metro uses cab signaling along with a centralised automatic train control system consisting of automatic train operation, Automatic Train Protection and automatic train signaling modules. A 380 MHz digital trunked TETRA radio communication system from Motorola is used on all 6 lines to carry both voice and data information. For Lines 3 and 4, Siemens Transportation Systems has supplied electronic interlocking, operation control systems, and automation control systems. An integrated system comprising an optical fibre cable, on-train radio, CCTV, and a centralised clock and public address system is used for telecommunication during train operations as well as in emergencies.

For Line-1 and Line-2, ALSTOM has supplied signaling system, and for Lines 5, 6, Bombardier has supplied signaling system.

9.29 Telecommunication

The telecommunication system acts as the communication backbone for Signaling Systems, such as SCADA and AFC, provide telecommunication services to meet the operational and administrative requirements of the metro network.

The proposed telecommunication facilities help meet the requirements.

1. Supplementing the Signaling system for efficient train operation.
2. Exchange of managerial information
3. Crisis management during emergencies
4. Passenger information system

The proposed telecom system will cater to the following requirements:

- Train Traffic Control
- Assistance to Train Traffic Control
- Maintenance Control
- Emergency Control
- Station-to-station dedicated communication
- Telephone Exchange
- Passenger Announcement System and Passenger Information and Display System within the station and from Central Control to each station.
- Centralised Clock System
- Train Destination Indicator
- Instant online Radio Communication between Central Control and Moving trains and maintenance personnel.
- Data Channels for Signaling, SCADA, Automatic Fare Collection, etc.

9.30 Automatic Fare Collection

Mass Rapid Transit Systems handle large numbers of passengers. Ticket issues and fare collection play a vital role in the efficient and proper operation of the system. To achieve this objective, the ticketing system should be simple, easy to use, easy to maintain, and minimise the burden on accounting facilities. It should be capable of issuing single/multiple journey tickets, amenable to quick fare changes, and require overall lesser manpower. In view of the above, a computer-based automatic fare collection system is proposed.

The System is based on the concept of a Contactless Smart Card for multiple-journey store value tickets and a contactless token for Single-Journey Tickets. The DMRC has also established a Central Clearing House System for connecting to other Operators in the City of Delhi.

Thales has successfully delivered the Automatic Fare Collection (AFC) systems on the first section of the Delhi Metro. Thales is a long-standing provider of Automatic Fare Collection (AFC) systems to the Delhi Metro and is now meeting the needs of Phase III with its innovative AFC solutions. Contactless ticketing technology provides fare-paying passengers with a smooth and rapid ticket validation process, enabling them to continue easily to their chosen destinations. The four Dream Gates (two entries at Lal Quila station, two exits at Jama Masjid) from Thales enable passengers to travel using a QR code on their mobile phones — a first in India.

9.31 Maintenance Depot

DMRC follows a mixed approach for Operation and Maintenance. The core operation and maintenance of major assets are conducted in-house, with specially recruited and trained staff assigned to this purpose. Non–core and offline activities are generally outsourced, though some of them are managed in-house. The outsourcing of activities is decided based on techno-economic considerations.

Delhi Metro follows O&M practices, where core functions (operations and maintenance of major assets) are performed In-house, while Housekeeping, ticket counter manning, elevator and escalator maintenance, Firefighting, Air Conditioning, and Building maintenance services are fully/ partly outsourced. A Maintenance regime comprising preventive and corrective maintenance, as prescribed by OEMs, is generally followed, reviewed, and revised periodically. DMRC has become a member of NOVA to have access to international O&M practices and follow Key Performance indices (KPIs) as per international norms.

Fig.9.107 includes aerial view of Mukundpur Depot and **Fig.9.108** shows the entry/exit of the same depot. **Fig.9.109** shows the Ghitorni Depot and **Fig.9.112** includes Khyber Pass Depot.

Monitoring equipment performance involves assessing key parameters. The concept is to progressively evolve the need-based maintenance regime, which can be suitably configured in the form of minor & major schedules. Unit replacement and getting essential repairs done by the OEMs is preferred. Labour-intensive procedures are kept to a minimum & emphasis is placed on automation with state-of-the-art machinery to ensure quality and reliability. Multi-skilled maintenance staff ensure quality and productivity in their performance.

Formulation of Disaster Management Plan and enforcement of periodical mock drills of various sub-systems to ensure preparedness in handling emergencies and disasters. xiv. Reliability, Availability, Maintainability (RAM) targets are mandatory. Presently, no Metro Rail system in India has outsourced the entire Operations and Maintenance activities.

Depot cum workshop for scheduled inspections, major schedules viz Periodical overhaul (POH) and major unscheduled repairs. The main depot generally houses Operation Control Centre (OCC), Administrative Building, maintenance facilities for Civil – track, buildings, water supply; Electrical – traction, E&M; Signaling & Telecomm.; etc. apart from necessary facilities viz stabling lines, scheduled inspection lines, workshop for overhaul, unscheduled maintenance including major repairs, wheel profiling, heavy interior/under frame/roof cleaning etc. for the rolling stock operational on the corridor. To start the morning services, some rakes will need to be kept at terminal stations, and stabling facilities will need to be provided at the depots.

9.31.1 Aspects of Depot Design

- Conceptual design of Stabling lines, Inspection Shed and Workshop to provide maintenance facilities and stabling facilities for the Rolling Stock.
- Operational and functional safety requirements
- Ancillary buildings for other maintenance facilities

Fig.9.107: Mukundpur Depot

Fig.9.108: Mukundpur Depot

Fig.9.109: View of Ghitorni Depot

9.31.2 Key Performance Indicators (KPIs) followed by DMRC

a. **Punctuality:** The Delhi Metro considers a trip as delayed if it incurs a delay of more than 2 minutes.
b. **Car KM:** All car km that were operated in revenue (passenger carrying) service.
c. **Percentage of cars used during peak hours**: Total number of cars in use in the weekday peak hour
d. **Ridership**: Total number of passengers travelled on the metro, including Token, Card and Group booking users.
e. **Passenger Earning**: Total revenue from Fare Box collection.
f. **Passenger KM**: Total distance travelled by each passenger on metro trains, summed up for all passengers.
g. **Car Hours:** Actual time in which each car is operating in revenue (passenger carrying) service, summed up for all cars.
h. **De-boarding**: Train withdrawn from service, and passengers need to be deboarded en route from the train due to failure.
i. **Trips run on time:** Number of trips run as per the scheduled time.

DMRC has started monitoring the following KPIs every month:

Key Performance Indicators are

1. Staff Cost
2. Cleaning & Housekeeping Cost
3. Electrical Cost
4. Traction Cost
5. non-traction 4 Operation
6. Electrical Maintenance
7. Operation
8. Rolling Stock
9. S&T
10. Others

9.31.3 Corrective Maintenance

The essence of corrective maintenance is to prevent failure by continuous monitoring of certain vital installations and taking timely corrective action to eliminate/reduce the impact of failure. The following activities are adopted to enable corrective maintenance.

a. Manning of vital installations like RSS, OCC, DCC and SCR round the clock.
b. SCADA for Remote monitoring and control of Traction installations and Telecom installations.
c. BMS for remote monitoring of E&M assets (Firefighting system, DG sets, UPS, lights, Escalators/ Lifts, etc.) at stations.
d. AFC, Station and Central Computers for monitoring AFC gates, TOMs, TVMs, etc. at stations. (e) Failure analysis through causative indications on the electronic equipment/sub-systems.

The maintenance activities are performed in dedicated areas (Buildings & space for stables (Open or Covered or in Combination), those included in a Depot Plan. For planning details of a Maintenance Depot, refer to the relevant section in Chapter 6, Volume 1.

9.32 Operation Control Center

(Ref. The New Indian Express, Hindustan Times & Economic Times)

To control Metro operation is the most critical activity. The OCC integrates the entire control and command of operations at one location. With almost 100 trains running on the network, the operations control center or OCC at Shastri Park, is the nerve center of the Delhi Metro.

From here, every aspect of the Delhi Metro system is monitored, including the timetable, speed, security, traction, electricity, trains, auxiliary equipment such as air-conditioning and ventilation systems in underground tunnels, and even the crowds in the station. Keeping the Delhi Metro running all day long requires not only regular maintenance of trains and tracks but also constant monitoring of the system during operational hours, a process that takes place through the Delhi Metro's operations control center and maintenance depots to find out how it achieves this feat daily. With almost 100 trains running on the network, the Operations Control Centre (OCC) at Shastri Park is the center of operations for the Delhi Metro. From here, every aspect of the Metro system is monitored—the timetable, speed, security, traction or electricity, the trains, auxiliary equipment like air-conditioning and ventilation systems in underground tunnels, and even the crowds in the station.

In a major step towards creating the necessary infrastructure for the commencement of operations on Phase-4 corridors of Delhi Metro, DMRC unveiled its new integrated Operation Control Centre (OCC) at Metro Bhawan (DMRC's headquarters). The new state-of-the-art control and command centre (OCC) was developed. The OCC provides remote visibility and manoeuvrability of the entire metro system (Rolling Stock, Stations, Signals, Tracks, etc.), with concourse and platform-level monitoring. Thus, both passengers and the system are under monitoring by the OCC.

Now, the entire DMRC train operation, which includes a total of 415 km of track length, around 388 trains, 301 stations, and 40 interchange stations, is being controlled from a single unified building in a more efficient manner, with better staff utilization **(Fig.9.110).**

The operating center has installed a giant screen **(Fig.9.111),** allowing one to be part of the system and keep a close eye on another screen with multiple CCTV images from different stations flashing. With a fully automated rail system, it's not surprising that Delhi Metro regards the OCC as its nerve center. Delhi Metro's trio of systems keeps the network running—ATP (automatic train protection), ATO (automatic train operation) and ATS (automatic train supervision)

Fig.9.110: Shatri Park Operations Control Centre (photo: Economic Times)

Fig.9.111: Control room at Shastri Park OCC – (photo: Economic Times)

The system operates by assigning a unique identity, known as the train ID, to each train. The first part is the destination code, while the second is the train number. The screen at the OCC tracks the movement of the train through the track circuits, which are like electrical nodes that send the information to the OCC. These are placed at intervals of 20m along the track. With no lag in the real-time location of the train and the OCC circuit, every minute movement of the train is monitored by the system. The ATC (Automatic Train Control) system, which controls the train's route, works in tandem with the ATS, which interfaces between the train and the line-side equipment, such as track circuits. Together, the system moves the train from one station to the next, ensuring that a minimum speed is maintained. **(Rumu Banerjee)**

Fig.9.112: Khyber Pass Depot – (photo: Economic Times)

9.33 Safety Engineering

a. Safety is of utmost importance for the Project, and a structured process of Safety Engineering shall be complied with by all stakeholders (Employer, OMC, Contractors, etc.) as per EN50126 and Yellow Book guidelines.
b. An Engineering Safety Case is developed for the Project by consultants. Similarly, OMC is to develop an Operations Safety Case and Safety Management System (SMS). The Project-wide ISA reviews these safety cases, and no objections must be raised before commercial opening.
c. OMC takes care of progressive updating of the SMS system during the Project life cycle.

9.33.1 Delhi Metro Safety and Security of Systems

a. The responsibility for the security of the DMRC system is with CISF and Delhi Police.
b. DMRC deploys CCTV for security purposes. However, the video analytic technique is not considered useful by DMRC. According to information available with DMRC through NOVA/ Comet, video analytics is not considered practical or feasible for a mass rapid transport system. London Metro, after the serial Bomb blasts in 2005, had planned the use of video analytics, which was discontinued in 2008 as it was not considered useful.
c. DMRC organises Mock drills related to safety preparedness and handling of disasters in association with DDMA/NDMA/NSG/Delhi Fire Service/Delhi Police. During these mock drills, passengers are evacuated from tunnels/viaducts under various scenarios involving different lines and stations. Normally, these drills are conducted during non-revenue hours or early revenue hours to avoid inconvenience to passengers.
d. CISF has prepared a contingency plan as per the directions of MHA detailing the role and responsibility of each stakeholder, i.e. DDMA/NDMA/NSG/Delhi Fire Service/Delhi Police/ DMRC. As per this contingency plan, each stakeholder has prepared its own contingency plan. Accordingly, DMRC has prepared its own contingency plan. In addition to this, DMRC has its own Disaster Management Plan, as required by the Disaster Management Act 2005. CISF also prepared a contingency for each station.

Drills for using the Rescue & Relief Train, attending lifts, and responding to escalator failures are conducted weekly on various lines and at different stations.

Mock drills are conducted to assess the preparedness of various sub-systems in the underground section.

DMRC has a safety organisation that functions under one of the heads of the O&M department. Additionally, part-time executives from various departments conduct regular safety checks.

9.34 Awards and Unique Approach

The metro has received awards for environmentally-friendly practices from organisations including the United Nations, RINA, and the International Organisation for Standardisation. It is the second metro system in the world, after the New York City Subway, to be ISO 14001 certified for environmentally friendly construction. By March 2023, 64 metro stations, four sections on the central verge between piers, and 12 other Phase I and II locations on the network will have rainwater harvesting systems for environmental protection. All 27 Phase-IV elevated stations will also harvest rainwater, and 52 recharge pits are being constructed for this purpose. It is the world's first railway project to earn carbon credits after being registered with the United Nations under the UN's Clean Development Mechanism, and has earned 400,000 carbon credits with the regenerative braking systems on its trains.

The DMRC installed the metro's first rooftop solar power plant at the Dwarka Sector 21 station in 2014. The network received 35 per cent of its energy from renewable sources by April 2023, which it intends to increase to 50 per cent by 2031. Of this, 30 per cent comes from the Rewa Ultra Mega Solar park in Madhya Pradesh, four per cent (50 MWp) comes from rooftop solar panels, and one per cent comes from a waste-to-energy plant in Ghazipur. The DMRC has installed solar panels at 142 locations, including 15 depots, 93 stations, and 34 other buildings.

The metro has been promoted as an integral part of community infrastructure, with artwork depicting the local way of life displayed at stations. Students at local art colleges have designed murals at metro stations, and the viaduct pillars of some elevated sections have been decorated with mosaic murals created by local schoolchildren. The metro station at INA Colony has a gallery of artwork and handicrafts from across India, and all stations on the Central Secretariat – Qutub Minar section of the Yellow Line have panels depicting Delhi's architectural heritage. The Nobel Memorial Wall at Rajiv Chowk features portraits of the seven Indian Nobel laureates: Rabindranath Tagore, CV Raman, Hargobind Khorana, Mother Teresa, Subrahmanyan Chandrasekhar, Amartya Sen, and Venkatraman Ramakrishnan.

9.35 Mishap During Execution

Several mishaps occurred during the execution of the Metro Network. A few of them have been mentioned below.

In October 2008, a launching gantry and part of the overhead Blue Line extension under construction in Laxmi Nagar collapsed and fell on a passing bus. Workers were using a crane to lift a 400-tonne concrete span of the bridge when the gantry and a 34-metre-long span of the bridge collapsed on the bus. The driver and a construction worker were killed.

In July 2009, a section of a bridge collapsed while it was being constructed at Zamrudpur, east of Kailash, on the Central Secretariat–Badarpur corridor. Six people died and 15 were injured. A crane removing the debris collapsed the following day, and two other nearby cranes collapsed, injuring six. In July 2009, a worker at the Ashok Park Metro station was killed when a steel beam fell on him.

Views after the Collapse in Zamrudpur

In a major accident that occurred in Zamrudpur on Sunday, July 2009, a chargesheet was filed before a local court. The Delhi Police have charged 24 people, including 11 senior officials, with negligence in connection with the 2009 Zamrudpur Metro pillar collapse incident. Several people had died. That charge sheet includes the then-chief engineer and the chief project officer of the site. The Director of the Consultant and the Engineer of the Contractor were also named in the chargesheet.

The Crime Branch of Delhi Police, in its chargesheet, has held all parties involved in the project responsible for the accident. The investigation opined that there was sufficient evidence that the work was planned, carried out and continued negligently and endangering human life and thus committed an offence causing hurt by an act endangering life or personal safety of others, and an act endangering life or personal safety of others, causing death by negligence.

A year after the Zamrudpur incident, in which several people died when the cantilevered bracket of Metro's pillar collapsed, crucial files related to the project have gone missing. The concerned authority stated that the files had been stolen. The contractor for the project was barred from participating in any Delhi Metro project for the next two years. The concerned Authority had attributed the accident to a "problem in the design" of the pillar and claimed that they "were trying to rectify it". Crime Branch has also pointed out that the crack in the pillar had been noticed in advance, but the officials still did not conduct a load test and continued with the construction work.

Accident Photographs

An upturned crane and launcher litter the accident scene of the Metro site at Zamrudpur on Monday, July 13, 2009. Cranes toppled on Monday while clearing the debris of Sunday's flyover collapse, in which more people were affected. The petition was filed by an NGO working on behalf of the affected labourers.

Several Labourer/s were critically injured when a girder fell at a metro construction site at Punjabi Bag

Views after the Accident

The magenta line of the metro network, designed for Driverless operation, crashed before the launching. An old man died, and four people were injured when the wall of the Gokul Puri elevated. The metro Station collapsed and fell on the road below.

An accident happened at the metro rail construction site when two cranes overturned at the Mehrauli Metro Station site, and several people were injured (2009)

Over a hundred people, including 93 workers, have died since work on the metro began in 1998.

In April 2018, several people were injured when an iron girder fell off the elevated section of a Metro structure under construction at the Mohan Nagar intersection in Ghaziabad. A car, an auto rickshaw, and a motorbike were also damaged in the incident.

CHAPTER 10

Mumbai Metro

10.1 General

Mumbai, India's financial capital, has experienced phenomenal population and employment growth over the past few decades and is one of the country's fastest-growing metropolitan regions. The Mumbai Metropolitan Region (MMR), spanning an area of 4,355 km^2, comprises eight municipal corporations, 11 municipal councils, and 996 villages. The population of MMR in 2005 was 20.8 million, which increased to 22 million by 2011 and is expected to continue growing, with the population expected to reach 34 million by 2031. Mumbai is a linear city, with the southern island city having a width of approximately 5km, and the northern section of the town is more expansive, varying from 11 to 17 km. The northern part of the city is dominated in the centre by the Sanjay Gandhi National Park, which leaves 5 to 6km wide strips of land on either side for development. Historically, the central business district of Mumbai was in the south at Nariman Point. Most of the population lived in the north and commuted to the business district in the south.

Mumbai is the largest metropolis, followed by Delhi and Kolkata. Mumbai is one of the most congested cities in the world. It needs urgent development in the transport sector. The dominant feature of passenger movements in Mumbai is the overwhelming reliance on public transport modes and walking. In MMR, public transport systems are overcrowded, and the road network is congested because of a significant gap between demand and supply.

Fig.10.1: Mumbai Metro During Construction (thecoreengineers.com)

In early 1996, the Mumbai Metro Planning Group (MMPG) was set up by signing a Memorandum of Understanding by six members, led by Tata Consultancy Services, to undertake a detailed feasibility study for the Metro Corridor Seven (Seventh Corridor) in Mumbai. The author was part of the study group on behalf of CES and Prof SS Chakraborty. The study was completed at the end of 1997 and submitted for consideration by the Governments of India and Maharashtra. The report was supported by seven Working papers, each focusing on the Travel Demand Forecasting Model and the Metro Engineering & Cost Study.

The final report included important aspects of a metro system, Concept, Alignment, and Construction Methodology, Travel Demand, Engineering and Cost, Environment, Innovative Resources, Financial, Institutional, Legal and Implementation aspects. A review of the 1974 techno-economic feasibility studies for the MTP (R) was conducted by Tata Consultancy Services in July 1994. The initial review established that both the sixth and seventh corridors are technically feasible. Furthermore, the MMRDA has since prepared and published a Draft Regional Plan for 1996-2011. The transportation sector proposals in the Regional Plan 1973 consisted of a list of transportation projects for inter-regional and intracity transport facilities, including rail facilities.

The revised Regional Plan supports this strategy that emphasises the need to improve rail capacities to address acute overcrowding in existing facilities. The strategy calls for integrating the two suburban systems and visualising the establishment of the Urban Metropolitan Transport Authority. To ensure long-term relief from the problem of overcrowding and to combat the chaotic situation, two World Bank projects, namely MUTP I & MUTP II, were executed. The World Bank also insisted on the participation of the Railways for a comprehensive solution.

In the Initial stage, TCS planned to associate with IL&FS & CES, but in April 1995, IL&FS expressed their unavailability for the project. During the same time, Dr MQ Dalvi, a transport economist based in London, offered his services. MMPG submitted their comprehensive report, recommending the underground system, utilising the Cut-and-cover method for stations and the Tunnel corridor using a TBM.

In Mumbai, public transport is the preferred mode of transport, but the capacity is grossly inadequate. It requires immediate improvement. Although introducing a metro in Mumbai to ease traffic congestion has been discussed for the last 50-60 years, concrete plans did not materialise until MMRDA prepared the Master Plan **(Fig.10.2)** of the Mumbai Metro network in 2003. The Master Plan totalled 146.5 km, comprising some corridors.

The Mumbai Metro map is a colour-coded diagram that visually represents all operational and planned lines, interchanges, and terminal stations. Metro is reshaping Mumbai's future with rapidly expanding routes and improved last-mile options. Interchange stations are DN Nagar - Lines 1 & 2A, Gundavali (Andheri East)–Lines 1 & 7, and Dahisar East - Lines 2A & 7. are given below.

Fig.10.2: Mumbai Metro Master Plan

Table 10.1: Status of Corridors

S. No.	Corridor	Length (Km) Status	
		Total	
1	Versova – Andheri – Ghatkopar	11.4	Operational
2A	Dahisar – DN Nagar	18.6	Operational
2B	DN Nagar- Mandale	23.6	Under Construction
3	Cuffe Parade - Seepz	32.5	Partially in Operation
4	Wadala-Kasarvavali	32.3	Under Construction
5	Thane-Bhiwandi-Kalyan	24.7	Under Construction

6	Swami Samarnath Nagar-Vikhroli	14.50	Under Construction
7	Andheri (E) – Dahisar (E)	16.5	Operational
8	CSMIA - NMIA	35	Planned
9	Dahisar East – Mira Bhyandar - Andheri	13.60	Under Construction
10	Gaimukh – Shivaji Chowk	11.4	Approved
11	Wadala – CSMT	12.7	Approved

CES prepared the DPR for Thane MRTS (Preliminary Report by RITES) and Line 1: Andheri – Ghatkopar, based on a Concept prepared by DE Consult of Germany. Later, the line extended to Versova, where DMRC prepared the Report in 2005. DMRC also prepared DPR for 2: Colaba – Bandra – Charkop – 2008, Line – 3: Bandra – Kurla - Mankhurd – 2006.

10.2 Mumbai Transport Scenario

The Suburban Railway, comprising the Central, Western, and Harbour lines, very efficiently transported commuters on the north-south axis from the northern suburbs to the southern business district. Over the last few decades, the growth of the Indian Economy has led to a significant increase in population and employment, resulting in the development of new business districts in central and northern Mumbai. The Suburban railway's efficiency and speed in ferrying commuters across the city have become its greatest enemy. Passengers experience a degree of overcrowding that is uncommon in rail systems worldwide. Each minute, trains arriving at Chhatrapati Shivaji Terminus and Churchgate Station inject 2,000 people into the city's historic core during rush hour. Nine-car trains designed to hold 1,700 passengers travel with up to 5,000 commuters, with an average of 13 people per day killed in rail-related accidents. The passenger density during peak hours reaches as high as 12 passengers per square meter, significantly exceeding international norms of 5 passengers per square meter. (US Federal Transit Administration Report, 1996)

Fig.10.3: Overcrowding and congestion on the transport network

Currently, the Suburban Railway System carries 7.4 million trips daily and is supported by the local buses operated by the municipal corporations. The Brihan Mumbai Electric Supply and Transport Undertaking (BEST) bus system operates primarily within the Municipal Corporation. (MCGM) Area is the largest operator with a fleet strength of 4,700 buses that carry 5 million trips daily within Greater Mumbai. Despite the severe overcrowding **(Fig.10.3)** and inhumane travel conditions, Mumbai's public transport share is approximately 70%, among the world's highest (Source: Comprehensive Transportation Study for MMR, 2008).

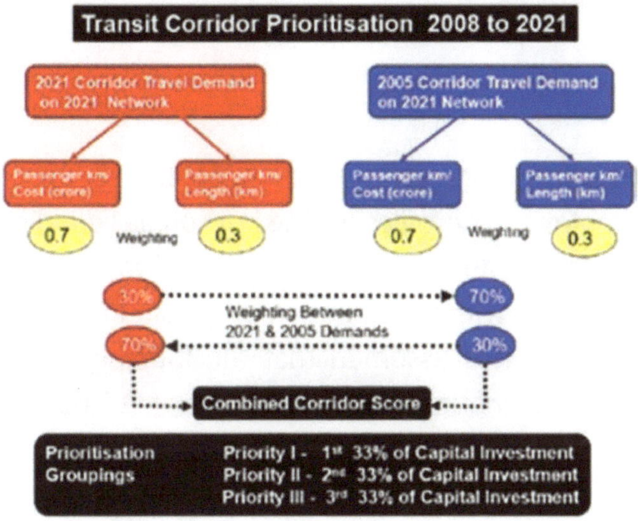

Fig.10.4: Transport share comparison

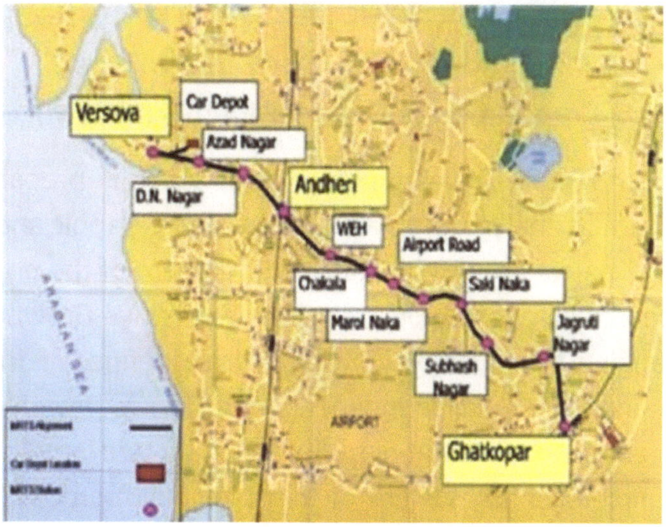

Fig.10.5: Line 1 Alignment

Over the last 50 years, the length of Mumbai's Road network has only doubled. To improve the travel conditions for commuters and increase the public transport mode share **(Fig.10.4)** in MMR, the Government has taken up several ambitious. Public transport projects include the Mumbai Urban Transport Project (MUTP) I and II, the implementation of the Mumbai Metro Master Plan, and the Mumbai Monorail. Under the MUTP projects, the Suburban Railway network has been expanded, and new rakes have been introduced. Of the Mumbai Metro Master Plan, Metro Line 1, **Fig.10.5,** (Versova – Andheri – Ghatkopar), an 11.40 km corridor with 12 stations, has been operational since 2014 and has a daily ridership of over 300,000 commuters. Metro Lines 2A, 3, and 7 are currently under implementation and are expected to be operational within a few years. To ensure a coordinated and holistic approach to improve the transportation scenario in the city, in 2008, MMRDA, with World Bank assistance under MUTP, appointed M/s LEA International Limited, Canada, in JV with LEA Associates South Asia Pvt. Limited, India, as consultants to prepare a comprehensive Transport Study and Business Plan for MMR. The key objectives of the study were as follows:

1. Identify the travel patterns of residents in MMR.
2. Select, develop, and operationalise an urban transport planning model using state-of-the-art modelling techniques and a software package appropriate to MMR's conditions and planning needs

3. Assess the relevance of the 1994 strategy and identify the consequences of pursuing alternative transport strategies and recommend/update a long-term comprehensive Transport strategy for MMR.
4. Identify for all modes a phased program of appropriate and affordable investments and policy proposals up to 2016 and
5. Help strengthen transport planning skills and transfer all data, planning model/tools and knowledge obtained through the study to MMRDA and other agencies such as Mumbai Rail Vikas Corporation (MRVC), City & Industrial Development Corporation (CIDCO) and Municipal Corporations.

10.3 Corporation of Greater Mumbai (MCGM)

CTS Scenario Development: The first step in preparing a transportation development strategy for the city is to analyse alternative future demographic and employment profiles for the horizon periods. Four alternative growth scenarios with a share of the population in Mumbai ranging from 40% to 60% have been envisaged. Separately, four Alternative Growth Scenarios with a share of employment in Mumbai ranging from 33%to 75% have been defined. Furthermore, it is assumed that the four population scenarios would be independent of the four employment scenarios, resulting in 4 × 4 = 16 possible scenarios (combinations of population and employment scenarios). The analytical process involved a first-level assessment of the sixteen Options, which led to a short-listing of six Options. A second-level screening process of these six Options led to the Technical Advisory Committee (TAC), comprising MMRDA officials and field experts, selecting three Options (P2E2, P3E3, and P3E4) that formed the basis for formulating long-term transport strategies for MMR.It is worth noting that these three population and employment scenarios represent quite diverse land-use or land-development phasing strategies

The population and employment for 2016 and 2021 have been assessed for the three growth scenarios, comprising two population scenarios and three employment scenarios. In deriving these figures, reasonable judgments were made in consultation with MMRDA officials and with the advice of TAC members. The estimated figures include two population distributions (P2 and P3) and three employment distributions (E2, E3 and E4). Considering Mumbai's forecasted growth in population and employment, transport trends, and the city's geographic limitations, it was concluded that the development is robust and multifaceted. Public transportation is crucial for meeting travel demand. This can be achieved by capacity enhancements to the existing suburban railway system, creating new metro corridors, connecting major existing and planned activity centres of the region, and providing exclusive bus lanes to reinforce rail-based transit with a higher-order road-based public transport system. Transit First is, therefore, considered the guiding principle in preparing the transportation plan for MMR.

10.4 Analysis of Existing Condition

A standard four-stage travel demand modelling approach has been adopted for the CTS. Trip Generation, Trip Distribution, Mode-Split and Assignment models have been developed after thoroughly validating the demand matrices assessed from other surveys/ studies. Six purposes (Home-Based Work Office, Home-Based Work Industry, Home-Based Work Others, Home-Based Education, Home-Based Others, and Non-Home-Based) and seven modes (Suburban train, metro, bus, auto, taxi, car, and two-wheeler) were considered. The study area has been divided into 1,030 Traffic Analysis Zones (TAZs) for travel demand analysis. The model used for travel demand modelling and network analysis is EMME (Equilibria Multimodal Equilibrium) software. Further, as part of the study, the following significant surveys have been carried out:

a. Home Interview Survey (66,000 sample size).
b. Road Network Inventory Surveys (2,300 km).
c. Outer Cordon/Sub-Regional Cordon/ Inner Cordon/ Mid-block count surveys (93 locations).
d. Rail Passenger surveys.
e. Speed-Flow surveys (for developing Volume-Delay functions for 16 carriageway types);
f. Terminal surveys.
g. Journey Speed Surveys (550 km).
h. IPT surveys (50 locations);
i. Pedestrian count surveys (50 locations); and
j. Workplace Surveys (5000 sample size)

In addition to the above, secondary information/data is collected on transport systems operating in MMR. GIS-based maps for the entire MMR have been prepared, incorporating the land use details and Development Plans. Based on these surveys, a database has been prepared for the region's household socio-economic and travel characteristics, enabling detailed travel demand modelling to study the future transport network requirements of the MMR. Following the collection of this data, desire lines were plotted for the existing traffic scenario and the P3E3 scenario, considering the existing transport scenario. The figures below show the intensity and direction of travel demand.

Fig.10.6: Special Steel Structure for Crossing WR Tracks

Fig.10.7: First Metro Corridor across WH

10.5 CTS Recommendations

As observed from the desired lines and trip assignments, the volume of users on the public transport and road network is expected to increase significantly, well above current levels. It is also observed that the east-west trips within the region are also growing. Several public transport options, including the Bus Rapid Transit System, Light Rail, and Mass Rapid Transit Systems, were considered to address the growing need for public transport. While the gross land area of MMR is extensive, encompassing some 4,355 sq. km, only about 12.5% of this land can be categorised as potentially developable. Approximately 1,134 sq. km are under the jurisdiction of various municipalities and their respective development plans, spread across multiple geographically diverse areas, ranging from wetlands to mountains. The existing built-up area is approximately 418 sq km, leaving a balance of 718 sq km. However, since this also includes water bodies and green zones, the actual developable land within municipal boundaries is approximately 168 square kilometres. Further potential urbanisable zones, identified in the Regional Plan for 1996-2011, fall under two categories, i.e., U1 and U25, contributing 182 and 180 square kilometres, respectively.

This means that approximately 530 sq km of land is deemed suitable for future development, or about 12.5% of the total MMR area. Within Greater Mumbai, 41% of the 438 sq km land area is designated as "No Development Zones" for several reasons, primarily related to environmental constraints or the preservation of parks and open spaces. Less than 10 sq.km of designated land is yet to be developed in Greater Mumbai. Considering MMR's geography, topography, scarcity of developable land and massive demand for public transport, a comprehensive metro network that enhances east-west connectivity and increases the capacity of north-south links is necessary. (Source: Comprehensive Transportation Study for MMR, 2008). A comprehensive investment plan was proposed, encompassing investments in the enhancement of existing transportation modes and the introduction of new modes, such as metro rail. A summary of the investments in public transport is as.

It is observed that, between 2008 and 2031, the total metro network required for MMR is 450 km, and the investment in metro rail systems accounts for over 50% of the total investment required in MMR's transportation network. For the development of the transit corridors (metro system and suburban railway), a prioritising plan that weighs travel demand and the cost of development has been developed. The travel demand parameter is measured in passenger-kilometres per length, and the development cost parameter is measured in passenger-kilometres per cost. The methodology followed is as described below. Following the prioritisation plan outlined above, the 24 metro corridors have been ranked, and a phasing plan has been prepared, as shown in **Fig.10.8** and Table 10.5.

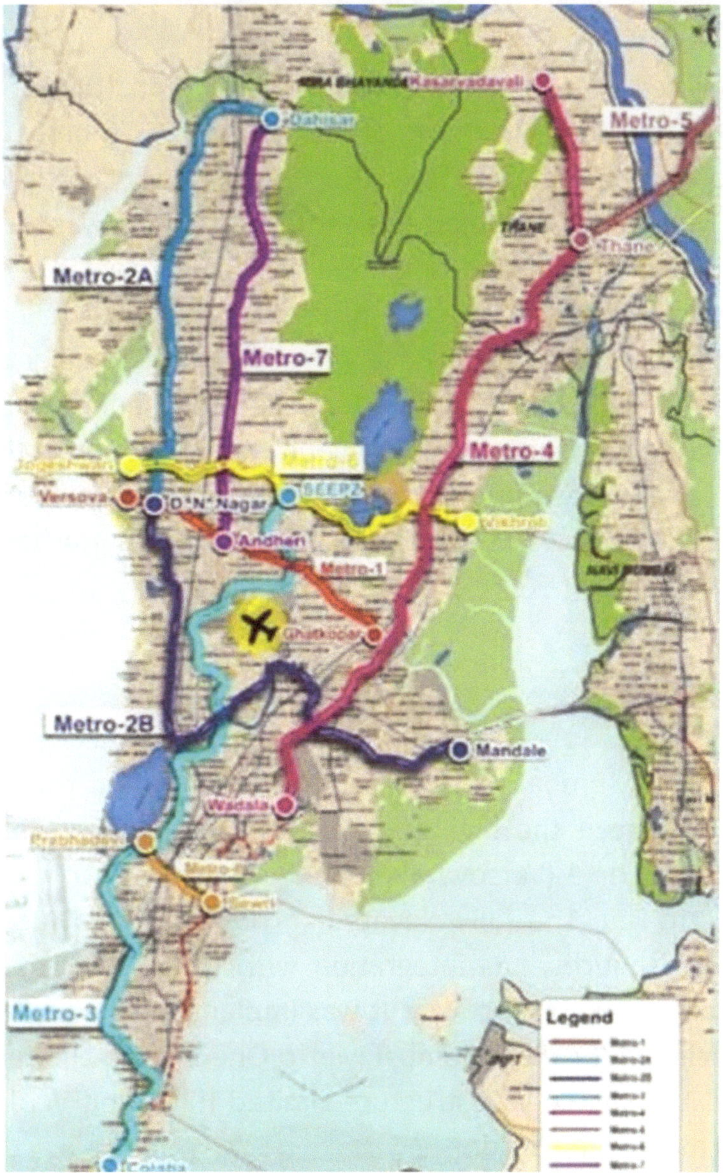

Fig.10.8: **Mumbai Metro Network Plan**

Fig.10.9: Narrow ROW, Metro Line 1 Fig.10.10: Utility Shifting before Metro Construction

10.6 Mumbai Metro Master Plan

In 2003, MMRDA appointed M/s. Delhi Metro Rail Corporation (DMRC) (was assisted by TATA Consultancy Services (TCS) and Indian Institute of Technology (IIT), Mumbai) to prepare the Master Plan for Mumbai Metro and the Detailed Project Report for priority corridors within the Municipal Corporation of Greater Mumbai limits. Objectives of the master plan was to first provide a rail-based mass transit system to areas that the existing Suburban Rail System does not currently connect, and secondly to ensure connectivity to mass transit stations within a half to one mile radius. The Master Plan proposes nine corridors covering a length of 146.5 km to be developed in three phases. Intermodal integration with the Suburban Rail System has been planned at Churchgate, Chhatrapati Shivaji Terminus, Mumbai Central and Bandra Terminus. Metro rail connectivity to the international and domestic airports has also been provided. Connections with areas outside MCGM limits have been planned through interchanges at Sewri (via Mumbai Trans Harbour Link) and Mankhurd suburban station.

10.6.1 Mumbai Metro Line 1

Mumbai's Suburban Railways provide rapid north-south connectivity; however, east-west connectivity is lacking. The Mumbai Metro Master Plan identified this failing of the transport network in the city and placed the highest priority on developing east-west connectivity in Phase 1 through the Metro Line 1 (Versova – Andheri – Ghatkopar) corridor and Line 3 (Mahim – Kurla – Mankhurd). Metro Line 1, a fully elevated corridor that provides connectivity between the eastern and western suburbs, has integration with the Western and Central Railways at Andheri and Ghatkopar stations, respectively. It was implemented on a PPP/BOOT basis in 2008. A Special Purpose Vehicle (SPV) called Mumbai Metro One Private Limited (MMOPL) was jointly incorporated by MMRDA, Reliance Infrastructure Limited (RIL), and Veolia Transport France for the implementation of the Metro corridor.

The corridor primarily runs along the median of the Andheri-Ghatkopar Link Road, spanning a length of 11.40 km and featuring 12 stations. Its alignment posed many engineering challenges, including crossing the Western Railway tracks at Andheri, crossing the Western Express Highway (WEH) at the second level, narrow rights-of-way, and numerous uncharted utilities

The first metro service was launched in June 2012; currently, the average daily ridership exceeds 0.3 million commuters. It is observed that the metro has reduced the travel time along the corridor from 71 minutes to 21 minutes, significantly alleviating traffic congestion. Further, a 35% modal shift from private vehicles and IPTs to the Metro has occurred. To ensure safe and efficient commuter dispersal at metro station areas, Station Area Traffic Improvement Schemes (SATIS) have been implemented at all stations on the corridor. The schemes include widening footpaths, installing railings, relocating bus shelters, and establishing intermediate public transport (IPT) pick-up/drop-off points to enhance intermodal connectivity.

10.7 Current Mumbai Metro Status

The CTS recommended extending the metro master plan from within the MCGM limits to include the neighbouring Thane District, which has witnessed rapid growth in recent years. Based on the study's recommendations, the master plan was extended to 172 km with connectivity to the Thane District. Since the development of the Mumbai Metro Master Plan, only Metro Line 1 has become operational. In 2015, to address the critical transportation issues facing Mumbai, the Government of Maharashtra, in consultation with the DMRC, identified 118 km of elevated Metro corridors as a high-priority project to be implemented over the next three to four years.

10.8 Mumbai Metro Development Challenges

10.8.1 Narrow ROW

As described above, Mumbai's road infrastructure is underdeveloped, narrow **(Fig.10.9),** and availability of alternate or parallel roads. The existing roads, other than the Western Express Highway and Eastern Express Highway, are typically 3+3 lane arterial roads. Elevated metros require a minimum of 7m of ground space for the piling and casting of pile caps and piers. This significantly reduces the capacity of the roads; furthermore, many of Mumbai's roads have commercial establishments along them, resulting in the curbside lane being used for short-term parking and pick-ups/drop-offs. Managing the movement of heavy construction equipment and carrying out the construction works while minimising the impact on vehicular traffic and commercial establishments abutting the road is hugely challenging. High levels of coordination and cooperation are required among the Traffic Police, contractors, the implementing authority, residents, and business owners in the vicinity of the metro construction.

10.8.2 Encroachments

Encroachments in the form of illegal hutments, religious structures, hawkers, and stalls can become significant hindrances to the progress of construction activities. Shifting these structures are time-consuming, cause substantial delays, and high cost.

10.8.3 Utility Shifting

Mumbai has numerous utilities **(Fig.10.10)** under the carriageway and footpaths, including water and sewage mains, stormwater drains, gas lines, telephone, and ICT cables. For metro construction, utilities along or across the alignment must be shifted, diverted or relocated. The relocation of mapped utilities can be planned for and included in the project's cost and schedule; however, many older utilities are uncharted or not accurately located according to available maps/drawings. Encountering uncharted utilities is a significant challenge and often becomes a critical path item on the project schedule, significantly impacting the project timeline.

10.8.4 Mumbai Metro Learnings and Key Initiatives

MMRDA has implemented Metro Lines 7 and 2A, and Metro Lines 2B and 4 are currently in the tendering stage. As learnt from the implementation of Metro Line 1, MMRDA has taken several steps to reduce project delays and ensure smoother project implementation.

10.8.5 Project Coordination

The MMRDA has established steering committees at the working and Chief Minister levels to ensure high coordination and expedite clearances from all departments involved in the project. The steering committees help ensure the collaboration of contractors, consultants, and government agencies.

10.8.6 Public Consultation

Public consultations have been conducted to sensitise residents and business owners along the metro alignments, and their opinions have been incorporated into the social and environmental management plans.

10.8.7 Traffic Diversion Plans

Following a thorough study of traffic and travel patterns along the road, detailed traffic diversion plans have been developed to minimise the impact. With the coordination and assistance of the Traffic Police, reversible lanes have been introduced where traffic is higher in one direction during peak hours. The Traffic Police have also provided additional field equipment, vehicles, and manpower.

10.9 Mumbai Metro in the Next Few Years

MMRDA has taken up the implementation of 18.6 km long Metro Line -2A (Dahisar (E) – D.N. Nagar) and 16.5 km long Metro Line – 7 (Dahisar (E)- Andheri (E)), which have a project cost of US$986 million (Rs. 6,410 cr) and US$955 million (Rs. 6,208 cr) respectively. The lines are expected to be completed by 2019, benefiting approximately 0.9 million commuters. Metro Line 2B (DN Nagar – Mandale) and Line 4 (Wadala – Thane), which have a project cost of US$1,690 million (Rs. 6,410 cr) and US$2,238 million (Rs. 6,208 cr), respectively, are currently in the tendering stage, and civil works shall commence shortly.

The Detailed Project Reports for Line-5 (Thane-Bhiwandi-Kalyan) and Line-6 (Swami Samarth Nagar-Jogeshwari-Kanjurmarg-Vikhroli) have been approved by the MMRDA Authority, chaired by the Hon'ble Chief Minister of Maharashtra. They are expected to receive approval from the State Government shortly. When all metro lines are completed, the Metro system will carry about 7 million commuters, equal to the Mumbai Suburban Railway system. This will double the capacity of the public transport system, alleviating Mumbai's traffic problems for 30 to 40 years.

With RDSO and CMRS clearances, it is reasonable to expect the project to be completed in six and a half years, making it a viable timeline for the first Metro Line in Mumbai. During the initial years, when the fare structure was Rs. 6, Rs. 9, and Rs. 11, the ridership per day on a few days reached nearly 500,000. After the steep fare increase by SPV, ridership remained consistently around 350,000 daily, and the line had an operational surplus despite overall losses due to debt servicing. During the COVID-19 Pandemic, the system suffered significant losses, like other mass transit systems worldwide, and is now slowly recovering to pre-pandemic levels.

10.10 Current Status

The construction commenced in February 2008, and the project achieved financial closure in October 2008. The Concessionaire estimated the project's completion date in mid-2011, which proved too optimistic. With complete handholding and cooperation of Mumbai Authorities and the Public at large, the project was opened to the Public in June 2014. The type of challenges, like Western Railway crossing north of Andheri Station, the presence of religious structures at more than three locations and completing all mandatory checks for its court litigations, arbitrations, and disputes, make the Metro Line 1 of Mumbai the most successful PPP project in terms of its efficiency and functional utility to the public.

10.11 Financing Information

The successful bidder estimated the total project cost at 2,356 crores. The project will be financed based on a Viability Grant of ₹650 crores, contributed by the Government of India (₹470 crores, equivalent to 20% of the project cost) and the Government of Maharashtra (₹180 crores, equivalent to 7.5% of the project cost).

The remainder will be financed by 70% debt and 30% equity. The private operator has also arranged a debt of Rs 1,147 crore for the project. This was tied up with a consortium of banks led by IDBI, Corporation Bank, Karur Vysya Bank, Canara Bank, Indian Bank, and Oriental Bank of Commerce. IIFCL (U.K.) is providing the foreign currency loan for the project.

Fig.10.11: Bridge over Jogeshwari Flyover - WEH

Fig.10.12: Construction Challenges – Underground Utilities

The loan has been secured for a moratorium period of 2 years and a total loan repayment period of 15 years. The project has also considered a service debt facility of approximately 70-80 crore as part of the project cost, ensuring that cost overruns are addressed during the project's tenure. Senior Lenders have also been notified of and approved these provisions. The financials estimated by the successful bidder are a Project EIRR of 8%, an Equity IRR of 15%, and a Debt-Equity ratio of 70:30.

10.12 Conceptualisation and Feasibility

The Government has been exploring the viability of various efficient, economically viable, and environmentally friendly mass transit systems. In this context, a detailed feasibility study was carried out.

Under the Indo-German Technical Cooperation, the consultancy work was entrusted to TEWET in association with **DE-Consult & TCS** from 1997 to 2000. Later, the Detailed Project Report (DPR) was prepared by **CES led Team** under the leadership of Dr BC Roy, the author.

After examining several alternative corridors and alignments, the study recommended a mass transit corridor from Andheri to Ghatkopar as potentially bankable and economically viable. The project was then decided to be bid out on a PPP basis based on DPR. To manage the transaction process, a consortium consisting **of Reliance, Louis Berger as a technical consultant, PricewaterhouseCoopers** (PwC), **and Masons and Economic Law Practices** was appointed in 2003-04 to assist the MMRDA.

10.13 Procurement

The Government of Maharashtra approved the project in August 2004, and global bids were invited in the same month through an Expression of Interest (EOI). Almost 150 bidders responded to the EOL, and a pre-bid meeting was held in November 2004.

Prospective bidders' suggestions were incorporated into the agreements being prepared for the project. The bid process was essentially two stages: technical and financial.

Only those consortia whose technical bids met the technical criteria were allowed to submit financial bids. Technical bids were invited for the project in May 2005. The consortia that submitted bids were:

- Hindustan Construction Company and RITES
- Reliance Energy Limited and Connex - France
- Shakti kumar Sacheti Limited and Lingkaran Metro
- Siemens, L&T, Gammon, BEML
- IL&FS and ITD Thailand, and Unity Infra projects

Only two financial proposals were received from two consortia of Reliance and IL&FS in January 2006.

After the bid process, negotiations commenced with the lowest financial bidder, Reliance Energy and Connex France. The consortium finally agreed on an Equity IRR of 15% on their investment, bringing the VGF to 650 crores from 1250 crores.

An application for VGF was submitted to the Government of Maharashtra in June 2006, following the selection of the successful bidder. The project faced delays in obtaining approval for Viability Gap Funding (VGF) as it was conceptualised before the model concession agreement was put in place by the Government of India. There was no Model Concession Agreement available at the time the bids were prepared in 2004-2005.

Moreover, the Public-Private Partnership Appraisal Committee (PPPAC) at the central government level was not constituted until then, and only tentative guidelines were in place for the PPP agreements. Therefore, various options to obtain grant funding were explored at that time, including securing grant funding through the JNURM scheme. However, the JNURM funds were capped at 10 % of the project cost.

The issue was finally resolved by a VGF grant in the form of a special one-time grant given to the state. The GoI agreed to provide a special grant of 20 % of the project cost.

10.14 Project Implementation

Project preparation commenced in 2007, following the consortium's signing of the Concession Agreement, although financial closure had not yet been achieved. The consortium was committed to ensuring the success of the first PPP project without delay.

Significant milestones achieved in the development phase are presented below:

The SPV was incorporated in December 2006.

- The Engineering and Project Management Consultants, a consortium of **Parsons Brinkerhoff (USA) and Systra SA (France)** joined the team on February 14, 2007.
- Signing of the Concession Agreement and Shareholders' Agreement occurred on March 7, 2007.
- MMOPL and the Government of Maharashtra entered the State Support Agreement on April 20, 2007.
- Construction commenced on February 8, 2008. Financial Closure for the project was completed on October 3, 2008.

All major contracts for the project were awarded promptly. Before construction began, 90 % of the right-of-way was handed over to MMOPL.

Some of the significant risks as perceived by the Private entities to provisions of the PPP document prepared by the MMRDA are explained below:

Table 10.2: Different Types of Risks

Risk Type	Sensitivity	Risk Period	Primary Risk Bearer	Comments
Financing Risk	Medium	0-5 years	Private Sector	Only 85% of the viability gap will be released during the construction phase of the project. These tranches would only be release the funds for the project. The remainder of the capital would be released 6 months after the project is operational.
Operational Phase Risk				
Technology Risk	Low	0-35 years	Private Sector / Government	The technology risk would vest with the both the private operator/government as the project would be executed in conformance with the Specification and Standards specified in the Schedules of the Agreement. (i.e. Performance Requirements, Performance standards and Technologies, Proposal submitted by private operator)
Operations & Maintenance Risk	Medium	0-35 years	Private Sector	The operations risk lies with the private sector with the private operator having to submit an operations and maintenance manual to MMRDA for approval. The private operator can mitigate this risk as he is allowed to appoint O&M contractors for the running of the system
Market Risk	High	0-30 years	Private Sector	The private operator would be allowed to levy and collect the fares from the users of the Metro and would not be entitled to revise these fares. All revenues accruing from the system would be deposited in an Escrow account from which payments would be disbursed. The private operator would be have access to any revenues in the account after all payments have been made. The fares would be revised at a rate of 11% every fourth year. The private operator can however approach the Government of Maharashtra for any upward revision of fares beyond those permitted in case of an increase in costs. There is no revenue guarantee from the government and the market risk is completely borne by the concessionaire.

Risk Type	Sensitivity	Risk Period	Primary Risk Bearer	Comments
Performance Risk	High	0-30 years	Private Sector	A lock-in condition exists where the private operator has to hold at least 51% equity during construction and in the 2 years after completion of the project. The lead consortium member will have to hold at least 26% equity stake in the project for a minimum period of 15 years after completion of project.
Handover Risk				
Handover Risk	Low	35th year	Private Sector	A joint inspection would be conducted by both the parties (Private Operator and MMRDA) 60 months prior to expiry of the concession period to gauge their compliance with the serviceability requirements defined in the agreement. This would be verified by an Independent Engineer, and the private operator would have to pay for the changes to comply with these serviceability requirements if found deficient.
Private Operator Event of Default	Low	0-35 years	Private Sector	Only leaders are protected, and so the equity holders bear the major risk. MMRDA takes over the assets and is liable to pay 90% of the debt due less insurance claims.
MMRDA Event of Default	Low	0-35 years	Government	MMRDA takes over the assets and is liable to pay a higher value of the assets. (110% of adjusted equity and 100% of debt due). Lenders are covered.
Pre-Operative Risks				
Delays in land acquisition	High	0-5 years	Government	The land will be handed over to the concessionaire as per the schedule submitted in the contract. If the MMRDA is not able to provide access to any part or parts of the site for reasons other than a Force Majeure then the MMRDA is liable to extend the Project Completion Date, Financial Period as determined by an Independent Engineer.

Risk Type	Sensitivity	Risk Period	Primary Risk Bearer	Comments
Financing Risk	Medium	0-5 years	Private Sector	The Private Operator has to achieve financial closure 180 days after the signing of the contract. MMRDA can extend the date for financial closure for a further 180 days in case the private operator cannot achieve financial closure.
Planning	Medium	0-5 years	Private Sector	The risk for the planning and execution of the project would vest with the private operator and would need to be executed in conformance with the Specification and Standards specified in the Schedules of the Agreement. (i.e. performance Requirements, Performance Standards and Technologies, Proposal submitted by private operator)
Regulatory, administrative & approval delays	Low	0-5 years	Private Sector	This risk is borne by the private operator who has to obtain all applicable permits/clearances from the GoI/GoM with regard to the implementation of the project. This is a Condition Precedent to the Concession Contract However, if the GoI/goM inordinately delay the project applicable permits so as to cause a Material Adverse Effect then MMRDA (i.e. Shareholder in MMOPL SPV) shall not terminate the agreement.
Construction Phase Risks				
Design Risk	Medium	0-5 years	Private Sector	The private operator would submit all the drawings and the schedule of the project to MMRDA. These would be reviewed by MMRDA and scrutinized by the Independed Engineer. MMRDA would not be responsible for any delays caused due to the drawings of the project.
Construction Risk	Medium	0-5 years	Private Sector	The private operator for the due and faithful performance of its obligations, shall provide a performance security of ₹ 14 Crore (USD 2.8 million) for the due and faithful performance of its obligations. This security would be renewed from time to time and would need to be replenished within 30 days. The private operator wuld also be liable to pay damages at the rate of ₹ 2 crores/ day if it fails to achieve any milestone.

Risk Type	Sensitivity	Risk Period	Primary Risk Bearer	Comments
				Furthermore, the private operator Must submit Monthly progress reports and allow the independent Engineer to inspect the progress of construction. The independent Engineer also would subject the MRTS to test and provide a provisional completion certificate for the project.
Change in Scope Risk	Low	0-5 years	Private Sector	It the government requires additional works and services which are beyond the scope of the project all such change in scope shall be made by MMRDA by an order and giving consideration to the operations and maintenance costs which would be spent by the private operator and then subsequently reimbursed by MMRDA
Interface Risk (with other metro corridors)	Medium	throughout	Private Sector	The concession agreement only touches upon the interface aspect. It states that as and when the other corridors become operational integration would be necessary which will call for co-operation and co-ordination amongst various SPVs / Operators. The risk would be with the private concessionaire with no financial involvement of MMRDA.
Force Majeure	Low	throughout	Shared (Depending on the type of event)	In case of a Force Majeure event occurring before the financial closure date, the project completion date and concession period would be extended. In case of a Force Majeure event arising due to an indirect and non-political event each party would bear their respective costs. In case of a Force Majeure arising out of a direct political event the MMRDA would reimburse the private operator to the tune of the costs incurred due to the Force Majeure. The compensation shall include O&M expenses, debt and increases in cost of contraction. In the event of termination there is compensation payable to the concessionaire till the extent of debt outstanding and varying level of equity compensation depending upon the nature of events.

Risk Type	Sensitivity	Risk Period	Primary Risk Bearer	Comments
Change in Law Risk	Low	throughout	Private Sector	No compensation from the government, although enabling provision to mutually discuss in good faith to suitably amend the terms of the concession if the financial effect of the change in law is greater than ₹ 1 crore. MMRDA is required to pay the amount of cost increase in case no settlement is reached.

10.15 Implementation Analysis

The only successful model in India or worldwide is the implementation of Metro Projects by the Government or wholly owned government entities. Risks perceived by private entities are generally not considered at that time. Delhi Metro Rail Corporation (DMRC), a public sector organisation, has successfully managed project timelines and costs for the Delhi Metro Network. On the other hand, there is no precedent of a similar institution in rail-based public transport in Maharashtra. Data on other public works contracts could not be adapted for the Mumbai Metro One project.

The following primary factors listed below were perceived to contribute toward creating value for money for the state government about this project:

1. **Reduction in financial burden on the state budget:** The PPP project was structured to ensure viability by providing viability gap funding. The financial bid parameter was the lowest viability gap quoted. The lowest financial bid quoted was with viability gap funding of ₹ 1,250 crores, which was reduced after negotiations to ₹ 650 crores. The government is also committed to compensating the concessionaire for expenses incurred in relocating public utilities along the right of way. This is estimated to be around 50-60 crores. The total estimated taxes for the project are approximately 300 crores. The asset would also transfer back to the government at the end of the concession period. In conclusion, the state government can cover the construction and operation of critical infrastructure through private sector participation for 30 years, with a reduced requirement for upfront capital.
2. **Substantial risk transfer:** The private sector has undertaken substantial project risks, such as financing, construction, operations and traffic/ revenue risks. Traffic risk, one of the significant risks associated with a user-fee-based model, is wholly vested in the private operator, with no clauses that provide for any compensation by the state government if the metro's ridership is low. The transfer of significant risks during the operational phase ensures that the private operator will make every effort to operate the system in a manner that promotes its sustainability.

While the above factors do point to the fact that there was substantial merit in adopting the PPP approach, we must acknowledge the implicit risks that the public sector would continue to retain;

there have been several international precedents where governments have taken back certain risks or bailed out ailing public infrastructure projects on the grounds of protecting public interest. For example, during the Mexican toll road crisis, the government had to assume the financial liabilities of ailing toll road operators, such as the bailout of the Thai BTS. Indeed, such financial support or bailout packages were, perhaps, necessary for those economies. However, these are implicit risks for the government due to the nature of public service or infrastructure facility being rendered, and the government's stated or implicit obligations to its citizens to provide the same need to be understood.

10.16 Key Learning and Observations

The pending bid process is critical to ensuring an excellent response to the proposal, as choosing the successful bidder took over two years. This resulted in a smaller number of bidders competing for the project. Similar hurdles were encountered during the bid process for Metro Line 2, as the concession agreement was based on the standard model concession agreement. However, this agreement had to be tailored to implement a metro system. These delays resulted in only one bidder finally submitting a bid for the project.

1. **Delay in Obtaining VGF approval:** There was a substantial delay in obtaining approval for VGF from the Government. While this was attributed to the model concession agreement not being in place, the PPP Appraisal Committee not being constituted, and only tentative guidelines concerning VGF approval being available at the time, this issue was a deterring factor for developers and is also likely to have impacted the level of interest in the Phase 2 bid.

Fig.10.13: Metro Line 7 in Operation

2. **Delays in approvals** can potentially derail the project. There was a delay in obtaining approvals from the railway authorities for the bridge that passed over the railway line. This had the potential of delaying the project schedule. This was due to the railways exploring the feasibility of another project that would invade the path of the metro line. However, a quick resolution of this issue ensured that work could continue. It is recommended that authorities be aware of all other upcoming infrastructure projects that may potentially affect the planned project's operations.
3. **The Land Acquisition** process can lead to issues in the project. The government committed to procuring the land for the project, which essentially consists of land allocated for the depot, in accordance with the land procurement schedule provided in the agreement. However, this land was under private ownership and was in dispute. The government was exposed to the risk of land being unable for the depot. The issue was finally resolved after the private owner of the land agreed to provide 75% of the land for the project's development, on the condition that the government grant him the right to the Floor Space Index (FSI) available over the entire plot of land for 25% of the land. This land has been rented on a nominal lease to the concessionaire for the concession period.
4. **Clear Specifications on Asset Transfer on Termination**: On the termination of the project through the efflux of time, five years before the expiry of the concession period, a survey of the assets would be carried out to determine whether they are in working condition as given in the agreement. The survey is to be conducted by an independent engineer, based on a schedule of specifications, to assess the condition of the assets. However, the schedule in the concession agreement lacks precise and robust specifications. There is a risk of a difference of opinion between the concessionaire and the government, which could potentially lead to a dispute. The government could manage this better by incorporating precise and robust specifications on the condition it would want the assets to be handed over to the government.
5. **Public Support for the Project**: For a project of this magnitude, it is crucial for the government agency to garner adequate public support to ensure a smooth implementation. MMRDA ensured sufficient public support for land acquisition and road expansion activities through a dialogue with the affected individuals. Despite these efforts, the project was susceptible to delays and similar difficulties were experienced in phase 2 of the project.
6. **Role of Good Project Preparation:** The viability gap funding used in the project (650 crores) makes up a significant component (27.5 per cent) of the project cost. The project cost has been shared between the central and state governments. The initial quote submitted by the successful bidder was subsequently revised to the current figure through negotiations. Thus, there is an increased need for good project preparation before the procurement process to ensure fair bids are received for the projects. This would eliminate private operators colluding with each other and/or speculative bids.
7. **Expediting the bid process** is critical to ensuring an excellent response to the proposal: The entire bid process for choosing the successful bidder took over two years. This led to a smaller number of bidders to bid for the project. Similar hurdles were experienced in the bid process

for Metro Line 2, as the concession agreement was based on the model concession agreement. However, this agreement had to be tailored for the implementation of a metro system. These delays resulted in only one bidder finally submitting a bid for the project.

8. **Delays in approvals** can potentially derail the project. There was a delay in obtaining approvals from the railway authorities for the bridge that passed over the railway line. This had the potential of delaying the project schedule. This was due to the railways exploring the feasibility of another project that would invade the path of the metro line. However, a quick resolution of this issue ensured that work could continue. It is recommended that authorities be aware of all other upcoming infrastructure projects that may impact the planned project's operations.

   The Land Acquisition process can lead to issues in the project. The government committed to procuring the land for the project, which essentially consists of land allocated for the depot, in accordance with the land procurement schedule provided in the agreement. However, this land was under private ownership and was in dispute. The government was exposed to the risk that the land would be unable to accommodate the depot. The issue was finally resolved after the private owner of the land agreed to provide 75% of the land for the project's development, on the condition that the government grant him the right to the Floor Space Index (FSI) available over the entire plot of land for 25% of the land. This land has been rented on a nominal lease to the concessionaire for the concession period.

9. **Clear Specifications on Asset Transfer on Termination:** On the termination of the project through the efflux of time, five years before the expiry of the concession period, a survey of the assets would be carried out to determine whether they are in working condition as given in the agreement. The survey is to be conducted by an independent engineer, based on a schedule of specifications, to assess the condition of the assets. However, the schedule in the concession agreement lacks precise and robust specifications. There is a risk of a difference of opinion between the concessionaire and the government, which could potentially lead to a dispute. The government could manage this better by incorporating precise and robust specifications, on the condition it would want the assets to be handed over to the government

Public Support for the Project. For a project of this magnitude, it is essential for the government agency to provide adequate public support to ensure smooth implementation. MMADA ensured sufficient public support for land acquisition and road expansion activities through a dialogue with the affected individuals. Despite these efforts, the project was susceptible to delays and similar difficulties were experienced in phase 2 of the project

The project initiated under the Tramways Act, after having gone through rigorous discussions at the State and Central Government levels, got into a legal tussle when the project was declared under the Metro Act 2008 during the fag end of the construction phase. The Act lacked adequate provisions to cover ongoing PPP projects, particularly regarding the Concession Agreement's provisions on changes in Law, including rights of development, Fare fixation, and Non-Fare Box revenues.

Fig.10.14: Mumbai Metro Line 10

10.17 Andheri-Ghatkopar-Versova Metro Corridor Line 1

10.17.1 Background

The Island City has experienced rapid growth in economic activity, and its suburbs are growing faster than the City itself in terms of population distribution and economic activity. The major challenge is in promoting growth through adequate infrastructure inputs, which would improve the quality of life. The MMRDA (Mumbai Metropolitan Region Development Authority) is a nodal agency for promoting planned development, which automatically makes it the nodal agency for implementing transport-related projects within the Mumbai Metropolitan Region. MMRDA prepared a Regional Plan for MMR for 1996 to 2011, which was approved by the GOM. As per the Regional Plan, a poly-nucleated land use.

Mumbai is served by two of India's zonal railways, the Western Railway (WR) and the Central Railway (CR). The Western Railway line runs northwards from Churchgate terminus station in the Mumbai CBD towards northern and western India. The Central Railway runs from Chhatrapati Shivaji Terminus (CST), located on the eastern side of the CBD and serves a large part of Central and Eastern Mumbai. Mumbai generally prefers the Public Transport System. Public bus services are provided by BEST and TMT in Thane and MSRTC elsewhere. The dispersal of rail commuters from the main railway terminals to their final destinations, such as Fort, Ballard Estate, Colaba, or Nariman Point, is primarily carried out by the bus system. The Public Transport System is an efficient user of space, with a reduced level of air and noise pollution. With the increase in population, the share of public transport, road or rail-based, should increase. With Mumbai's population crossing 12 million, the share of public transport at 88% is good.

While upgrading the existing suburban system through the MUTP, additional mass transit corridors were needed to meet the expanding demand. Metro systems are superior to other modes of transportation as they offer a higher carrying capacity, provide faster, smoother, and safer travel, occupy less space, are non-polluting, and are energy-efficient.

While the suburban rail system and road-based transport system are being managed by MUIP and MUTP, the metro system proposed in the past was implemented in an ad hoc manner. It was necessary to have a master plan for the metro system. Based on the survey and detailed study, probable corridors were discussed with local authorities, and a network comprising 146.5 km was selected as the Master Plan. And the Versova-Andheri-Ghatkopar Metro corridor was selected as Line 1, and it was decided to implement. To start with, **DE Consult** prepared the study report for the Ghatkopar-Andheri metro corridor. Later, MMRDA engaged **CES** to prepare a comprehensive DPR for the Ghatkopar-Andheri corridor, the first corridor in Mumbai, named Line 1. After the report was submitted and the tendering process began, MMRDA decided to extend the corridor to Versova. The CES Report was open for selection of the location for the Depot, whether at Godrej, with a flyover approach to vacant land, which is not a preferred solution and would require acquisition, or at the Andheri site. Near Andheri Suburban Railway Station, the Depot must be elevated, which is very costly. The alternative was to go beyond the suburban railway tracks. MMRDA appointed DMRC to prepare the DPR for the extended corridor and decided to implement it in a PPP mode.

Fig.10.15: Blue line- Line 1 Mumbai Metro

Fig.10.16: Azad Nagar Station during Construction

Fig.10.17: Marol Naka Station during Construction 2012

Fig.10.18: First Metro in Mumbai at DN Nagar in 2013

10.17.2 Salient Features of Line 1

The Versova-Andheri-Ghatkopar corridor is an 11.40 km elevated corridor. It provided connectivity to the eastern and western suburbs of the Western & Central Railway. The facility facilitates smooth and efficient interchange between the suburban rail system and Mumbai Metro Line 1.

Gauge -- 1435 mm

Route Length -- Elevated 11.07 Km

Number of stations -- Elevated 12 Nos.

Traffic Forecast --

	Passenger/day (Lakh)	PHPDT
Year 2011	4.28	15,565
Year 2021	6.65	23,590
Year 2031	8.83	30,550

Designed PHPDT	23560 (2031)
Designed Train headway	3 minutes
Operational Headway	5 / 3 minutes
Train Composition	4 Coaches
Design speed	80 kmph
Traction system voltage	25 KV ac
Current Collection	Overhead Catenary
Sub Station	2 No.
Rolling Stock	3.20 m wide, stainless-steel body
Axle load	17 t
Seating arrangement	Longitudinal
Capacity	1178 Passengers
Maintenance Depot	D N Nagar

Signaling, Telecommunication & Train Control

Type of Signaling: Cab signaling with ATC & ATP

Telecommunication: Integrated System with Fibre Optic cable, SCADA,

Train Radio, PA system, etc.

Train information system-- telephones and

Centralised Clock System.

Fare Collection: AFC with BOM and POM, Smart card, etc.

Construction Methodology: Elevated viaduct consisting of prestressed concrete 'U'-shaped, Girders on a Single pier with pile/ Open foundations

Financial Indices EIRR 21.53 %

FIRR

(i) Govt. Financing 7.56 %

(ii) PPP 11.06 %

10.17.3 Alignment

The project alignment has two sections: i) Andheri–Ghatkopar, the eastern segment, and Andheri-Versova, the western segment. **MMRDA appointed CES** to prepare a Detailed Project Report based on the study report done by DE Consult. The corridor was planned as an elevated corridor. The total length of the corridor is 11.07 Km.

The area for the maintenance depot was not readily available. During the preparation of the DPR, the Godrej site needs to be connected by a flyover and acquired at the Ghatkopar end. The option was studied as one of the possible sites. At the other end, the depot, if proposed, had to be elevated, as it was close to the existing Andheri Railway Station and the Suburban Railway tracks, and the final location for the Depot could not be frozen. Later, during further study, the depot site was proposed to be located on Jai Prakash Road near DN Road, where a plot of 12.25 hectares was nearly vacant. And the corridor was extended to Versova.

The corridor was planned mostly over the existing road. Since the roads were not wide enough, the MMRDA decided to widen them before starting the Metro construction. The stations were in the middle of the road. In total, twelve stations were planned along the entire corridor from Ghatkopar to Versova with a Depot near DN Road.

10.17.4 Geotechnical Investigation

Geotechnical investigations were conducted along both corridors to a depth of 30 m in soil, 10 m in soft rock, and 5 m in hard rock. Soil and rock samples were collected and tested in the laboratory. The subsoil strata at the proposed site comprise various types of layers, including silty sand, sandy clay, silty sandy clay with gravel, slightly weathered to fresh breccia/shale/limestone, highly to moderately weathered sandstone/breccia/basalt, and completely weathered rock. For the elevated section pile foundation with 1.2m diameter piles socketed in rock, is recommended. The bearing capacity of the soil is unlikely to cause any problems for the foundations.

10.17.5 Utilities

The proposed Metro alignment passes along major arterial roads of the city's road network, which serve Institutional, commercial and residential areas. Many surface and subsurface utility services, such as sewers, water mains, stormwater drains, telephone cables, electric poles, and traffic signals, exist along the proposed alignment. Details of the existing utility services along the proposed alignment were collected from the relevant authorities, namely the Municipal Corporation of Greater Mumbai (MCGM) and the Public Works Department. (PWD), Irrigation and Flood Department, MCGM, BEST (Bombay Electric Supply & Transportation), Mumbai Traffic Police, BSES (Bombay Sub-urb Electric Supply)/Reliance Energy, TATA Power, Mahanagar Gas Ltd., MMRDA & MAHADA, TATA Tele Services, Reliance Info. Ltd, Western & Central Railway, and others, as required. The affected portions of the services, with reference to the proposed alignment, were identified, and temporary diversion & relocation proposals for the affected services have been indicated.

10.17.6 Land Requirement

Every effort had been made to keep land requirements to the barest minimum, and the acquisition of private property is minimal. Land was mainly required for depots, route alignment on sharp bends, station buildings, platforms, entry/exit structures, traffic integration, power substations, administrative buildings, and construction depots/work sites, among other purposes. The land requirement on the Versova–Andheri–Ghatkopar corridor was approximately 12.64 hectares, out of which 0.031 hectares belonged to the government, while the remaining 12.61 hectares was private land.

10.17.7 Viaduct Structure

For fast and economical construction, precast segmental construction was proposed. Two types i) Pre-cast prestressed segmental Superstructure & ii) Pre-cast prestressed segmental concrete U-sections were proposed. Pre-cast segmental construction offers additional advantages, including superstructure segments that can be cast off-site and transported to the site as needed, thereby minimising disruptions to existing traffic. Casting of segments can be done simultaneously, when the foundations and substructure are in progress, and it is easy to incorporate last-minute changes, particularly to adjust span length, if required, to ensure an aesthetically pleasing outcome. However, precast box segments were preferred due to longer durability, though the use of U-segments lowers the Rail Level.

Later, in the DPR prepared by DMRC for the full corridor, i.e., Versova-Andheri-Ghatkopar, the use of U Girder was recommended.

a) Superstructure

The superstructure of a large part of the viaduct comprises simply supported spans. However, at major crossings/over or along existing bridges, special steel or a continuous unit was proposed.

However, both the DPR recommended precast prestressed concrete segments. The maximum number of spans of c/c of piers of standard simply supported constructed by the precast segmental construction technique had been designed as 25.0m. The usual segments were 3.0m in length, except the Diaphragm segments, which shall be 2.0m each.

The other spans comprise 28.0m, 31.0m, 22.0m,19.0m & 16.0m, which were proposed by removing/ adding usual segments of 3.0m each from the center of the span

For major crossings exceeding 40.0 m, a special unit of 3–span construction was envisioned. Continuous units (in cases provided at the obligatory location) were to be constructed using the cast-in-situ balanced cantilever construction technique. Alternatively steel structure could be adopted.

b) Substructure

The viaduct superstructure was planned to be supported on a single cast-in-place RC pier. The shape of the pier follows the flow of forces. For the standard spans, the pier gradually widens at the top to support the bearing under the box webs. The size of the pier was recommended to be 1.6m in diameter, of circular shape at the central verge.

To prevent the direct collision of a vehicle with the pier, a Jersey-shaped crash barrier of 1.0m height above the existing road level was provided all around the pier. The longitudinal center-to-center spacing of the elastomeric/pot bearing over a pier would be about 1.8m. The transverse spacing between bearings would be 3.2m, or as specified in the bearings' design.

c) Foundations

The soil substrata along the corridor comprise completely weathered rock, followed by highly to moderately weathered and slightly weathered to fresh brown, grey Brecia/Shale/Limestone/ Basalt at varying depths of 3.5m to 8m, 4.5m to 9m, and 7m to 14.8m, respectively. Accordingly, pile foundations with varying pile depths depending on the depth of rock from ground level were recommended.

d) Stations

The stations were placed in the median of the road. The total length of the station was planned as 135 m. All the stations were proposed at two levels. The 90 m concourse was proposed in the middle of the station, with staircases leading from either side of the road. Passenger facilities, including ticketing, information, and operational areas, were located at the concourse. Like other metros in India, the concourse included both paid and unpaid areas, i.e., public and non-public zones. The unpaid area included operational facilities such as Station Control Room, Station Master's Office, Waiting Room, Meeting Room, UPS & Battery Room, Signalling Room, Train Crew Room & Supervisor's Office, Security Room, Station Storeroom, Staff Toilets, etc.

Minimum vertical clearance of 5.5 m provided from road level to under the concourse. Accordingly, platforms were at a level of between 12.5 m & 14m. Platform roofs had been proposed to be of steel structure with aluminium cladding to achieve a light look. Portals across the road were proposed over which the station structure would rest, and over the road. The rest of the structure was supported on a single column, on the Central Verge

The **Ticketing Gates** were provided at the unpaid area, with numbers calculated based on a gate capacity of 45 persons per minute and an estimated traffic horizon year. At least two ticketing gates were planned at any station.

Platforms were proposed on both sides and from either end to provide direct access from both sides of the road through the concourse. A uniform platform width of a minimum of 3 m, excluding staircases and escalators. This would also help to locate the tracks at the centre throughout. The platform width was also checked to ensure a sufficient holding area in the event of an emergency.

Parking and circulation areas were planned for each station based on prevailing norms, considering the facilities required for the type of linkages planned in each station. Traffic and transport integration facilities are provided for two different types of linkages:

i. Feeder links to provide integration between various metro corridors and road-based transport modes, i.e. public, hired, and private vehicles.
ii. Walk links provide access to pedestrians.

10.17.8 Construction

Construction activities, as recommended, are to be planned with a mechanised approach, with the minimum area for the construction equipment, without storing the construction materials, and a minimum labour force, at the in-situ construction site, such as for substructure and foundations, Requires minimal hindrance to existing traffic, prioritising safety and health aspects for both personnel and machinery & with minimal or no environmental impact.

For precast construction, precast segments are to be cast off-site. The pre-casting site should be within a reasonable distance, and the safe transport of segments should be feasible through the urban road network. If a batching plant if erected at the site, then care should be taken to ensure the supply of concrete of the required strength around the clock. Procurement of materials should be conducted in accordance with the specification and design basis. Stacking of precast segments in the yard and at the site (temporary stacking during erection) should adhere to the design basis. Launching a girder, if used for the erection of segments, both design and safety norms must be approved by the Engineer before it is put into operation.

10.17.9 Epoxy Bonded Joints and Shear Keys.

A minimum compressive stress of 3 kg/sqcm should be provided uniformly over the cross-section for the closure stress on the epoxied joint until the epoxy has set. The curing period for the application of the compressive stress, the method of mixing, and application of epoxy and all related aspects, including surface preparation, must be as per approved manufacturer's specifications. The temporary compressive stress during the curing period must be applied using an approved external temporary bar prestressing system (such as Macalloy or Diwidag bar systems or an approved equivalent).

10.17.10 Environmental Impact Assessment

The aim is to ascertain the existing baseline conditions and assess the impacts of all factors resulting from the proposed corridor during its operation and construction in phases. The changes likely to occur in different components of the environment, viz. Natural Physical Resources, Natural Ecological (or Biological) Resources, Human/economic development resources (Human use values), and Quality of life values (socioeconomics) would be studied and assessed to a high degree of accuracy. The assessment includes Hydrology, Surface Water Quality, Air Quality, Soils, Noise, Geology, Socioeconomics, archaeological /historical monuments, etc.

Based on project particulars and existing environmental conditions, potential impacts result from the proposed MRTS. The positive environmental effects are i) Traffic congestion reduction, ii) Quick service and safety, iii) Less fuel consumption, iv) Reduction in Air Pollution, v) Better roads, and vi) Employment opportunities. Potential negative impacts likely to result from i) Impacts due to project location, ii) Impacts due to construction works, and iii) Impacts due to project operation.

10.17.11 Mitigation Measures

Based on the project requirement, Environmental Baseline Data and Environmental Impacts, the Environmental Management Plan may be planned for

i. Compensation for Loss of Land,
ii. Compensation for Loss of Trees,
iii. Compensatory Afforestation and Fencing,
iv. Compensation for Relocation/Resettlement,
v. Water Supply & Sanitation,
vi. Oil Pollution Control
vii. Noise Control
viii. Vibration Control

10.17.12 Permanent Way

i) Choice of Gauge

Standard Gauge (1435mm) was recommended for Line 1. The Versova-Andheri-Ghatkopar corridor, although in India, uses the national gauge, which is Broad Gauge (1676mm).

ii) Track Structure

Track on Metro Systems is subjected to intensive usage with very little time for day-to-day maintenance. Thus, it is imperative that the track structure selected for Metro Systems be long-lasting and require minimal or no maintenance, while ensuring the highest level of safety, reliability, and comfort, with minimal noise and vibrations. Ballastless track with long, welded, and hardened rails has been recommended, including the depot, as the track is planned to be located on an elevated deck. At the Godrej depot, the track structure is ballasted. Rail Section: In view of the proposed axle load, it is proposed to adopt a UIC-60 (60 kg/m) rail section. Since sharp curves and steep gradients would be present on main lines, the grade of rail on main lines was planned to be 1080 Hard Head.

Briefs on depot design and maintenance facilities have been included in Section 10.20.

System

10.17.13 Rolling stock

Six international firms—Siemens, Bombardier, Alstom, Rotem-Hyundai, CNR Changchun, and Nippon Sharyo—were shortlisted to provide rolling stock for the line, but CSR Puzhen was ultimately chosen to supply the rolling stock. CSR Puzhen was awarded a contract in May 2008 to supply 16 trains, each consisting of 4 cars. The first rake was shipped from Shanghai in March 2010, and the last rake arrived at Mumbai port by the end of February 2014.

Blue Line was designed with 64 coaches. Fourteen trains (of four cars) are in service during rush hour, and seven/eight in non-peak hours. All coaches are air-conditioned and have humidity control, and are designed to reduce noise and vibration. Each coach is approximately 2.9 wide and has 48 seats. A coach has a capacity of 375 passengers, and a single four-car train has a total capacity of 1500 passengers. The coach body is made of lightweight stainless steel, with fire-resistant metal doors. Coaches have metallic silver exteriors and are covered with vinyl sheets to display advertisements. The interior features anti-skid floors and longitudinal seats with dedicated space to accommodate wheelchairs. Trains on the Blue Line are fitted with the VTS Fire tide 7010 video transmission system. Coaches are also equipped with LED displays showing a dynamic route map, as well as LCD TVs for entertainment, information, and advertising. Windows in coaches are made of double-glazed laminated glass to shut out noise. Each coach has eight externally hung, sliding bi-parting doors, except the pilot cabin, which has only 2. Doors are broad to enable wheelchair access. The maximum gap between the station platform and the doors is 85 mm.

a) Headway

According to the study on line 1 ridership, the train composition consists of four coaches, which satisfies the requirement until 2031 with a headway of 5 minutes. And estimated as 4 mts. In 2021 and 3 mts in 2031.

Fig.10.19: Internal View of Metro Coach (Indianexpress.com)

10.17.14 Power

A power supply is required for the operation of the Metro system – for running trains, providing station services, workshops, depots & other maintenance infrastructure. Broad estimation of auxiliary and traction power demand is made based on: - (i) Specific energy consumption of rolling stock – 70KWh/1000 GTKM (ii) Regeneration by rolling stock – 30% (iii) Elevated/at –grade station load – initially 250KW, which will increase to 350 KW in the year 2031 (iv) Depot auxiliary load - initially 2000KW, which will increase to 2500 KW in the year 2031.

The ABB Group was awarded the contract for supplying power systems to Blue Line in July 2008. The line is electrified using 25 kV AC at 50 Hz via an overhead catenary system.

In early 2017, MMOPL completed the installation of light-emitting diode (LED) lighting at all metro stations, the depot, and administrative buildings to reduce electricity consumption. The MMOPL also began a project under the Renewable Energy Service Company (RESCO) model to install rooftop solar panels at all stations and the metro depot. A total of 2.30 MWp is expected to be generated by the system, with approximately 2 MWp generated by the station's rooftop solar panels and 0.30 MWp generated by the depot's solar panels. The LED lighting system and solar panels are estimated to reduce Blue Line's carbon emissions by 2,430 tons per year.

i) Standby Diesel Generator (DG) sets

In the unlikely event of simultaneous tripping of all input power sources or grid failure, the power supply to stations, as well as to trains, may be interrupted, and it was proposed to provide a standby DG set of 100 KVA capacity at the stations to cater to the following essential services: (i) Essential lighting (ii) Signaling & telecommunications (iii) Firefighting system (iv) Lift operation (v) Fare collection system

10.17.15 Signaling and Train Control

Blue Line features are an advanced signaling system, including an automatic train protection system (ATPS) and automated signaling to control train movements. Siemens supplied the signalling and train control systems required for the project, while Thales supplied the communication systems. The network's signaling and train control systems are based on LZB 700M technology. The requirements were planned to be achieved by adopting automatic signalling in the section and a Blaise-based Automatic Train Protection (ATP), a state-of-the-art automatic train supervision system.

10.17.16 Telecommunication

The telecom system was designed to cater.

- Train Traffic Control
- Assistance to Train Traffic Control
- Maintenance Control

- Emergency Control
- Station-to-station dedicated communication
- Exchange Telephone
- Passenger Announcement System within the station and from Central Control to each station.
- Centralised Clock System
- Train Destination Indicator
- Instant online Radio Communication between Central Control and Moving Cars and maintenance personnel.
- Data Channels for Signalling, SCADA, Automatic Fare Collection, etc.

10.17.17 Automatic Fare Collection

The ticketing system shall be simple, easy to use/operate, and maintain, with minimal impact on accounting facilities. It shall be capable of issuing single/multiple journey tickets, adaptable to quick fare changes, and require overall manpower. In view of the above, a computer-based automatic fare collection system is proposed. The AFC system proves to be cheaper than the semi-automatic (manual system) in the long run due to reduced manpower cost for ticketing staff, reduced

10.17.18 Accidents

Eight accidents occurred during the construction of the Blue Line. Two people were killed and 21 were injured during the construction of the line. The first accident occurred in May 2008 when 1 person was killed, and another was injured when a pile rig collapsed at a construction site in Andheri (West). In 2009, 4 people were injured when a steel reinforcement cage and temporary scaffolding of a concrete pillar caved in, in Andheri (East). In April 2012, a crane at a construction site in Ghatkopar veered off a truck and crashed into a portion of the nearby Sarvodaya Hospital. No one was injured as the affected portion of the hospital building was empty at the time.

On 5 September 2012, a slab collapsed at the under-construction Subhash Nagar metro station in Andheri, killing one construction worker and injuring 16 people. Following the incident, construction work on the metro was suspended. MMOPL fined the contractor. On 11 September 2012, MMOPL appointed SGS Consultants, based in Geneva, as independent safety consultants for the construction of the Blue Line. According to the additional metropolitan commissioner, "The accident was basically due to voids in the support. The support weakened due to rain, and the utilities underneath created voids. The weakening of the support was primarily due to the loosening of the soil caused by construction activity and its subsequent erosion from heavy rain. If, for any reason, a part of the support settles, stress increases on the staging and causes it to fail." A similar incident occurred in September 1979, when a slab slipped at the start of concreting on the roof of an underground substation, which was two meters thick, near Park Street during the construction of the Blue Line (Line 1) of the first Indian Metro in Kolkata. Staging was completed before the monsoon, and concreting was carried out after the monsoon, despite thorough checks of the staging. The reason was that the silt penetrated at the base of the timber log (part of the staging system), which was supported on a mud mat made of 1:3:6 concrete, but the location of this mat could not be determined.

During the initial stage of concreting drag of green concrete was the cause of staging collapse. Mishaps in other sections in line 3 are included in the last of this section.

10.18 Metro Line 3 (MML 3)

10.18.1 Background

Mumbai Metro Line 3, starting at Cuffe Parade, passing through BKC & Seepz, terminates at Aarey Colony, covering approximately 32.5 km. The entire length of the corridor is underground, comprising 28 stations connected by an underground tunnel corridor, except for Depot and Station AAREY, which were added later and are proposed to be at grade. At the beginning of the project planning, due to the uncertainty of land availability, the depot was initially planned at one or two levels, depending on the land made available before the finalisation of the Depot Design.

The main tunnels consist of a twin-tube configuration with an internal diameter of 5.8 meters each, excavated using an Earth Pressure Balance (EPB) machine with a 6.35-meter boring diameter. A total of 42 Tunnel Boring Machine (TBM) breakthroughs is planned. As of December 2020, 35 TBM breakthroughs have been recorded. In Mumbai, suburban trains are overcrowded, carrying 5000 passengers against the design capacity of 1750. MML-3 is expected to substantially reduce the passenger load. Traffic congestion would be reduced, and significant waiting time would be saved. Once implemented, it is estimated to reduce traffic on the corridor by about 35% or by 4.5 lakh vehicles, leading to possible decongestion of roads.

Mumbai Metro Rail Corporation Limited (MMRC) is the nodal agency responsible for implementing the Mumbai Metro Line–3 (MML-3) project. It has been constituted as a JV of the Govt. of India (GOI) and the Government of Maharashtra (GOM) on a 50:50 sharing basis. It aims to provide a Mass Rapid Transit System to supplement Mumbai's inadequate suburban railway system by bringing the metro closer to the people's doorsteps. With the full-scale operation of MML-3, the city is expected to witness a significant reduction in traffic volumes, and the local transit facility is anticipated to receive a boost. The MML-3 would bring multiple benefits, including comfortable Air-conditioned travel for over 16 Lakh commuters daily, reduced travel time, decreased air & noise pollution, and improved safety & security for citizens. Upon completion, stations would connect six business districts, 30 educational institutes, and 30 recreational areas.

Services provided for MML 3 by **MAPLE Consortium** (AECOM-Lead, PADECO, LBG & Egis Rail, the Author was part of AECOM Team) as **GC**, **DMRC for Operation & Maintenance** & **Independent Safety Assessor – Burea Veritas Italia SpA**.

The civil works were divided into seven underground contract packages (UGC) and one depot package. **UGC-01 (Cuffe Parade – CST Metro):** Larsen & Toubro – Shanghai Tunnel Engineering

Co (L&T-STEC) JV **UGC-02 (CST Metro – Mumbai Central):** Hindustan Construction Co – Moscow Metrostroy JV (HCC-MMS JV), **UGC-03 (Mumbai Central – Worli):** Doğuş – Soma JV, **UGC-04 (Worli – Dharavi):** Continental Engineering Corp – ITD Cementation – Tata Projects JV,**UGC-05 (Dharavi – CSIA T1):** J Kumar Infraprojects – China Railway Tunnel Group JV, **UGC-06 (CSIA T1 – Marol Naka):** J Kumar Infraprojects – China Railway Tunnel Group JV,**UGC-07 (Marol Naka – Aarey Ramp):** Larsen & Toubro – Shanghai Tunnel Engineering Co (L&T-STEC) JV, **Package 8 (Aarey Colony Depot):** Sam (India) Builtwell Pvt. Ltd.

Companies were awarded contracts for specific system works, rolling stock, and equipment: **Rolling Stock (Trains):** Alstom Transport – Alstom Transport India Ltd. JV, **Signaling & Train Control, PSD, Telecom:** Alstom Transport – Alstom Transport India Ltd. JV, **Trackworks (Mainline & Depot):** Larsen & Toubro Ltd.**Traction & Power Supply Systems:** Alstom Transport – Alstom Transport India Ltd. JV (Traction) and Larsen & Toubro Ltd. (Power Supply), **Automatic Fare Collection (AFC):** Kalindee Rail Nirman (Texmaco) – Asis Elektronik VE Billisim (Kalindee-Asis JV), **Tunnel Ventilation Systems (TVS) & ECS:** Shanghai Tunnel Engineering Co. Ltd (Phase 1), Blue Star Ltd. (Phase 2 Lot 1), and Voltas Ltd. (Phase 2 Lot 2),**Escalators & Lifts:** Yuanda – Royal JV, Johnson Lifts – SJEC JV, and IFE – CRANEX JV, **Rail Supply:** Mitsui & Co. Limited

Fig.10.20: Metro Car Line 3 (news9live.com)

The Underground Corridor portion was planned to be made of Precast Concrete Segmental Tunnel, while Stations & shafts are planned to be constructed using the cut–and–cover approach, either using Top–Down or Bottom-Up, depending upon site conditions. In some areas, NATM had also been proposed. Most existing structures were within the zone of influence of tunnelling.

Accordingly, it was anticipated that the selection and proper use of a Tunnelling method appropriate to the sub-surface conditions encountered is good enough to control the ground movements. As required, additional protection would be applied to avoid damage to the existing structures.

This corridor has been divided into seven Packages, which include many stations and connecting corridors. A brief detail of the Packages is included in Table 10.3, given below.

MML-3 had been envisioned as the modern transport of the 21st century. State-of-the-art technologies would be harnessed during construction to protect existing building structures. The stations would be equipped with Platform Screen Doors to prevent trespassing on the tracks. The stations would also be with automatic fare collection machines for contactless revenue collection.

Sr No	Contract Packages	Chainages	Station
1	UGC—1	Ch. – 0.488 to Ch. 3767	Cuffe Parade, Vidhan Bhavan Churchgate, Hutatma Chowk
2	UGC--2	Ch. 3767 to Ch. 7818	CSTMetro, Kalbadevi Girgaon, Grant Road
3	UGC—3	Ch. 7818 to Ch. 12592	Mumbai Central, Mahalaxmi Science Museum, Acharya Atray Station
4	UGC—4	Ch. 12592 to Ch. 18959	Worli, Sidhi Vinayak, Dadar, Shitaladevi
5	UGC—5	Ch. 18959 to Ch. 25199	Dharavi, BKC, Vidyanagri, Santacruz
6	UGC—6	Ch. 25199 to Ch. 28842	CSIA Domestic, Sahar, CSIA International
7	UGC—7	Ch. 28842 to Ch. 32488	CSIA Domestic, Sahar, CSIA International

10.18.2 Main features of MML 3 - Table 10.3 Brief Details of Packages

- India's first fully 33 km underground metro line
- Stations- 27 (26 U/G+1)
- Completion cost: Equivalent INR 33,406 Cr
- JICA Loan Component: 57.2 % rest by GOI/GOM/other stakeholders
- Ventilation and ECS in stations and tunnels
- PSD doors at all stations.

Fig.10.21: Hon'ble PM of India inaugurating the Metro Line 3 (hindustantimes.com)

10.18.3 Alignment

The consultants developed various alignment options between Cuffe Parade - Seepz and presented them to MMRC during various discussions/meetings, as well as in the "Options Report". The corridor is proposed to have a double-line track, with a capacity to run 8 coach trains and fully underground, either on the edge of the existing road or along the median of the road, and care was taken to cause the least disruption to existing services/ traffic movement. Underground structures are planned with consideration for Mumbai's aggressive marine environment and high corrosion proneness, incorporating suitable anti-corrosive treatments for reinforcement/structural steel, as well as concrete surfaces in contact with the earth and those exposed to the atmosphere.

Fig.10.22: Platform of Worli UG Station

The alignment begins at the southern tip of Mumbai City, near Cuffe Parade, at the WTC junction on Captain Prashant Pethe Marg, and moves northwards. After crossing Badhwar Park, it turns left along Jagannath Bhosle Road, moves off the road and runs under the police barracks at Nariman Point to connect Vidhan Bhawan and Mantra Laya. Furthermore, it aligns with Jamshedji Tata Road to connect to Churchgate Station. After crossing Oval Maidan along Veer Nariman Road, the alignment continues along Dr D.N. Road, heading west of CST Railway Station, along Mahapalika Marg. Further, the alignment runs along JSS Road, Lamington Road and Dr Anandrao Nair Marg. After crossing the WR tracks at Mahalaxmi, the alignment runs along Dr E Moses Marg, Dr Annie Basant Road, N. Gokhale Road and Lady Jamshedji Road. At Mahim, the alignment turns right along R. Hospital Road, moves along Mahim Sion Link/Station Road, crosses Mahim creek (Mithi River) just east of the Sion-Bandra Link Road, and further crosses Metro line II (Charkop – Bandra – Mankhurd corridor) at Bandra Metro Station (ITO). The alignment passes through Bandra Kurla Complex, Bharat Nagar, and Valmiki Nagar slums, as well as Kalina University. Further North, the alignment passes parallel to Santacruz–Chembur Link Road and the eastern service road of Western Express Highway. The alignment enters Mumbai Airport's premises and covers the Domestic terminal, Sahar Road and the

International terminal. The alignment crosses Metro Line I (Versova – Andheri – Ghatkopar corridor) at Marol Naka. Further, the alignment passes through the MIDC and SEEPZ area along Krantiveer Lakhuji Salve Marg with a terminal station at SEEPZ opposite the SEEPZ Bus Depot. The total length of the alignment between C/L of Cuffe Parade Station to SEEPZ Station is 32.546 km, and the total alignment length is 33.508 km. A total of 27 underground stations has been proposed along the entire length of the alignment. 3 underground stations are proposed in Mumbai Airport'

Fig.10.23: Alignment Line 3

10.18.4 Technologies Used

Table 10.4: Technologies Used

Sr. No.	Category	Parameters	Description
1	Tunnelling	• Tunnelling Boring Machine (TBM) • Tunnel ventilation system (TVS) to maintain temperature during normal operations and remove smoke during emergencies.	• A highly mechanised device that is used as an alternative to drilling and blasting methods to construct tunnels with circular cross-section
		• New Austrian Tunnelling Method (NATM)	• Modern tunnel construction method where the surrounding rock or soil formation of a tunnel is integrated into an overall ring-like support structure • Mobilization of the strength of rock mass
2	Station	• Civil construction	• By Cut & Cover Method Or • By NATM Method Or • Combination of both
		• Station Finishes	• Modern finishes with extensive use of granite, glass and stainless steel.
		• Station Lighting	• Modern energy friendly smart lighting is planned for station areas.
		• Ventilation and Air-conditioning (VAC) System	• Advanced Air-conditioning systems for station areas along with Smoke Management System and Tunnel Ventilation System
		• Automatic Fare Collection System (AFC)	• Efficient operations/quicker fare changes • Zero chances of revenue loss • Faster evacuation
		• Platform Screen Doors	• Technology to be used for public safety to prevent any fall, mishap, suicidal attempts etc accidents at metro stations
3	Traction System	• 25kV, Single Phase AC Traction	• Rigid Overhead Conductor System (ROCS) in tunnel • Flexible OHE in Car Depot • GIS Technology Main / Aux. Subsystem • State of act SCADA System

4	Rolling Stock	• Car Body	• Light weight and cost-efficient car body • Bolster less/ light weight bogies.
		• Width/Axle Load	• 3.2 meters/17 T
		• Braking System	• Electro-pneumatic (EP) service brake • Fail-safe, pneumatic emergency brake • Spring applied parking brake • Regenerative Braking & blending with EP.
		• Passenger Doors	• Remote-controlled door from the driver's cab
		• Air Conditioning	• 2 air-conditioning units in each coach
		• Seating Arrangement	• Longitudinal seats for maximum seating and standing arrangements
5	Signalling and Communication	• Train Control	• Automatic Train Protection and Automatic Train Supervision • Automatic application of brakes in case of disregard of signal • Continuous Tracking of train position
		• Centralized Train Supervision	• Controlling and monitoring of train operation remotely • Operation Control Centre (OCC) to monitor the track status, points, signals, operation of vehicles and also traction power control.
		• Telecommunication Features	• Traffic management • Mobile Radio Communication • LED/LCD based display boards • Passenger Announcement System • Centralized Clock System • Passenger Information Display • CCTV Surveillance System

10.18.5 Challenges Faced

* Administrative

* Social Safety

* Instrumentation & Monitoring

*. Utility Shifting

* Traffic

* Engineering, Engineering & Construction-Engineering and Construction Challenges, being addressed, best techniques followed around the World.

10.18.6 Civil Work Packages

Table 10.3: Civil Work Packages

<table>
<tr><th rowspan="2">Contract No.</th><th rowspan="2">Contractor/JV Address & Contact Details</th><th rowspan="2">Length (km)</th><th colspan="2">Station Names</th><th rowspan="2">Total in Rs. Cr.</th></tr>
<tr><th>Cut & Cover</th><th>NATM</th></tr>
<tr><td>MM3-CBS-UGC-01</td><td>L&T – STEC JV MUMBAI</td><td>4.2</td><td>Cuffe Parade
Vidhan Bhavan
Churchgate</td><td>Hutatma Chowk</td><td>2,988.53</td></tr>
<tr><td>MM3-CBS-UGC-02</td><td>HCC - MMS JV</td><td>4.05</td><td>CST</td><td>Kalbadevi
Girgaon
Grant Road</td><td>2,521.89</td></tr>
<tr><td>MM3-CBS-UGC-03</td><td>Dogus - Soma JV</td><td>5.06</td><td>Mumbai Central
Mahalakshmi
Science Museum
Acharya Atre Chowk
Worli</td><td></td><td>2,557.84</td></tr>
<tr><td>MM3-CBS-UGC-04</td><td>CEC - ITDCEM-TPL JV</td><td>6.81</td><td>Siddhi Vinayak
Dadar</td><td>Shitaladevi</td><td>2,830.10</td></tr>
<tr><td>MM3-CBS-UGC-05</td><td rowspan="2">J.Kumar – CRTG (JV)</td><td>4.95</td><td>Dharavi,
B.K.C.
Vidyanagari,</td><td>Santacruz</td><td>2,817.02</td></tr>
<tr><td>MM3-CBS-UGC- 06</td><td>4.45</td><td>CSIA Domestic
Sahar Road
CSIA International</td><td></td><td>2,118.40</td></tr>
<tr><td>MM3-CBS-UGC – 07</td><td>L&T – STEC JV MUMBAI</td><td>4.15</td><td>MIDC
SEEPZ</td><td>Marol Naka</td><td>2,281.45</td></tr>
</table>

10.18.7 Interchanges

Passenger interchange facilities, connecting to other railway lines, will be provided at the following stations:

- Churchgate (connects to the Western suburban line towards Virar and beyond)
- Grant Road (connects to the Western Suburban Line towards Virar and beyond)
- Mumbai Central (connects to two lines: the Western suburban line towards Virar and beyond; also to the National Rail Network for travel to cities across India)
- Mahalaxmi (connects to two lines: the Western suburban line and the Mumbai Monorail)
- Mumbai CSMT (connects to three lines: the Central suburban line towards Thane and beyond; the Harbour suburban line towards Panvel via Vashi; and to the National Rail Network, because this is a major terminus for trains going to other cities across India)
- Marol Naka (connects to Line-1 of Mumbai Metro, which runs between Ghatkopar and Versova via this station, Marol)

- Santacruz (connects to Santacruz through the skywalk, which connects the Western suburban line)
- Bandra Kurla Complex (connects to Line-2 of Mumbai Metro, which runs between Dahisar and Mankhurd via this station)

10.18.8 Station Area Planning

(Interim Consultant for Mumbai Metro Line III Colaba-Bandra-SEEPZ)

This Section describes the station-specific structural form, station planning, entrance locations, and the sizing of station facilities. Station planning for Mumbai Metro Line 3 was conducted based on the following criteria.

- Adequacy of road width where cut & cover construction is proposed
- Land for cut & cover access shafts is available, where NATM construction is proposed.
- The availability, adequacy, and suitability of land are earmarked in the DPR for Station Typology Configuration
- Various factors, such as land availability, construction methodology, and road widths, could divide the stations into the following categories.
- Cut-and-Cover Station: A typical cut-and-cover station box is about 285 m long, 24.5 m wide and 17 m deep. This typology is applied to stations where a road width of more than 28 m is available for the construction of the station.
- This prototype is a two-level structure with a concourse at the upper level and an island platform, 185 m long and 12 m wide at the lower level. Space beyond the platform on either side is used for housing the auxiliary substations, one at either end and two tunnel ventilation fans at either end, along with the plenums for draft relief and tunnel ventilation system shafts.
- The concourse level has a centrally paid area long enough to accommodate vertical circulation elements (lifts, escalators, and stai
- This paid concourse is flanked by unpaid concourses on either side, separated by fare collection barriers. These two unpaid concourse areas have ticket vending windows and space for future TVMs (Ticket Vending Machines).
- ECS (Environmental Controlled Systems) plant rooms, trackway exhaust, and under-Platform supply fans are in the zone towards the end of the station box at the concourse level. Other station facilities, such as S&T (Signaling &Telecom) rooms, SCR (Station Control Room), and staff areas, are located at the concourse level beyond passenger areas.
- At least four exits/entries are provided from the station box to the ground, two from each unpaid concourse.
- Elongated Cut & Cover Station: Where there is a crossover proposed next to the station,
- Making the station box about 400 m longer results in additional space at the concourse level. This extra space accommodates ancillary services like water tanks, chiller plants, DG sets, etc., usually accommodated in the ancillary structure above ground in

- The standard cut-and-cover station type.
- Street-level facilities such as entrance structures, ventilation shafts, and ancillary buildings.

10.18.9 Station Typology & Configuration

Various factors, such as land availability, construction methodology, and road widths, could divide the stations into the following categories.

Fig.10.24: Station Platform- Mumbai Metro Line 3 (livemint.com)

Fig.10.25: Concourse level- Station Mumbai Metro Line 3

10.18.10 Cut & Cover Station

The construction of the underground metros requires deploying various design and construction techniques to accommodate the availability of materials (mostly indigenous, due to cost considerations), site conditions, and to ensure future provisions. The cut and cover stretch has a box configuration built either with Diaphragm Walls (DW), heavy-duty sheet piles, contiguous bore piles, soldier piles (H-pile) and timber lagging. A few of these walls, such as DW, contiguous bore piles, or continuous secant piles, can be used both as temporary and permanent structures. In Metro line 3, continuous secant piles were used for the cut-and-cover method in station areas.

10.18.10.1 Secant Piles

A secant pile wall can be a temporary or permanent structure in underground metro works. Secant pile walls are formed by constructing intersecting reinforced concrete piles. The piles are reinforced with either steel rebar or steel beams and are constructed by drilling through strata. Primary piles are installed first, with secondary piles constructed between them once the latter gain sufficient strength. Pile overlap typically depends on verticality tolerance, rigidity, and depth of pile.

The main advantages of secant or tangent pile walls are:

1. Increased construction alignment flexibility.
2. Increased wall stiffness compared to sheet piles.
3. Can be installed in difficult ground (cobbles/boulders).

4. Less noisy construction.
5. No need to install timber lagging.
6. Reduced water leakage.

The main disadvantages of secant pile walls are:

1. Verticality tolerances may be hard to achieve for deep piles.
2. Total waterproofing is very difficult to obtain.
3. Increased cost compared to sheet pile walls.

Installation

Secant piles are being installed at the metro stations to stabilise the structure by bearing the load and ensuring it does not collapse in the future. Soft piles (without steel reinforcement and M20 grade concrete used) and hard piles (with steel reinforcement and M40 grade concrete used) have been placed alternately. The pile has a diameter of 800 mm with 40 mm casing on both sides, which makes the total width 880 mm. The equipment used for the piling is the rig machines (70 to 80 tonne, 55 to 60 tonne, and 60 to 70 tonne). Installation of piles begins by marking the survey point using the total stations. Secondly, augers of 600 mm diameter are used for hard pile to cut the ground till the desired length. Next, the casing is fixed to prevent the interruption of the mud flow. After that, the boring operation begins, and at the same time, the auger continues cutting the ground up to the desired length. Once the desired length is reached, the sounding chain is used to verify the pile's length, known as the terminating length. Then the steel reinforcement cage is inserted into the pile. After that, tremie pipes are inserted into the pile, along with the hopper at the top, which allows the concrete to pass through, and the pipes aid in the compaction of concrete. At the end, the concrete is added to the pile. Once it reaches the top, the tremies are slowly removed, compaction is done, and then the casing is removed.

Anchoring

Anchoring is done to prevent the collapse of the pile, which occurs due to the lateral earth pressure. At the beginning, a protective cover known as the waler beam is constructed. Next, the drilling of the hole for the anchor is done by the head cutter. The anchor consists of simple wire. 3 PVC pipes (for protection from corrosion) and the grout pipe. An anchor with a diameter of 150 mm is placed at an inclination of 380 using the angle definer at a depth of 20-25 metres. This anchor contains a slotter at the mouth, which is the first component to enter the drilled hole and begins cutting the inside portion whenever the simple wire gets stuck inside. Anchoring is done manually and takes around 50 minutes to get over. Once the entire anchor is inserted into the hole and reaches the desired depth, then the grouting takes place. Here, two types of grouting are performed: primary and post-grouting. Primary grouting is done first, and every void is filled whenever possible. Post grouting is done by allowing the grout to overflow a bit to ensure that all the voids are filled. Installation of Piles

10.18.10.2 Station Box

A typical cut-and-cover station box is about 285 m long, 24.5 m wide and about 17 m deep. This typology is applied to stations where a road width of more than 28 m is available for the construction of the station. This prototype features a two-level structure, with a concourse at the upper level and an island platform, measuring 185 m in length and 12 m in width at the lower level. Space beyond the platform on either side is used for housing auxiliary substations, one at each end, and two tunnel ventilation fans at each end, along with plenums for draft relief and tunnel ventilation system shafts.

The concourse level has a centrally paid area long enough to accommodate vertical circulation elements (lifts, escalators, and stairs). This paid concourse is flanked by unpaid concourses on either side, separated by fare collection barriers. These two unpaid concourse areas feature ticket vending windows and space for future Ticket Vending Machines (TVMs).

ECS (Environmental Controlled Systems) plant rooms, trackway exhaust, and under-platform supply fans are in the zone towards the end of the station box at the concourse level. Other station facilities, such as S&T (Signaling & Telecom) rooms, SCR (Station Control Room), and staff areas, are located at the concourse level beyond passenger areas. At least four exits/entries are provided from the station box to the ground, two from each unpaid concourse.

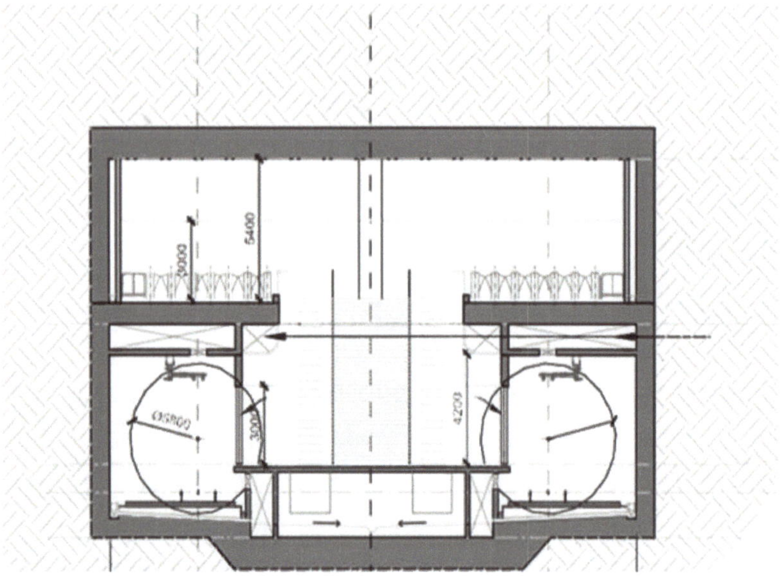

Fig.10.26: Typical Cut-and-Cover Station

10.18.10.3 Elongated Cut & Cover Station

Where there is a crossover proposed next to the station, the station box is about 400 m longer, which results in additional space at the concourse level. This extra space accommodates ancillary services like water tanks, chiller plants, DG sets, etc., usually accommodated in the ancillary structure above ground in the standard cut-and-cover station type.

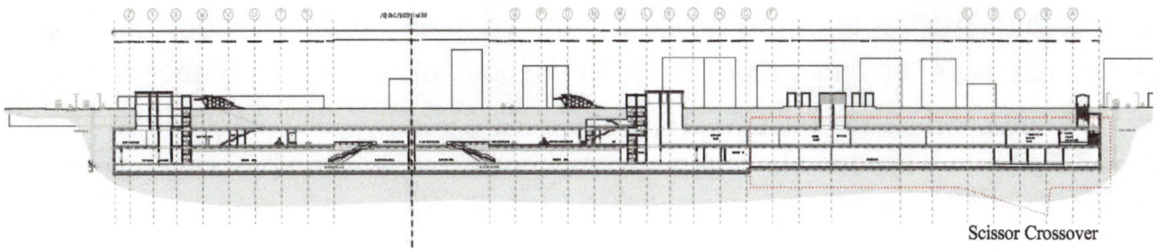

Fig.10.27: Typical Long Section of a Cut-and-Cover Station

Conditions as per NFPA 130 have been considered in the design for the 'worst-case scenario, which:

1. All passengers accumulated on the incident platform planned to be evacuated within 4 minutes, requiring the combined escape capacity of stairs and escalators serving that platform to be sufficient to evacuate 'the platform occupant load' in 4 minutes.
2. The platform occupant load includes-
    - Passengers waiting at the platform, factored to include a service disruption (between 5 to 10 minutes for train services of short intervals,
    - Train loads on all the tracks serving that platform mean the section loads of trains arriving from both directions.
3. Escape capacity includes the combined capacity of all stairs and escalators serving that platform.
4. Escalators moving in the direction of escape are considered as moving and as a means of egress, subject to operational conditions such that those moving in the opposite direction can be stopped and restarted in the direction of escape.
5. One escalator is considered non-available in egress capacity calculations which is on account of maintenance.
6. The maximum distance between any two means of escape is limited to 100 m, and the maximum distance of an escape stair from a dead end is not more than 25 m.

Two scenarios exist for escape requirements: the 2031 ridership figures and the ultimate capacity.

Egress Capacity:

The stations are sized and designed to achieve egress criteria as stipulated in NFPA-130:

1. Provide sufficient egress width to evacuate the platform occupant load in 4 minutes or less.
2. To facilitate the evacuation of the farthest passenger at the platform to the point of safety in 6 minutes or less.

Required Inputs:

1. **Platform Occupant Load (POL):** POL is calculated from PHPD (Peak Hour Peak Direction) ridership data, considering system design criteria like headway[2], disruption time[3], surge factor, ultimate load[4], etc.

2. **Station Layout**: The layout of the station is required to determine horizontal and vertical travel distances, passenger holding capacity beyond the point of safety, the number of exits, gates, etc.
3. **Egress element capacities & travel speeds:** The egress capacity of the staircase and escalator, and passenger travel speeds on the platform, concourse, and stairs should be considered per NFPA-130.

Assumptions:

1. Concourse paid/unpaid area is considered a point of safety as per NFPA-130.
2. The escalators considered should have the provision reversed in the direction of egress in a fully loaded condition, as per NFPA-130.
3. One escalator at each level is discounted while calculating egress capacity as per NFPA-130 clause 5.5.6.3.2.6
4. According to NFPA-130, the discounted escalator shall have the most adverse effect on egress capacity.
5. Elevators are not accounted for as a means of egress, even though their use for this purpose, it is permitted as per NFPA-130.
6. In the event of a fire, AFC gates shall remain open to facilitate evacuation as per NFPA-130.

The construction of the typical underground metro station is based on site conditions and alignment.

10.18.11 NATM Stations

Where making an open cut and constructing a station is impossible due to constricted road widths or difficulty in traffic diversion, other structural options are proposed. These options involve platforms in bored tunnels enlarged by NATM and station rooms at concourse level in open cut and cover boxes, varying in shape and location depending on specific site conditions.

Within this typology, variations have been evolved as follows:

Type 1. Cut-and-cover boxes (around 75m X 32.2m) are proposed at either end of the station, connected by two tunnels constructed by NATM to accommodate the platforms. These tunnels are connected at two to three locations with cross passages for inter-platform transfer.

The cut & cover boxes at the two ends are 3 level structures that accommodate Tunnel ventilation Fan rooms and Auxiliary Substations at the platform level. ECS plant rooms, Trackway exhaust & platform supply Fans, staff rooms and S&T rooms at an intermediate level, a concourse containing ticketing booths, AFC gates, and a Station control room at the uppermost level (-1 Level). DG sets, fuel tanks, fire water tanks, and pump rooms are also located at this level, suitably at either end, The station developed for such sites aims to reduce the cut-and-cover footprint and ease traffic management and R&R disruptions.

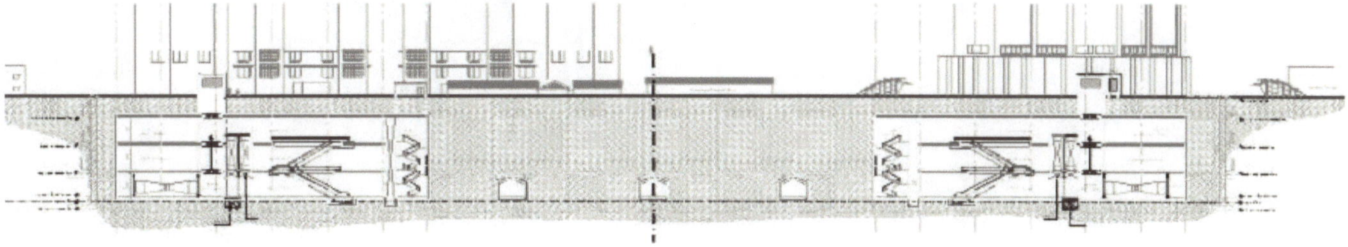

Fig.10.28: Standard NATM Longitudinal Section

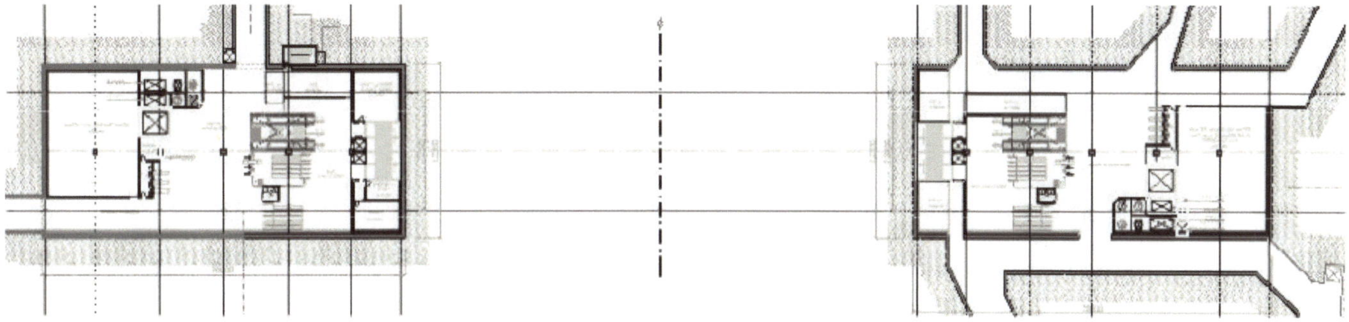

Fig.10.29: Standard NATM Concourse Level Plan

10.18.12 Special NATM Station

This type (Type 2) features a narrow, elongated cut-and-cover box at the center of the station, with rail lines and platforms on either side in NATM Tunnels. The width and alignment of this cut-and-cover box are derived from the availability of road space at street level and the minimum structural dimensions of the box walls, while the rail alignment is governed by the minimum clearances between the C&C box and tunnels as specified by NATM.

This C&C box, which essentially contains vertical movement elements in the central part and other station equipment and accommodation (such as tunnel ventilation fans, track way exhaust fans, under platform supply fans, ASS, S&T rooms at the two ends), is connected to the two tunnels containing platforms at several locations, for the movement of passengers as well as air.

Depending on specific site conditions, the box dimensions are determined, and in turn, the space available for accommodating ancillary facilities is determined. The rest are proposed at street level on land that is slated for acquisition for this purpose.

The stations following this type of configuration are:

1. Kalbadevi
2. Girgaon
3. Marol Naka (cut & cover construction is not continuous on account of the substructure of Metro Line 1 piers crossing the alignment.)

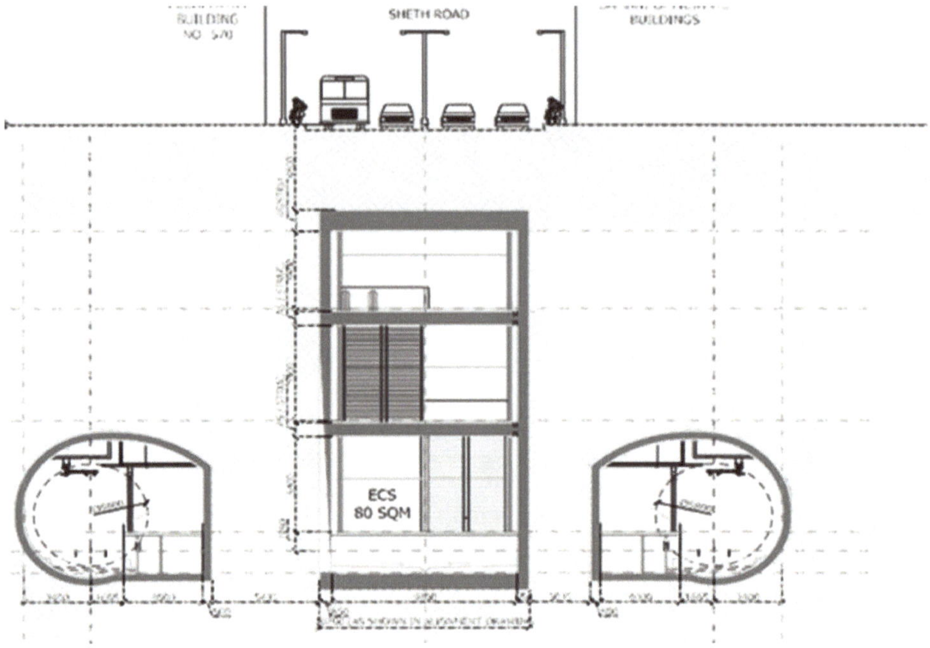

Fig.10.30: Special NATM Cross-Section

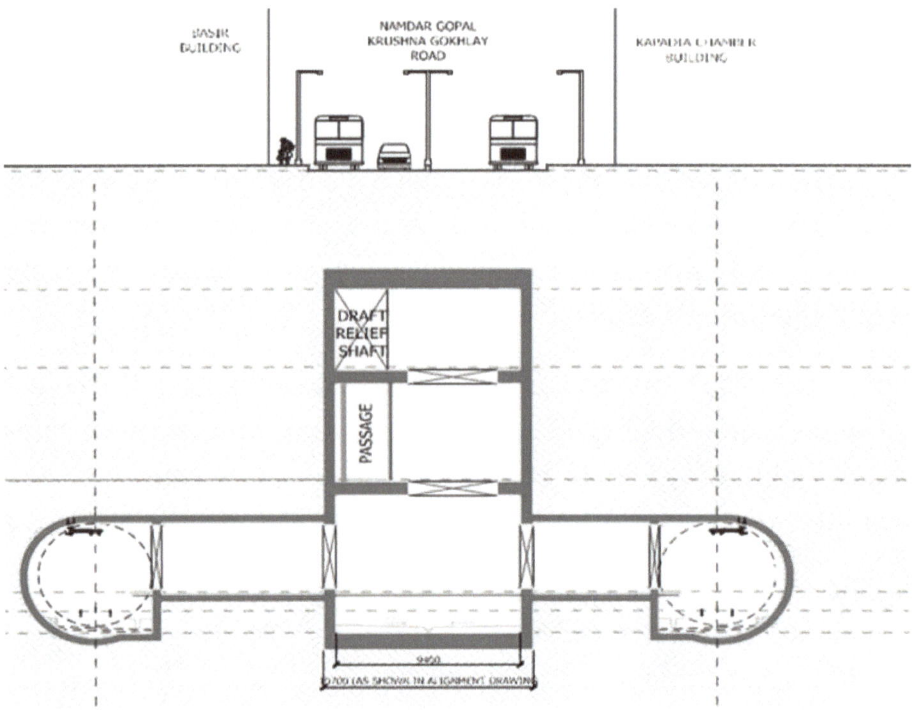

Fig.10.31: Special NATM Cross-Section through cross-passage

10.18.13 Side NATM Station

This type (Type 3) is applied where the road space is not too narrow but still not wide enough for conventional construction (C&C). Here, one of the platforms is proposed inside the C&C box, while the other is in a tunnel by NATM.

At this NATM station, a narrow, elongated cut-and-cover box is proposed above one of the rail lines to accommodate all station facilities, including ticket halls, station platforms, passenger access and emergency egress, ventilation and smoke control, staff rooms, and equipment rooms.

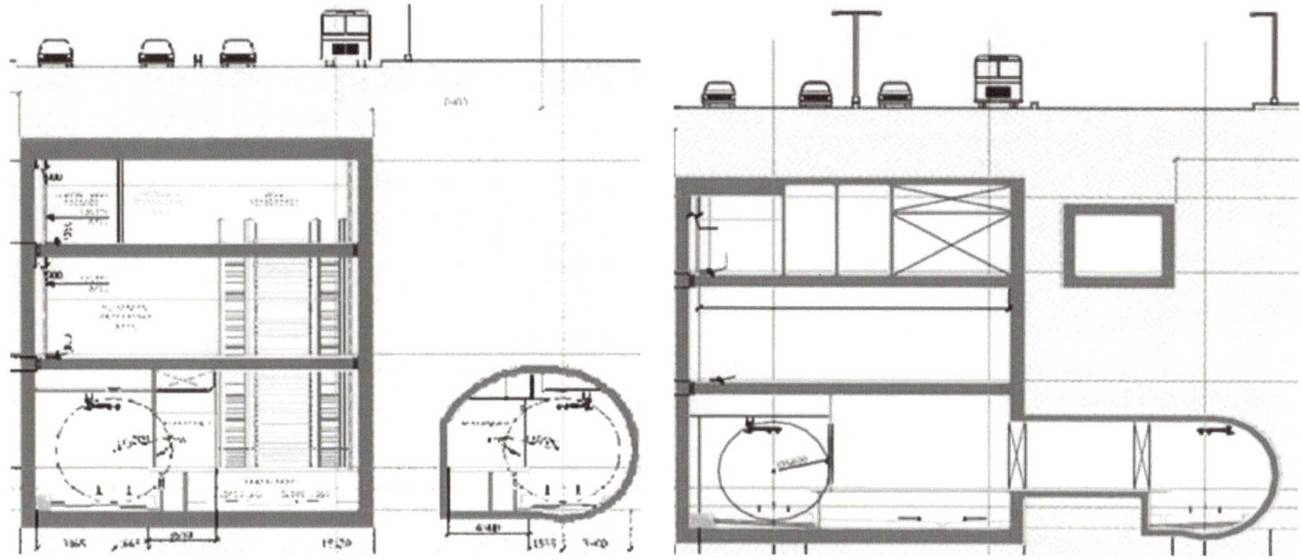

Fig.10.32: Side NATM Cross-Section through cross passages

10.18.14 Grant Road Station

Grant Road underground station is in a densely populated area of Mumbai, characterised by a built-up area of old and dilapidated buildings. Due to space constraints, a conventional cut-and-cover approach to building a station was not possible. A part of the station was designed below buildings using the Austrian Tunnelling Method (NATM), and for the downline platform narrow cut-and-cover box was adopted below the available road width. The geology of this area consists of 8-10 m of alluvial soil and slightly weathered breccia rock underneath, with Uniaxial Compressive Strength (UCS) below 10 MPa. Excavation in soil was proposed using a secant pile as the water table is near to ground. Due to constricted space, only a 4m rock column was available between the NATM and the station box outer wall. This narrow rock column was to carry the overburden in the tunnel and the surcharge load from buildings at the surface. Two-Dimensional Finite Element Method (2D FEM) analysis indicates that simultaneous excavation for NATM and cut–and–cover resulted in high stress concentration in the rock-column, lower Factor of Safety (FOS) and higher ground settlement. Different construction options for the tunnel and station were studied. It was found that higher FOS and lower ground settlement could be achieved when NATM excavation starts after completion of the cut-and-cover section.

The Grant Road Station box of cut-and-cover is proposed in the open land adjacent to the road. It would house part of the length of both tracks and include two NATM track platform tunnels in the remaining station length. In this NATM station typology, a shorter cut and cover box is proposed above both the rail lines on the road and most of the station facilities, such as the ticket hall, passenger access and emergency egress, ventilation and smoke control, staff rooms, and equipment rooms, etc., proposed in the open land adjacent to the road.

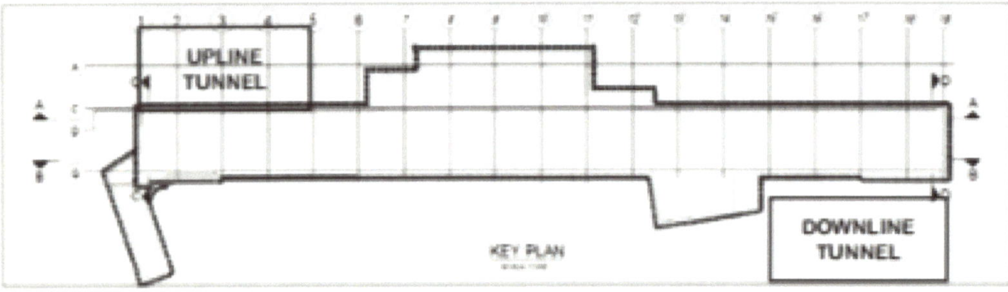

Fig.10.33: Grant Road Station- Key Plan

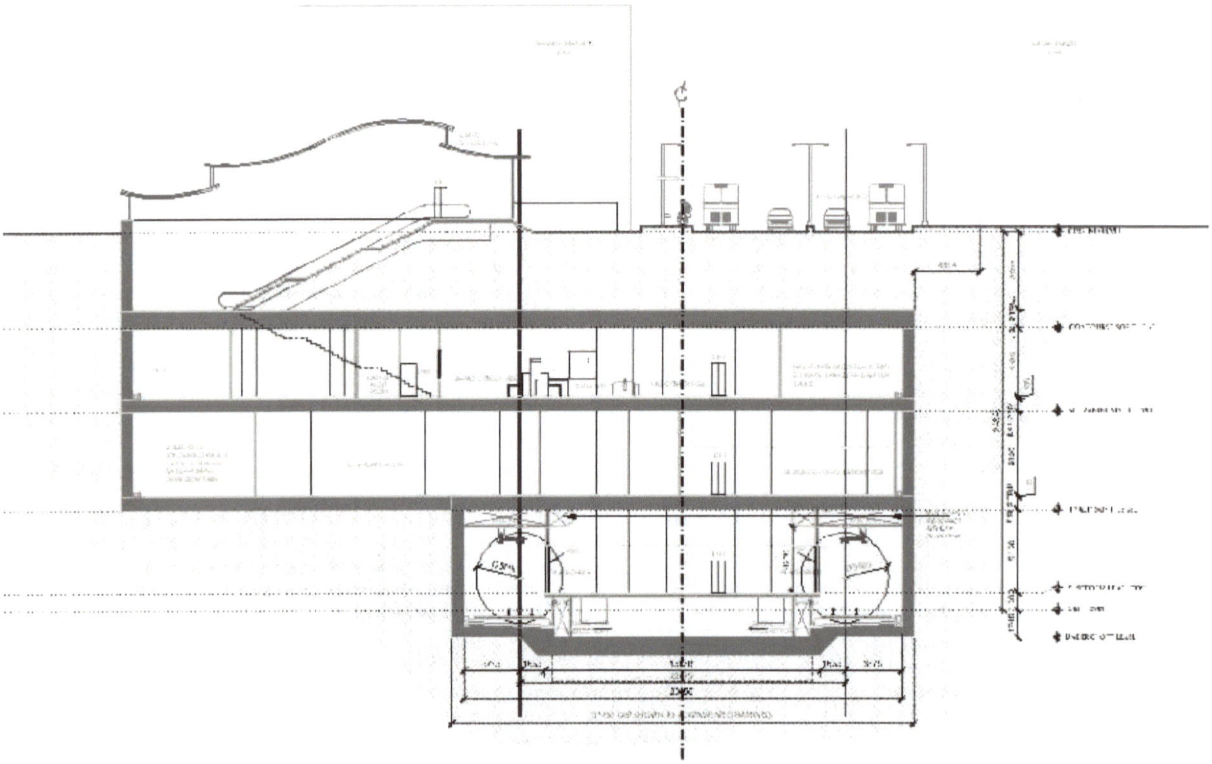

Fig.10.34: Grant Road Station Section

FE analyses were carried out using PLAXIS 2D for options 1, 2, and 3 to compare the principal stresses, surface deformation (settlement), and achieve a global factor of safety (FOS) during the critical stages of the construction sequence. Option 1: High stresses in the rock column were observed from the comparative results summary of FEM analysis. This could be due to the total overburden load transferred onto the unconfined rock column. Ground surface deformation (35mm) was observed just above the NATM location. This high surface deformation was predicted to be very critical for the existing building structures above and could not be allowed unless the buildings were evacuated and supported during construction.

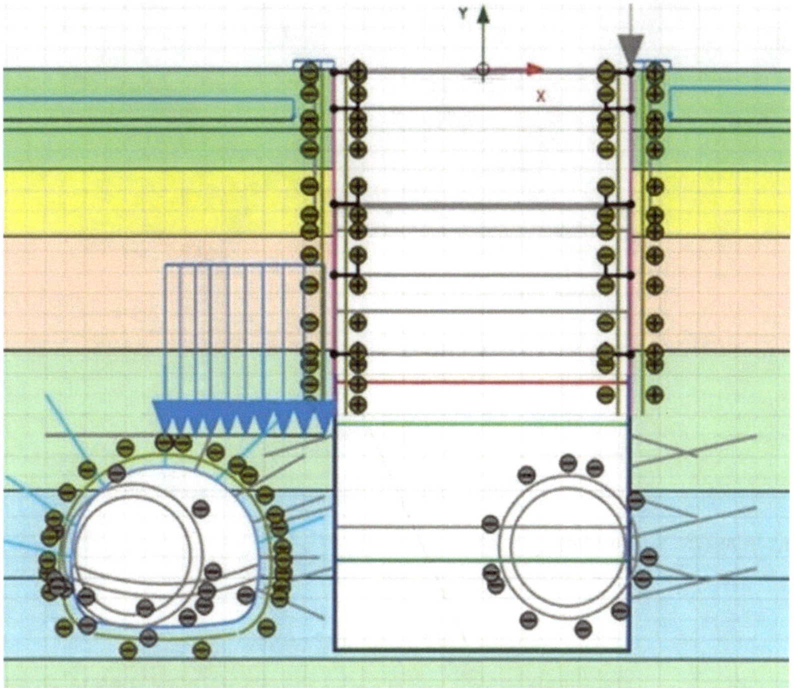

Fig.10.35: Completed NATM following Option-3

Construction Sequence Rock column stress (kN/m2) Global FOS

Option-1 5170 1.69

Option-2 2488 1.379

Option-3 3047 1.99

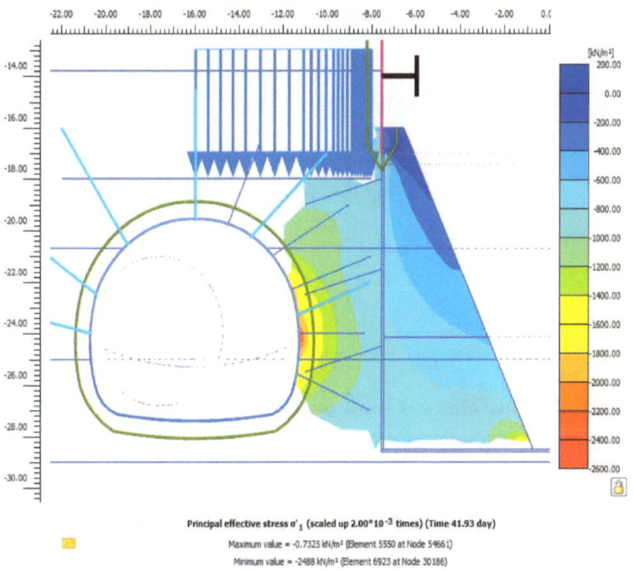

Fig.10.36: Rock column stress (Option-2)

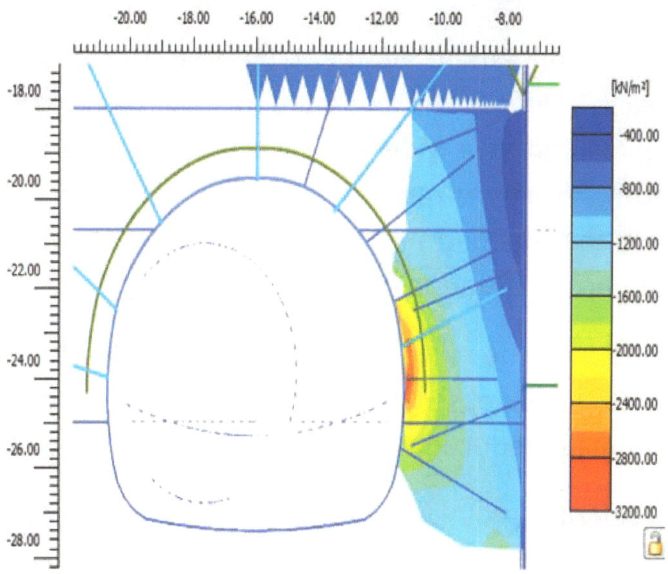

Fig.10.37: Rock column stress (Option-3)

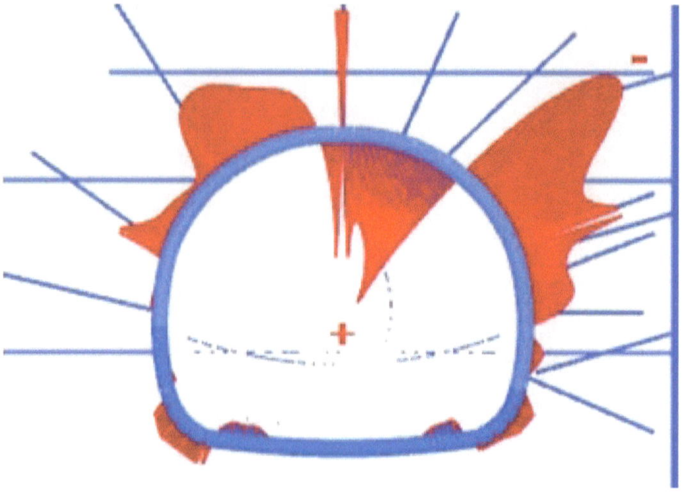

Fig.10.38: Bending moment in NATM tunnel lining

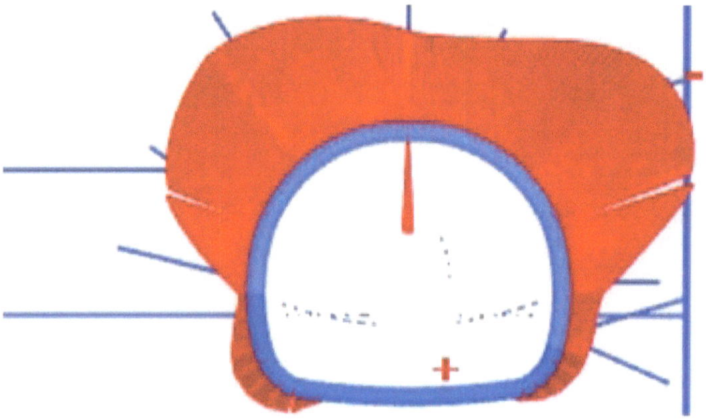

Fig.10.39: Axial force in NATM tunnel lining

As the confinement was available from the sloped berm in option 2, the rock column's allowable capacity bearing the overburden load was increased, and lower stresses were induced in the rock column as a wider area was available for dispersion. This resulted in controlling the surface deformation by as much as 17mm compared to option 1. During the construction sequences, it was noted that the berm and sloped excavation towards the NATM side reduces the speed of the work sequence and station box casting.

In option 3, the rock column was also confined at the station side by the station wall. From the FE analysis, it was observed that the rock column benefited from increased FOS during NATM excavation, resulting in reduced stress. Surface deformation was reduced by 12 mm compared with option 1.

Considering ground behaviour, FOS of excavation, lower ground deformation and construction timeline, option 3 was selected for detail design and execution. A section capacity check was done by applying a suitable ultimate limit state factor. The maximum ground settlement was estimated to be 23 mm, accounting for the effects of station construction, TBM tunnelling, NATM construction, and the drawdown of the water table during construction.

10.18.15: Station Sizing & Egress Requirements-

To determine tentative sizes of station boxes for NATM or cut-and-cover stations, it is essential to establish the sizes of basic passenger facilities that govern the sizing of the station's civil structure. These are essentially the widths of the platform and the combined widths of all vertical circulation elements (stairs and escalators) required for emergency evacuation of passengers. The two are, in turn, calculated to satisfy the provisions of the National Fire Protection Association (NFPA) 130, 2010, a standard followed worldwide for the design of rail-based mass transit systems in the event of an emergency at the station requiring passenger evacuation.

Conditions as per NFPA 130 (mentioned earlier) are considered for the calculation of egress. A few conditions are reproduced for ready reference.

1. Escape capacity will include the combined capacity of all stairs and escalators serving that platform.
2. Escalators moving in the direction of escape can be considered as moving and as a means of egress, subject to operational conditions such that those moving in the opposite direction can be stopped and restarted in the direction of escape.
3. One escalator will not be considered available in egress capacity calculations on account of maintenance.
4. The maximum distance between any two means of escape will not be more than 100 m, and the maximum distance of an escape stair from a dead end will not exceed 25 m.

Two scenarios exist for escape requirements: the 2031 ridership figures and the ultimate capacity.

a) Egress Capacity

The stations are sized and designed to achieve egress criteria as stipulated in NFPA-130:

1. Provide sufficient egress width to evacuate the platform occupant load in 4 minutes or less.
2. To facilitate the evacuation of the farthest passenger at the platform to the point of safety in 6 minutes or less.

b) Required Inputs

1. Platform Occupant Load (POL): POL is calculated from PHPD (Peak Hour Peak Direction) ridership data, considering system design criteria like headway2, disruption time3, surge factor, ultimate load4, etc.
2. Station Layout: The layout of the station is required to determine horizontal and vertical travel distances, passenger holding capacity beyond the point of safety, the number of exits, gates, etc.
3. Egress element capacities & travel speeds: The egress capacity of the staircase and escalator and passenger travel speeds on the platform, concourse, and stairs should be considered NFPA-130.

c) Plan and Section

1. The combined width of entry structures leading from the Concourse to the Ground level should be generally equal to the combined width of staircases provided from the Platform to the Concourse to the extent possible.
2. Wherever feasible, the entry structure has been provided with a combined staircase and Escalator bank. In cases where adequate space is not available, a single staircase has been provided.
3. Lifts have also been integrated with the combined Staircase and Escalator bank wherever space is available.

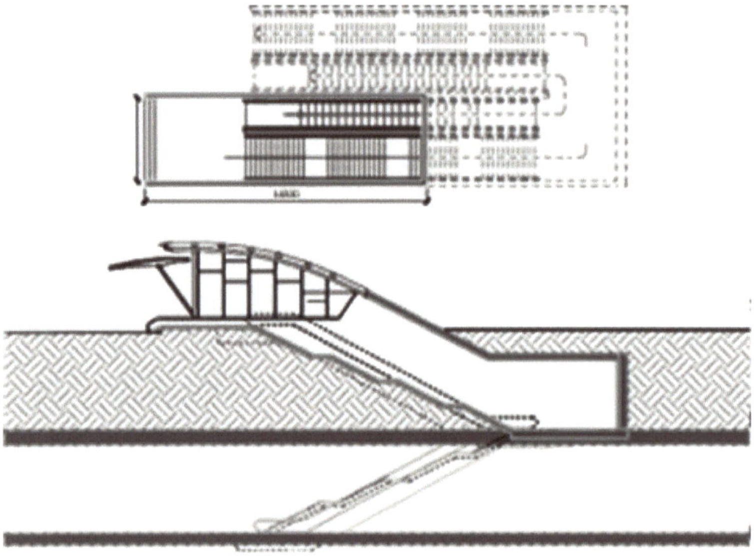

Layout 1

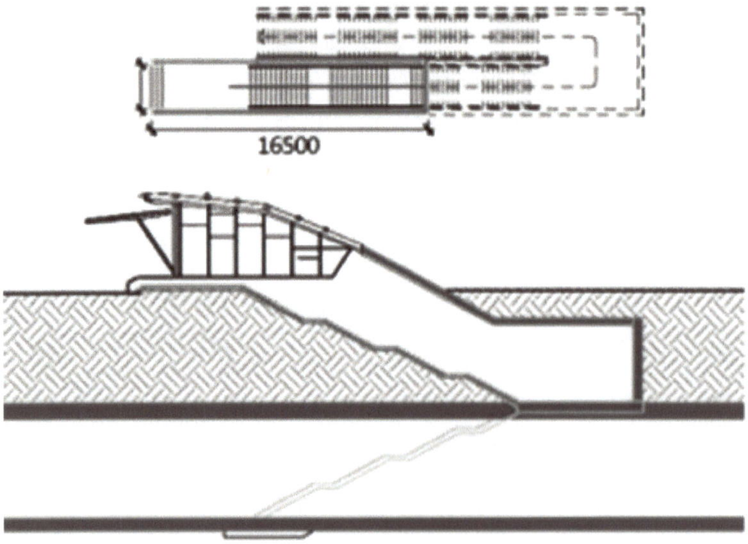

Layout 2

Fig.10.40: Integrated Staircase and Escalator bank Single

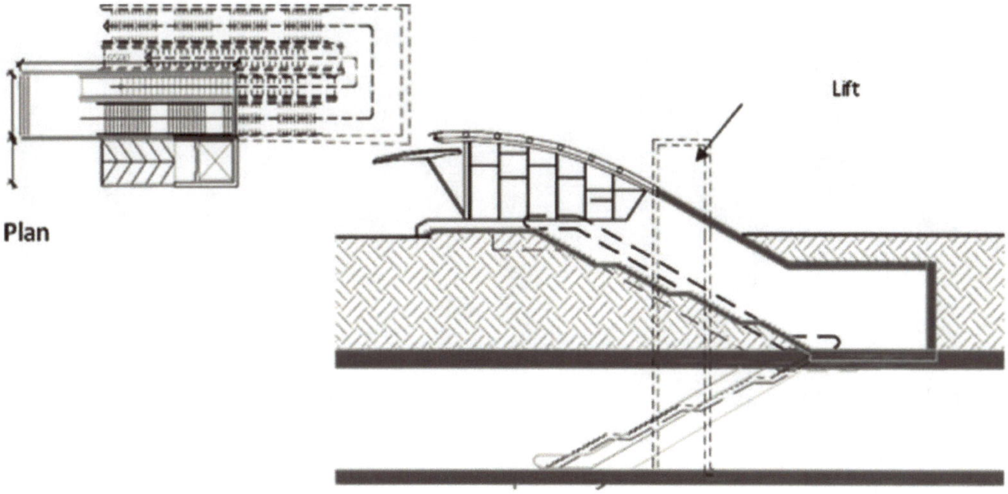

Fig.10.41: Integrated Staircase, Escalator and Lift

Lifts have also been proposed where adequate space to accommodate the required length of stairs and escalators is unavailable at street level. At least 1 lift is provided on either side of the station for disabled persons / public access.

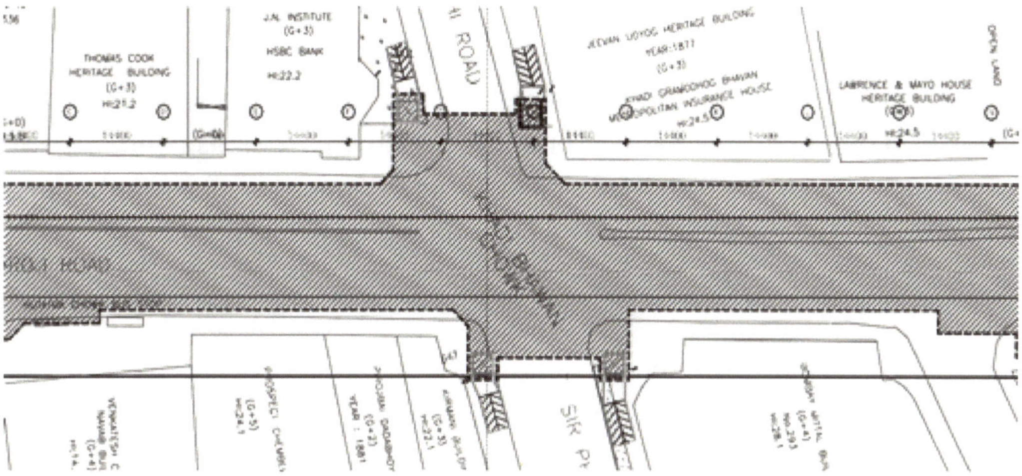

Fig.10.42: Station Plan showing Lift Locations

10.18.16 Cuffee Parade

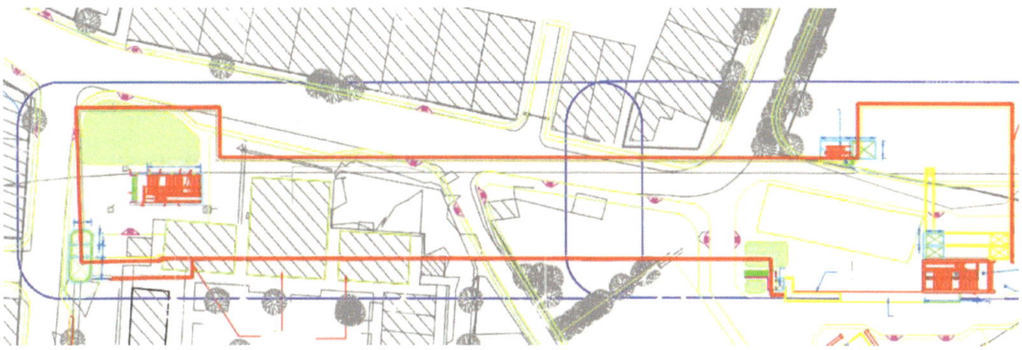

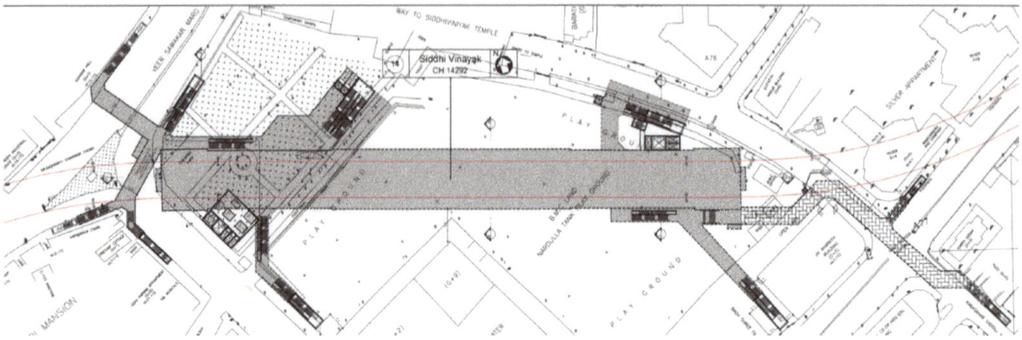

Fig.10.43: Cuffee Parade – Cut & Cover Station

a) Location

Cuffe Parade Station is the starting point of the proposed metro corridor. The station is situated in a BMC park, between the MVIRDC World Trade Centre and high-end residential buildings. The station location is convenient from a construction perspective, as the entire cut-and-cover activity will be off the road, causing little to no traffic disruption during construction.

b) Entrances

One major entrance is proposed adjacent to the Cuffe Parade Police Station, which also serves the catchment area from Ambedkar Nagar, Dhobi Chawl, World Trade Centre & Navy Nagar. Other entrances have been proposed in the open BMC Park to serve the catchment from residential quarters and other high-rise towers such as the Taj President Hotel and Dalmal Park Apartments. Another major entry/exit structure is planned on the northern end of the station box (near Julelal Mandir chowk) in the open land of the Parsi Dharamshala to get catchment from the surrounding residential areas.

c) Typology

Since the station is off the road, an open cut is feasible; hence, the Cuffe Parade station is planned to be constructed using the cut-and-cover technique. The cut-and-cover station construction is also necessary as a scissor crossover precedes the station. The station box is 24.5m X 405m, with track centers at 15.35 meters.

d) Station Planning

The station box will be elongated at track level beyond a typical cut-and-cover station due to the presence of a crossover. The additional space created at the concourse level to utilise for ASS (Auxiliary Sub Station). In addition, ancillary facilities such as chiller plants, water tanks, and pump rooms, which would otherwise be in an independent building at ground level, are planned to be accommodated in the extended box above the crossover.

The platform centres are planned such that the entire cut-and-cover zone is off the road, thus freeing the Julelal Mandir chowk road junction from being blocked during the construction of the station box. Most entrances, shafts, and ancillary buildings are located directly above the station box, within the BMC Park.

Fig.10.44: Siddhivinak UG Station during Construction

10.18.17 SEEPZ

a) Location

This station was planned adjacent to the SEEPZ campus on the SEEPZ road. It is situated on the central bridge along the 23-metre-wide road. The station box begins at the Kotak Mahindra Bank Road intersection and extends to the WRPC office building.

b) Entrances

One major entrance was planned inside the SEEPZ compound along the boundary wall, which would serve as a catchment to the SEEPZ campus as well as other office buildings on the Colaba end. Another major entrance was provided along the BEST Bus Stand compound and was clubbed with 4 vent shafts. This would require the reinstatement of Sulab Shouchalaya (public toilets) along with the vent shafts. This entry serves catchment for the Bus stand.

A single staircase located in front of the ERIT building serves as a catchment for the surrounding office buildings. This staircase is combined with the fireman's staircase, which opens onto the concourse level in the SCR. Another major entry is provided at the end of the box inside the boundary wall of the WRPC office campus. Four fire escape stairs had been provided at regular intervals to avoid clogging the station during an emergency. These exits are located on the footpath along the external side of the SEEPZ campus boundary wall.

c) Typology

The SEEPZ station was planned to be constructed using the cut-and-cover technique. The cut-and-cover station construction was necessary as a scissor crossover succeeds the station. The station box size is 24.5 m X 405 m, with track centres at 15.35 meters.

d) Station Planning

SEEPZ station is a typical cut-and-cover station with ample entrances abutting the station box from adjacent footpaths and connecting roads. The station box has been extended over the scissor crossover portion, adding a significant area at the concourse level to accommodate both paid and unpaid facilities, as well as other station amenities. Two banks of stairs and escalators connect the paid concourse to the platform. The area over the crossover houses retail areas, water tanks, and a DG set. The crossover also houses another set of water tanks at the platform level. Additional water tanks are provided to support the huge retail area offered at this station. The tunnels beyond the cross over terminate at the Depot. To achieve permissible gradients on tracks, the effective height of the concourse was reduced to 4.2 m. The required height for the ECS plant rooms is 5.4 m; therefore, they were separated from the concourse level and housed outside the station footprint. Two ECS plant rooms, each with a minimum area of 600 sq m, have been provided. One ECS was provided below the BEST Bus stand site, which also accommodates retail at the top of the box level. The other has been provided below the Water Department site, housing the chiller plant room at the top of the box level. Passages from the station concourse separately connect the ECS plant rooms.

Most of the ancillary block facilities were accommodated within the station box. Three cooling towers, each 6 m in height, four station ventilation shafts, two ventilation shafts from the chiller plant room and one from the DG set, were consolidated in the vacant part of the Water Department site.

Fig.10. 45: Station at Kalbadevi Under Construction

Fig.10.46: Worli Station (freepressjournal.in)

10.18.18 Tunnel Constructions in Line 3

The Colaba-Bandra-SEEPZ underground corridor of the Mumbai Metro has completed 100% of the tunnelling. With this, the overall project progress stands at 76.6 per cent. The project deployed 17 Tunnel Boring Machines (TBMs) to carry out 55 km of tunnels with a finished diameter of 5.8 metres. The first section of the Mumbai Metro Line 3, between Aarey Colony and Bandra-Kurla Complex (BKC), was expected to be operational by December 2023.

The tunnel, 5.8m in diameter, was designed using a similar approach to that employed in other metros in India, utilising precast concrete segments.

Following the initial geotechnical and hydrogeological studies, the chosen alignment and anticipated difficulties in TBM boring under the Mumbai soil led to the final selection of TBMs in discussions with the design-build contractors, which were of three different types. The choices were Slurry TBM, EPB TBM, and Dual Mode TBM.

The **Slurry machines** excavated part of Line 3, using twin-bore running, supplied by NFM/ Robbins Limited. In total, two slurries exist under conditions in congested areas and abutting sensitive structures. Sufficient space was available for locating the desilting plant near the launching shaft.

EPB TBM was used for excavating the major portion of the alignment. In total, seven TBMs, three of them supplied by Herren Knecht, two from Terratec and two from Shanghai Tunnelling Engineering Company. EPB machines were used along the sections, primarily in the largely varying geological conditions, surface features such as water streams, railway tracks, and the reduced cover over sensitive structures.

TBM manufacturers concluded that the geological risk existed at various sections in the alignment, which were

* Unstable face- weak rock mass and water
* Material erosion at the face, which had a risk of being washed out by the water flow
* Weak joint fillings, which can be washed out by water flow
* Localised/Concentrated water inflow, which might have eroded materials
* Small block fall at the face, resulting in local instability.

A **dual-mode TBM is** used for excavating a large section like EPB. The machines supplied by Robbins and Terratec. It was used in areas with geotechnical conditions, including Fresh Basalt with Volcanic rock, Basalt comprising infillings, slices, and zeolites. a mixed face with Volcanic tuff, fresh Basalt, and moderately weathered rock, as well as fresh greyish basalt.

10.18.18.1 Initial Drive of Tunneling Activities

i) Mining and Shovelling

The first TBMs were lowered using a 150-tonne Gantry crane, and the process of mining and shovelling commenced. The TBM had been pushed forward with the help of hydraulic jacks. There is a total of 16 pairs of hydraulic jacks numbered as 2, 4, 6, 8, 10, 12, 14, and 16 [8, pp.1135-1136]. The cutter head of TBM has a speed of 2.8 R.P.M (rotations per minute) with a face pressure of 1.8 atm. The excavator carries out the mining activity while the TBM continuously cuts through the ground. As the amount of mining is completed, chainages and the shovelling length are all monitored continuously using German Software known as VMT. If the TBM encounters technical issues and the cutterhead stops rotating, the VMT software detects this and sends a signal to the person monitoring the process. These processes are carried out under the initial drive, which is up to a length of 110 m.

ii) Mucking

Once the mining commenced and the cutter head of the TBM started excavating the tunnel, a lot of loose soil and mud, generally termed as "Muck," began to collect. Since the activity is still in its initial phase, the bucket used for mucking is smaller: 4 m3 capacity. Also, a small muck pit is constructed to store the muck. Mucking requires much better transportation facilities to increase the speed of the activity. The locomotive, which has been checked and declared fit to use by the third party, is being used to move the bucket containing muck from the inside portion of the tunnel to the outside part. The average estimated time for one cycle of mucking is approximately 18-20 minutes.

iii) Ring Building

Ring Building is dependent on Mining and Shoving length (Length when the hydraulic jack is extended outward). A precast ring made of concrete consists of 6 segments; each named a Tunnel Segment.

To facilitate erection, in sequence, segments are named as Ordinary 1, Ordinary 2, Ordinary 3 (O1, O2, and O3), Tapered 1, Tapered 2 (T1 and T2) and Key (K) and erected by the erector of the TBM. The key is the smallest component that locks or unlocks the entire ring. The key position is the one that determines the direction of segment erection. The basic rule of ring building is to place the key between two tapered segments (T1 and T2). The required shoving length for ring building is 1850 mm. Depending on the strata, it normally takes 7 to 8 hours to achieve the mining length as per design.

iv) Grouting

Pressure grouting is done to fill the voids left after the completion of ring building. If the ring building is completed for position P33, then grouting must begin for ring P31, which is the alternate position. Cement, sand and bentonite are mixed and are transferred through grout pipes into the voids. Sodium Silicate is also added to that grout (12% of the total grout in quantity) for rapid hardening, which increases the initial setting time. The grout tests are performed before the grout is decided safe to use.

10.18.19 Tunneling across the Mithi River and the Mangrove area

a) Challenges of Tunneling in the Mithi River Section

Mumbai is made up of basaltic lava flows. The main rocks are Basalt, Volcanic breccia, rhyolite, trachyte, and a few shale bands with a westward direction foliation, belonging to the upper Cretaceous to the lower Eocene (60 to 50 million years) according to the geological time scale. According to geo-mechanical In terms of properties, the in-situ rock strength varies from very hard to poor (grade 1 to IV).

b) Components under the Mithi River

i. Twin tunnel operated with TBM,
ii. Return to the BKC south NATM tunnel, located on the south side of the Bandra Kurla Complex station. It is a horseshoe-shaped structure with a base width of 5.5 meters and a clear height of 6 meters. The length of this tunnel is 152m, falling 100m under the Mithi River, and the remaining 52m is under the mangrove area.
iii. Cross Passage: Four passages falling under the Mithi River. One cross passage connects the down lines of the main tunnel, and the other three connect the turn back BKC south NATM tunnel with the up and down lines tunnel. The cross passage is D-shaped, measuring 2.35 meters in width and 2.83 meters in height.
iv. Cross-passage with sump includes the construction challenges encountered during tunnelling underneath the Mithi River. As the tunnel alignment is curved underneath the Mithi River, TBM mining became more challenging. Additionally, there were challenges during investigation, tunnelling, segment installation, construction of NATM cross-passage, high water seepage,

and the installation and observation of geotechnical instrumentation. Due to marshy land, mangrove areas, tidal impacts, and poor geology, these challenges were encountered during the pre-investigation along the Tunnel Alignment.

Fig.10.47: Track Laid in Tunnel Corridor Up/Down (railanlusis.in)

c) Investigation and Construction Challenges

Tunnelling of a 1.80 km-long section consisting of 980m of mangrove and 200m of the Mithi river. The tidal water level in the river and mangrove area varies from 3 m high to 1.5m low, compared to the normal river water level, during high tide and low tide, respectively. The twin tube tunnel underneath the river has a radius of 230 m and consists of four cross-passage sections, each approximately 10.60m in length. Engineers faced challenges of i). The overburden thickness between the river bottom levels and the tunnel top is 5 m, consisting of highly weathered basaltic breccia rock. ii). Chances of high-water seepage, iii). The tunnel lining segment should be of high strength and waterproof. iv) Construction of turn back tunnel (NATM Tunnel) on the south side of BKC station, v). Construction of a cross-passage underneath the Mithi River.

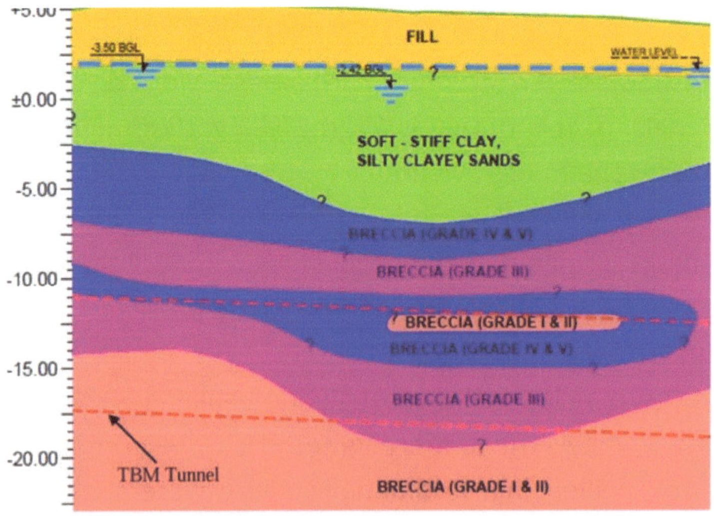

Fig.10.48: Geological Conditions below Mithi River

The tidal water level in the river and mangrove area varies from 3 m high to 1.5m low, compared to the normal river water level, during high tide and low tide, respectively, and the challenges were

i. Po,or geology- As the overburden thickness between the river bottom levels and the tunnel top is only 5 m, with highly weathered basaltic breccia rock.
ii. Chances of high-water seepage,
iii. The tunnel lining segment must be of high strength and waterproof. Construction of the turn-back tunnel (NATM Tunnel) on the south side of BKC station,
iv. TBM tunnel alignment deflection from the permissible limit.

During the site geological investigation stage, only three boreholes were possible along the Mithi River tunnel alignment. The core recovery was very poor. The rock condition was very poor, classified as grades IV and V, and the bedrock alignment varied drastically. Engineers used the best engineering measures based on similar projects worldwide. It was recommended that the EPB TBM operate in close mode to counterbalance water pressure at the tunnel face and periphery, thereby controlling gravity-driven failure and settlement. TBM face pressure is recommended at 1.8 bar for nominal conditions and 2.3 bar if soil formation is encountered at the tunnel face. Additionally, geotechnical instrumentation (surface settlement monitoring, vibration wire piezometer, standpipe piezometer, deep settlement marker, and optical target) was recommended to measure ground and tunnel settlement. The EPB TBM had to operate in a close mode to counterbalance water pressure at the tunnel face and periphery, thereby controlling gravity-driven failures and settlements. TBM face pressure is recommended at 1.8 bar for nominal conditions and 2.3 bar if soil formation is encountered at the tunnel face.

The Engineers had to ensure segment strength, tunnel segment gasket, tunnel alignment deflection, construction of NATM tunnel cross-passage, instrumentation monitoring, safety, and environmental aspects.

Fig.10.49: Twin Tunnel under Driving Line 3

10.18.20 Tunnelling in Other than the Mithi River

The scope of Line 3 includes 33.5 kilometres of twin tube tunnels constructed by TBM, 19 cut-and-cover stations, and 7 Stations, including cross passages and crossovers constructed using NATM. The shallow urban tunnelling is being undertaken for the first time in Mumbai's geology. This section includes, among other things, the ground and groundwater response to TBM tunnelling, which would be of value to ongoing and future projects of a similar nature in similar hydrogeological conditions. The superposition of two single tunnels were used to calculate the ground settlement induced by twin tunnelling. The back analysis of settlement data is carried out at Site 1 and Site 2.

Site 1: The TBM tunnelling was completed through a mixed ground comprising Volcanic Breccia rock of Grade III with Rock Quality Designation (RQD) of 40% - 60% and Grade IV with RQD of 10% - 40%, with a small rock cover of about only 0.5 tunnel diameter (minimum value), overlain by a layer of alluvial soils. The tunnel depth (measured from the tunnel crown) ranges from 15.5m to 16.5m. The groundwater is close to the ground surface.

Site 2: Site 2 represents mixed ground conditions comprising Basalt rock of Grades I (with RQD 90%-100%) to V (with RQD of 0% to 10%), overlain by loose to medium dense silty sands. Intertape Shale of Grades IV (10%-40%) to Grade III (40% to 75%) was observed below the tunnel axis level. The weak Shale, having a compressive strength of 5 MPa to 15 MPa, combined with the high-strength Basalt of approximately 50 MPa compressive strength, presents challenging ground conditions for tunnelling.

The rock earth pressure balance (EPB) machines were chosen primarily due to the largely varying geological conditions, surface features such as rivers, and the reduced cover over sensitive structures. The design features were:

- Large opening ratio of between 28 and 35%,
- Additional injection ports in the cutter head to enhance soil conditioning,
- Closer spacing than normal for forward probing and ground enhancements,
- Bespoke cutter head design with closely spaced disk cutters, scrapers and peripheral cutters to accommodate the large opening ratio.

The TBM was operated in open mode. The ground movements were monitored using arrays of ground settlement markers at regular intervals. The observed maximum ground settlement after completion of the twin tube tunnels was 7.6 mm, using 4 instrumentation arrays with settlement markers installed transverse to the direction of the twin tunnel drives. The Gaussian distribution curve is fitted for the observed ground settlement data from Site 1. The back analysed value of the trough width parameter (k) is 0.7, and the volume loss is 0.3%. These values correspond to a range of k values from 0.5 to 0.7 and volume losses of 0.4 to 0.6%.

A dual-mode TBM carried out the tunnelling at Site 2, with the geology in mind. The machine features are

- Heavy-duty main bearing and modified cutter head to cater for predominantly hard rock,
- Change over from "Open" to "Closed" mode for mixed face conditions,
- Capable of high excavation rates in the predominantly hard rock conditions,
- Ability to "close off" the machine should soft flowing ground or high-water inflows be encountered.

The settlement data in varying geological conditions was analysed. The data was fitted with a Gaussian model to estimate the volume loss and trough width parameters. The study indicated that volume loss ranges from 0.3% to 0.5% and the trough width parameter ranges from 0.5 to 0.7. However, the absence of groundwater drawdown control measures together with unfavourable geological conditions could increase the magnitude of volume loss to as high as 1.2%.

10.18.21 Design and Construction of Tunnels

i) TBM Tunnels

The stations relate to twin tunnels with an internal finished diameter of 5.6 m. The type, grade, and strength of rock, as well as the possibilities of mixed ground conditions and water inflows, govern the choice of TBM.

Anticipated average excavation rates from 8 to 12m/day. TBM tunnels in the Deccan trap rocks indicate that the cutter abrasion in basalts and breccia may be less due to a lower percentage of quartz and silica.

The alignment is designed to follow roads to the extent possible. This requires consideration of site-specific issues, such as the impact on existing buildings. There are numerous buildings classified in the moderate to severe damage category, as assessed during the Building Condition Survey. At two locations, tunnels cross the existing railway lines. Tunnelling below live railway lines needs stringent control on volume loss and ground movement. As such, total movements should not exceed 15 millimetres, and angular distortion should be limited to 1 in 2000 longitudinally and 1 in 2500 transversely.

Tunnelling below the Mithi River was discussed in the earlier part.to the diversity and complexity of the hydrogeological conditions and the sensitive nature of buildings, heritage, and religious sites, tunnelling in dilapidated structures becomes a critical issue and requires approval from the relevant authorities. The final choice of Tunnel type from the three different types — Slurry TBM, EPB TBM & Dual Mode TBM — discussed earlier must be commensurate with the site conditions.

The basic requirements of TBM design were established across the entire corridor, which was designed and built by different Contractors. Requirements were:

i. Geometric parameters, alignment, diameter, and segmental lining (in this case, precast concrete lining), including minimum radius of curvature along the corridor.
ii. Geotechnical & Hydro geotechnical parameters, including water table, type of rocks, specific gravity, cohesion, angle of friction, elasticity and permeability
iii. Assessment of key mechanical requirements of the TBM, such as thrust pressure, torque requirements, cutter head design layout, number and size of disc cutters, scraper bucket with their respective spacing, geometrical configuration of thrust pads and annular grouting pressures.
iv. Obstructions like rivers, railway lines, high-rise buildings, heritage structures, overburden & available land on each launching shaft. This type was used for approximately 15 km in length.

Fig.10.50: Terratec Dual Mode TBM

Cutterhead design, thrust, and torque requirements are fundamental and critical. The dual-mode TBM was the choice for favourable geological conditions, allowing for greater excavation rates to be obtained in open-mode operation with the use of a belt conveyor, with reduced risks associated.

Design features of the dual-mode type of the following geological conditions,

i. Variations in ground type
ii. Possibility of running ground and/or face collapse
iii. Flowing and faulted ground
iv. Water flow.

Fig.10.51: Slurry type TBM

Slurry TBM used for 3.6 km. Slurry machines were chosen due to the varied geological conditions along the section, the proximity of the tunnel alignment to heavily congested areas and sensitive structures, the condition of the buildings, and the availability of the required land.

Slurry TBM is suitable for larger variations in ground type over a short distance, with land consisting of hard rock to soft soils. Location with possible running groundwater, and/or with face collapse under sensitive and EPB TBM was also used for around 15 km. EPB machines were chosen for these sections, which have largely varied geotechnical conditions, surface features such as rivers and railway tracks, and reduced cover over sensitive structures. It is used at locations with geological risks, which are congested areas. Ground with flowing and faulted ground and with possible water inflow

i. Unstable face – weak rock mass and water
ii. Material erosion at the face
iii. Weak joint fillings
iv. Localised / Concentrated water inflow
v. Small block fall at the face, resulting in instability

Fig.10.52: Herren Knecht EPB TBM

10.18.22 Condition Assessment Monitoring and Basic Approach for Rehabilitation

a) Introduction

One of the primary lines in the Mumbai Metro Network is Line 3, which starts from Cuffe Parade and extends to Seepz, with a Depot Located at Aarey Colony. The last station is at Aarey Colony. As this corridor passes through thickly built-up areas, it has been decided to build the Metro system Underground, with Aarey Station at grade. This Corridor is full of existing buildings/ structures, as well as many Heritage structures. This brief report covers the Condition of existing buildings and Heritage Structures within the influence zone of the Line

The structures' protection activities, before and during the construction, were:

* To identify existing structures within the zone of influence.

- Compile records of design and as-built condition.
- To carry out appropriate analysis in evaluating the need for protection, such as underpinning, jacking, isolation, grouting, freezing or other means.
- Comparison of viable alternatives with respect to cost and constructability.
- To inspect and survey existing buildings, to find out the feasibility of underpinning, if required.

b) Project Geology

A geotechnical assessment of the alignment corridor was conducted. The geophysical profile comprises soil followed by rock, and the rock head varies non-uniformly between 3m and 10.5. The soil deposit is predominantly silty sand and clayey sand, including clay layers. Clay layers are of medium to high plasticity. For the soil type in the corridor, data were correlated with strength and stiffness parameters to N values. Rock is mainly Basalt with rock cover.

The groundwater level along the corridor varied between 1.9 m and 5.6 m below ground. The proposed tunnel was generally embedded in Basalt with a minimum rock cover of 4.3m (UGC -01). In some sections, it was noted that the face fractured the Breccia. These details were important for calculating volume loss and predicting settlement.

The geotechnical investigation revealed mixed face ground conditions. During construction, the interface needed to be established as accurately as possible to take greater precautions during tunnelling and excavation, thereby avoiding excessive ground movement/collapse.

Topsoil varying between 1.0 and 14.1 M decomposed rock varying between 0 and 16m BRECCIA, BASALT, TUFF. Ground pre-treatment in these areas was deemed necessary.

c) Existing Buildings and Heritage Structures

The structures were diverse and included historic buildings, monuments, gardens, cemeteries, landscapes, archaeological sites, remains, ruins, monuments, public spaces and natural features. The historic environment was not only a matter of material remains, but also the identity of individuals, communities, and the nation.

The importance of a property to the history, architecture, archaeology, engineering, culture and community. For selecting heritage, attention was:

i. Associat,ions with events, activities and patterns.
ii. Association with Important Persons
iii. Physical Characteristics of design, construction and forms.

iv. Important information, illustrating social, economic history, like terminals, town halls, clubs &markets.

v. Technological innovation

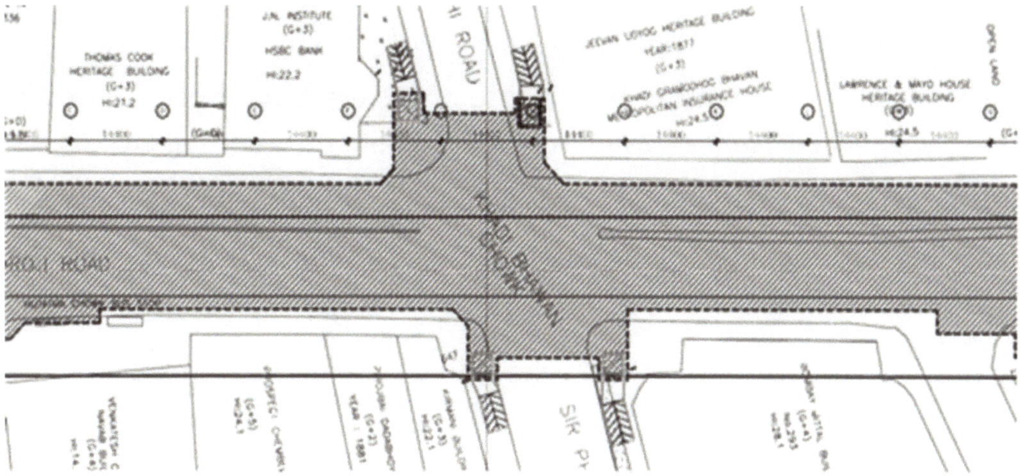

Fig.10.53: Location of Lifts in Station

A Visionary team of experts has listed a record of 'Heritage Structures', available to the public. This charter recommended heritage buildings and sites, classified as Grade I, II, and III depending on order of Importance. Buildings /Sites classified as Grades I and II were considered in accordance with the provisions of the Officials & Legal Manuals.

Buildings identified along the 32.5 KM corridor, spanning from Cuffe Parade Station to Seepz Station, within a 60 M bandwidth, encompass both heritage structures and non-heritage buildings. A list of the structures identified was included in the package-wise. A separate list of Heritage Structures has been prepared package-wise. However, all the heritage structures were located within the first three Packages. A package-wise list has been included in Table.10.6

Information on the Buildings identified is:

a. Year of Construction
b. Plans and Details of Buildings/ Structures
c. Type of Structures (Foundation/ Superstructure)
d. Major Repairs & Rehabilitation carried out in past

Since drawings of most buildings and structures were unavailable, Survey Team collected site data to the extent possible. MCGM certified structural engineers recorded the existing crack patterns. The list of Buildings included in this Report is for reference. It is the responsibility of the Contractors to review the list and match it with actuals at the site. Contractors need to inspect and re-inspect the condition of Buildings and Categorise Them based on existing damage. In categorising the building, conditions to be surveyed based on the following:

a. Distance from the centre of the Alignment.
b. Method adopted during Construction.
c. Condition at the time of execution.
d. Detailed structural examination.
e. Existing Condition

It is common and natural for residents of buildings to observe and attribute pre-existing conditions in buildings to the effect of adjacent excavation of Stations/Shafts/ Tunnels, etc.

Preconstruction photographs and Video are important in obtaining a history of building movements due to seasonal changes. And a logical evaluation of ground movements and building distortion due to excavation or tunnelling.

Understanding the chain of relationships and behaviour extending from tunnelling or excavation

i. Ground movement sources and ground loss
ii. Ground movement patterns and volume changes
iii. Building structural Characteristics and finishes – how it was built, maintained and repaired.

Effects may superimpose on pre-existing building conditions, such as cracks, distortion and deterioration.

Data available at present is mostly inadequate. And so, package-wise, Contractors had to carry out detailed building condition assessments, collect the drawings from the concerned Authorities, and investigate the foundation levels, details, and type of foundations, adopting proven technology without endangering the existing foundation. Given the large number of buildings and structures along the Corridor width, it is crucial to identify them easily. Accordingly, six-digit building identification numbers have been planned. A Building Condition Survey is crucial for the successful completion of the project, potentially reducing the risk of litigation. In this context, the Contractor may use the BCS report, prepared by the Interim Consultant, as a reference.

While carrying out BCS, the Contractor may follow the items outlined during the preliminary survey and, in addition, carry out any other items that are required, depending on site-specific requirements. Items carried out during the survey have been included here for ready reference, which are:

i. Visible Surface Cracks were recorded as non-structural cracks, and through cracks are structural.
ii. Surrounding Ground Condition to detect subsidence or upheaval.
iii. Verticality of Walls, particularly in the case of load bearing, in checking bulging, deformations or any other form of distress.
iv. Cracks at the interfaces.
v. Critical inspection of major overhangs to detect cracks, deformation, deflection, etc.
vi. Spalling in RCC Structures, particularly for slab, beams & columns.

vii. Indication of foundation settlement.
viii. Verticality / Plumb of the Building to check tilt/twist.
ix. Damage to non-structural members.

Table 10.6: Package-wise List of Heritage Structures

	PACKAGE:1				
S No	Building No	Name of Building	Building Address	Category	Type of Distress
1	6W007	Kitab Mahal	Dr DN Road	III	5
2	6W008	National Insurance Building	Dr DN Road	IIA	4
3	6W009	Taj Building	Dr DN Road	IIA	3
4	6W010	Empire House	Dr DN Road	IIA	2
5	6W011	Kodak House	Dr DN Road	IIA	2
6	6.00E+12	Capital Cinema	Dr DN Road	IIB (A)	3
7	6.00E+17	167 Terminal View Bldg, Co-op Society	Dr DN Road	IIA	3
8	6.00E+25	DDC BANK	Dr DN Road	III	3
9	6.00E+26	Sir Jamset JJ Jeebhoy Parsee Benevolent	Dr DN Road	IIA	2
10	6.00E+27	Badri Mahal	Dr DN Road	IIA	2
11	6.00E+28	Fort House	Dr DN Road	IIA	1
12	5W003	Narsari Building	Dr DN Road	IIA	2
13	5W004	Cox King Holiday Building	Dr DN Road	IIA	3
14	5W006	Lawrence & Mayo House	Dr DN Road	IIA	2
15	5W007	Jeevan Udyog Building	Dr DN Road	IIB	4
16	5w008	JN Patel Institute	Dr DN Road	IIA	3
17	5W009	Cooks Building	Dr DN Road	IIA	2
18	5W011	Standard Building	Dr DN Road	IIA	2
19	5W012	Albert Building (Siddarth College)	Dr DN Road	IIA(B)	1
20	5W013	Oriental Building	Dr DN Road	IIA	3
21	5.00E+01	Commissariat Bldg.	Dr DN Road	IIA	3
22	5.00E+03	Peninsula House	Dr DN Road	III	3
23	5.00E+04	Erichson bldg.	Dr DN Road	IIA	2
24	5.00E+10	Bombay Mutual Building	Dr DN Road	IIA	3
25	5.00E+12	Pirojbai Dadabhoy	Dr DN Road	IIA	3
26	5.00E+16	Dadysett	Dr DN Road	IIB(A)	2

27	5.00E+22	High Court	Veer Nariman Road	IIA(I)	2
28	4W002	Astoria Hotel	Jamshedji Tata Road	IIA	2
29	4W007	Moti Mahal	Jamshediji Tata Road	III	3
30	4E004	The Ritz Hotel	Jamshedji Tata Road	IIB	3
31	4E010	Industry House	Jamshedji Tata Road	III	1
32	3E012	Western Rly Officers Transit Flats	Captain Prakash Pethe Marg	I	NA 2
33	3E015	Panthaki	Captain Prakash Pethe Marg	III	2
34	3E017	Damani House	Captain Prakash Pethe Marg	III(0)	1
35	3E023	Baug-E-Abbas	21A Cuffe Parade	IIB	2
36	3E025	Taraporvala Mansion	Captain Prakash Pethe Marg	IIB(A)	NA2

	PACKAGE:2				
S No	Building No	Name of Building	Building Address	Category	Type of Distress
1	9W004,005,006	Pannalal Terrace	Dr Bhalkamkar Marg	III	3
2	9W030	Bhaskar Mansion, Opera House	Grant Road, Sardar Ballabhbhi Patel Road	III	2
3	9W033	Nagindas Mansion, Opera House	Do	III(IIA)	2
4	9W034	Do	Do	III(IIA)	3
5	9W035	Do	JSS Road	III(IIA)	3
6	9W036	13B Nagindas Mansion	Do	III(IIA)	2
7	4E062	St. Teresas Church	RR Road, Prathana Samaj Road	IIE(III)	1
8	9E 064	Shyam Sadan	Do	IIB	4
9	9E065	Kshtriya Dnyati Niwas	JSS Road	III	2
10	8W033	Goan Institute Building	JSS Road	IIB (II)	4
11	8E020	Zaoba Shreeram Building	JSS Road	III	2
12	7W029	Parsee Fire Temple Chirabazar	JSS Road	IIA (II)	NA 3
13	7W031	Bhiwandiwala Terrace	JSS Road	III	3
14	7W040	Aidum Building	JSS Road	IIB (III)	4
15	7W041	People's Free Reading room & Library	JSS ROAD	IIA(III)	1

16	7W042/043	Metro Big Cinema	JSS Road	IIA	1
17	7W051	659 Jer Mahal	JSS Road	IIB (III)	4
18	14E020	Podar Medical College	Dr J Palkar Marg	III	3
19	14E021 &022	Podar Ayurvedic Medical Centre	Annie Besant Road	III	4

d) Existing Structures Protection Program

The major portion of the proposed Metro alignment passes through densely populated/thickly built-up areas. Metro Construction has two major components. Corridors, as underground Tunnels and Stations, are being built using the Cut-and-Cover system, employing a top-down/bottom-up construction approach. Shafts are proposed to be built using a bottom-up approach.

For most structures within the zone of influence of tunnelling, it can be anticipated that the selection and proper use of a tunnelling method appropriate to the surface conditions encountered will be sufficient to control ground movements; however, if this is not the case, protection may be necessary.

The induced ground movement and groundwater drawdown during tunnel excavation are major considerations. Accordingly, the selection of the Tunnelling method is important. In this case, Shield Tunnel Boring Machine (TBM) is one of the preferred options, and precast reinforced concrete linings have been planned to give the final shape of the Tunnel Corridor.

Accordingly, issues that need to be studied for this type of structure are:

1. Nature and variability of ground conditions
2. Construction impact, like noise, vibration and ground movement, particularly to satisfy stringent control for Heritage Structures
3. Impact on all structures:

   Effect of ground movement due to tunnelling operation.

   Monitoring actual ground movements, including settlement, angular distortion, and crack width measurement within the influence zone

   Damage assessment based on the type of foundation, type of Structures, and condition of building/structure.

The cut-and-cover method is cost-effective, safe, easy to control and implement. But the disadvantage:

i. PUO pipes need to be removed/ diverted/ or kept hanging in position, keeping the system operative
ii. Archaeological issues need to be taken care of
iii. Availability of sufficient land both for permanent structures & required construction area.
iv. Appropriate traffic diversion with minimum hindrances to movement

To investigate the performance of deep excavation and the associated effect on the adjacent buildings and existing utilities. field observations include deflection of Diaphragm Walls (DW), vertical movements at the wall top, ground settlement and settlement of surrounding buildings and utilities.

The over excavation and long construction duration cause the DW s to develop substantial deflections. The base and middle floor slabs play a prominent role in reducing post-excavation deflection and settlement.

Precautions:

1. Verification of appropriateness of the selected Tunnelling method
2. Method of control of settlement at the surface, requiring additional protection like freezing, grouting or underpinning
3. Possible control of the development of stress
4. Progressive deformations
5. Supplement if any important data of existing buildings is missing
6. Protect structures by underpinning, jacking, isolation, grouting or other means

10.18.23 Condition Assessment of Existing Structures

There are 1478 existing Structures along the 60m Corridor between Cuffe Parade and Seepz, and an additional 111 structures, including the Aarey station area. Structures are listed in 7 Packages. These include Heritage structures. However, all heritage structures are in the first three packages.

For heritage structures, two different categories have been indicated. One is a type of heritage category. Category I, II and III. Category II has been further divided into II A & II B. The condition of the buildings/structures is categorised into six types: Type 0, 1, 2, 3, 4, and 5, based on their existing condition. Brief details of Heritage Category and Building/ Structures are:

1. **Category I:**
2. **Category IIA:**
3. **Category IIB:**
4. **Category III:**

Heritage Grade I buildings and precincts are of national or historical importance, characterised by excellence in Architectural style, Personality movement, or Institution.

Heritage Grade II (A & B) includes buildings of regional or local importance, special architectural or aesthetic, cultural or historical value of a lower grade.

Heritage Grade III are structures and precincts of importance in town space, aesthetic or sociological interest.

For Existing conditions of Structure types:

Sr. No.	Type	Condition	Criteria
1.	Type 0	Negligible	Minor cracks, which are sporadic and are Non-structural, Presence of a Water stain
2.	Type 1	Very Slight	Area and non-structural. Presence of water stains
3.	Type 2	Slight to Moderate	non-structural cracks Several non-structural cracks Presence of water seepage
4.	Type 3	Moderate	Minor, sporadic spalled concrete Presence of Water seepage
5.	Type 4	Severe	Major extensive spalled concrete Bulging of structural & load bearing Elements Reinforcement bars of columns and Beams Junction Presence of seepage
6.	Type 5	Very severe	Deteriorated structural/load bearing Elements Heavily exposed reinforcements In columns and beams of framed Structures Structural elements where deflection/ Tilt is visible Propped structural elements

The list of Buildings included in the Report is for reference only. It is the responsibility of the Contractor to review the list and match it with the actual ones existing at the site. Contractors need to check the condition & categorise the buildings based on the degree of existing damage. To evaluate the condition of the building, taking into consideration the following factors:

a. Distance from the centre of the Alignment
b. Methodology of Construction
c. Condition at the time of taking up Construction
d. Detailed Structural examination.

As per the preliminary Survey, out of the total 1478 buildings, between Packages I to VII, there are 250 buildings of Severe (Category 4) and 16 are of very severe in nature. Package-wise break-up is as follows:

UGC:01	Severe: 12 nos
	Very Severe: 4 nos
UGC:02	Severe: 92 nos
	Very Severe: 8 nos
UGC:03	Severe: 21 nos
	Very Severe: Nil
UGC:04	Severe: 73 nos
	Very Severe: 4 nos
UGC:05	Severe: 20 nos
	Very Severe: Nil
UGC:06	Severe: 11 nos
	Very Severe: Nil
UGC: 07	Severe: 21 nos
	Very Severe: Nil
Total:	Severe: 250 nos
	Very Severe: 16
Heritage:	Total: 56 nos
	Severe: 7
	Very Severe: 1

10.18.24 Effect of Transit Structure on Existing Structures During Execution and Operation

The effect of the Underground Metro System on the existing buildings/structures is two folds. First, this occurs during execution, and the other occurs during the system's operation. To determine the impact of executing the underground metro system, which includes tunnels, shafts, stations, and depots, it is necessary to monitor the distress caused during execution and compare it with any existing distress. During operation, in addition to monitoring new distress, one must monitor comfort and serviceability.

As per Skempton and McDonald, damage can be classified into three categories

- Architectural Damage – Affects the appearance of structures, panel walls, floors and finishes.
- Functional Damages -- Damages like jammed doors, windows, falling plaster, dis function utilities.
- Structural Damage -- Includes the effects on Structural, Stability & Safety.

In situ monitoring during the tunnelling operation is an integral part of the design, not only for checking structural safety and design, but also for verifying the basic concept of the ground's response to tunnelling and the effectiveness of the structural support. In situ monitoring will include

i. Control deformations of the Tunnel to ensure the tunnel profile.
ii. Appropriateness of selected tunnelling method.
iii. Control the Settlements at the surface and obtain the formation pattern of the ground.
iv. Monitor settlement caused by lowering the water level.
v. Development of stresses in Structural members.
vi. Measure progressive deformation, which may need ground and support strengthening.

10.18.25 Ground Improvement

The ground control achieved through the tunnelling method may not be sufficient to prevent settlements from reaching levels that cause functional or structural damage to structures along the alignment. For protecting buildings subject to settlements that are likely to cause damage, techniques such as Permeation grouting, Compensation grouting, Compaction grouting, Jet grouting and Underpinning are used.

Permeation grouting fills pore spaces in soil with grout.

Compaction grouting involves the controlled injection of stiff mortar into subsurface soils.

Jet grouting injects fluids at high pressure and flow rates in the ground.

Compensation grouting involves controlled injection of grout between excavations and structures that need protection.

Underpinning is the process of installing new foundations beneath an existing building's foundation.

Grouting or injections Of Ground may improve ground characteristics. Though in most cases Grouting is applied to close discontinuities in rock or to strengthen soft ground.

In addition, the ground is also stabilised by dewatering and by avoiding inflows of water. Ground failure may be avoided by minimising pore water pressure.

Improvement is also possible through ground freezing, which changes the property. Time-dependent stress-strain behaviour of frozen ground can be significantly different. Freezing causes an increase in water volume and heave at the surface.

10.18.26 Construction Approach

The construction approach of the major elements of the Project is:

In most Transit Tunnel Projects, significant portions of the alignments are constructed beneath buildings. Station boxes have been proposed to be a cast-in-situ reinforced concrete structure. Traffic management during construction is very important. Types of construction configurations are proposed based on site conditions, with reference to the Station Location and its footprint. Types of Construction planned for Stations and Shafts, including both launching and retrieval, as well as Tunnels and NATM, are required for cross passages and station areas, depending on site conditions.

In general, station boxes are planned to be made of cast-in-situ reinforced Concrete. Based on the Station configuration, land available or to be acquired & planned traffic management plan during construction, with approval of concerned authorities.

There are two basic cut-and-cover construction methods, namely Top-Down and Bottom-up. Both methods are suitable for station construction. The system, in principle, consists of temporary or permanent Retaining walls (Diaphragm wall / Contiguous bored piles, or Secant Piles), including a strutting system as per design requirements. However, considering the presence of high-water levels and the presence of Heritage Structures, Top-Down Construction will have an advantage over Bottom-Up.

The types of construction are as follows.

1. Cut-and-Cover Stations in open space, either applying top-down or bottom-up, depending upon a compatible excavation procedure appropriate for existing geophysical conditions (In general, Rock with Soil overburden).
2. Cut & cover Station box underneath the road-wide carriageway, sufficient to accommodate the Station footprint, may be cast in situ, top-down construction. However, the traffic flow on the surface and with underground utilities will require appropriate diversion, either temporary or permanent, depending on the existing site conditions and plan. In some areas, it may be necessary to provide a temporary deck for traffic and temporary supports for utilities.
3. When the type of Station is such that both items 1) & item 2) are not feasible to accommodate the full station, then the method using NATM may be planned.

10.18.27 Construction Impact

Impact during Construction, within the influenced area

i. Ground movement due to the Tunnelling process.
ii. Ground movements, including settlement, angular distortions, and measurement of crack widths within the influence zone of tunnelling operation.
iii. Effect of Ground loss.
iv. Effect of structure, condition of Building/ Structure.

10.18.28 Heritage Structures – Settlement & Damage Assessment: Approach Methodology

In assessing and protecting existing structures, both existing & potential damages need to be identified and addressed. This can be done based on ground movement, using Both Empirical and numerical analysis. For cut-and-cover sections, both vertical and horizontal settlement/movement were calculated. The design of the Hutatma Chowk Metro Station has been carefully planned to respect the heritage character of DN Road and its iconic landmarks, including Flora Fountain and the Hutatma Chowk Memorial.

Fig.10.54: Escalators were designed without a Roof not to intrude in the heritage environment

Chhatrapati Shivaji Maharaj Terminus (CSMT) Metro Station – a Heritage Structures **Revolutionising Connectivity** and designed considering the followings.

- Sleek modern design with arched entrance & glass facade
- Direct connectivity from CSMT Metro Station to CSMT Suburban Railway via subway
- Multiple entry & exit points for easy access
- Connects key business, heritage landmarks, and transport hubs in South Mumbai
- Reduces commute time and cuts surface traffic congestion

Based on experience from other metropolitan areas and empirical methods, the settlement due to an overburden of approximately 7m is estimated at 0.1%, or 7mm, and the maximum horizontal movement is expected to be around 0.15%, or 10mm. The settlement due to rock excavation is negligible.

Movement due to Tunnel excavation can be expected to be within the range of 5mm or less, using the empirical approaches of MAIR (1993) and PECK (1969).

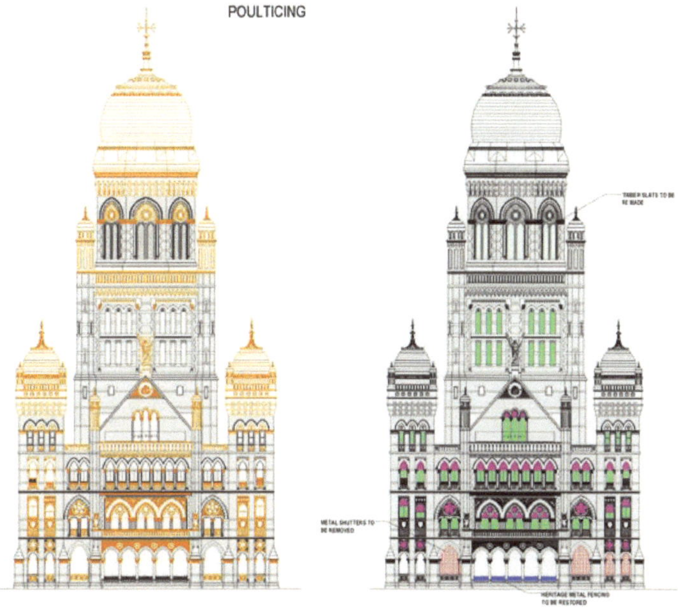

Fig.10.55: Restoration of Municipal Head Office, Mumbai: Roof, Exterior & Lobby (Heritage Building)

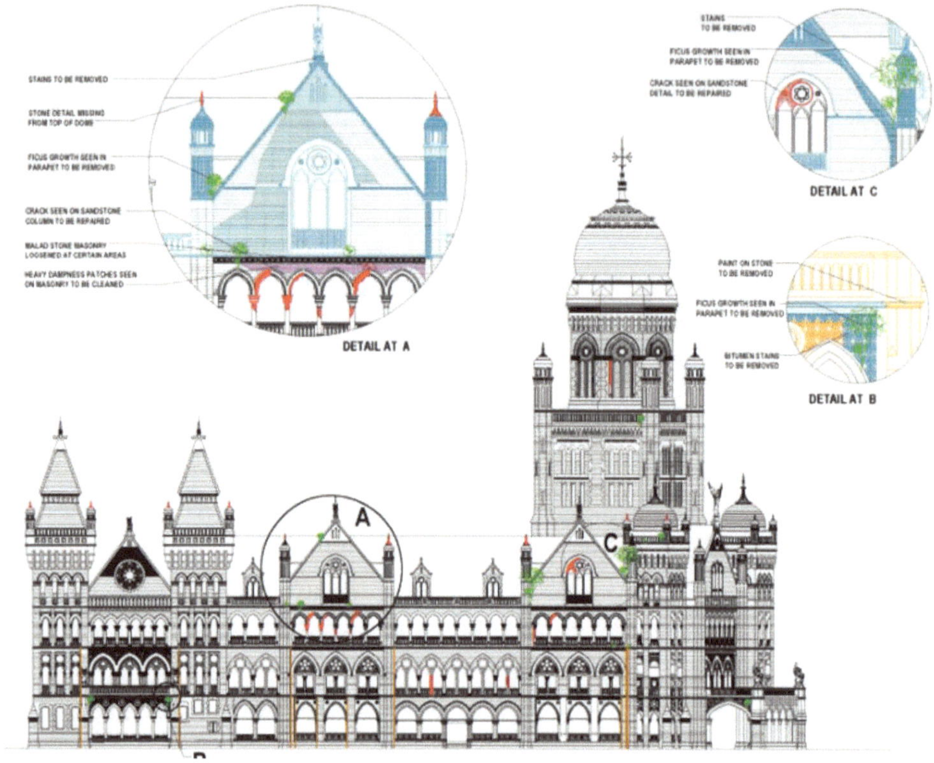

Fig.10.56: Restoration of Municipal Head Office, Mumbai

10.18.29 Instrumentation and Monitoring

Need for Instrumentation

(Ref. Dr MG Khare)

Instrumentation and monitoring need to be carried out both during implementation and operation.

To start with, one must establish a baseline for movement, distortion, groundwater level & its fluctuation, noise and vibration, and indications of excessive and undue movement of the ground surrounding existing Buildings.

To monitor movement and deformation for design verification, both for temporary and Permanent structures. Additionally, to ensure that maximum allowable tolerances are not exceeded for structures within the influence zone. Water drawdown outside the excavation must be controlled within the natural fluctuations of groundwater.

10.18.29.1 Instruments and purpose:

Types of Instruments Required

Parameters	Reason for Monitoring	Instruments Used
Pore Water Pressure	• Design for lateral earth pressures • Design for Uplift pressures	• VW Piezometer • Standpipe Piezometer
Lateral Deformation	• Verify the performance and safety of tunnels, temporary and permanent structures	• Inclinometer • Tape Extensometer
Vertical Deformation	• Verify the performance of structures adjacent to construction activities	• Settlement Markers • Rod / Borehole Extensometer
Load in temporary / permanent structural members	• Verify performance and safety of cut and cover excavations	• Load Cells, Strain Gauges
Vibrations	• Monitor vibrations during construction and operation	• Accelerometer

(Dr MG Khare)

A list of instruments, including their purposes, used is provided below.

Temporary Benchmark : to monitor vertical movement.

Precise Levelling Point : monitoring of monuments

Precise Levelling Studs : for installation on pavement at right angle to tunnel.

Piezometers : monitoring of groundwater level.

Inclinometers : monitoring of lateral movement, including that of the Diaphragm wall

Extensometer : to monitor vertical deformation or heave.
Strain gauges : to monitor strains of struts, anchors, etc.
Load cells : monitoring of loads at struts, anchors, etc.
Tilt meter : to monitor the tilt of buildings.
Crack meter : monitoring cracks.
Vibration : to measure the vibration of existing structures

Samples of Damage and types of Instrumentation used

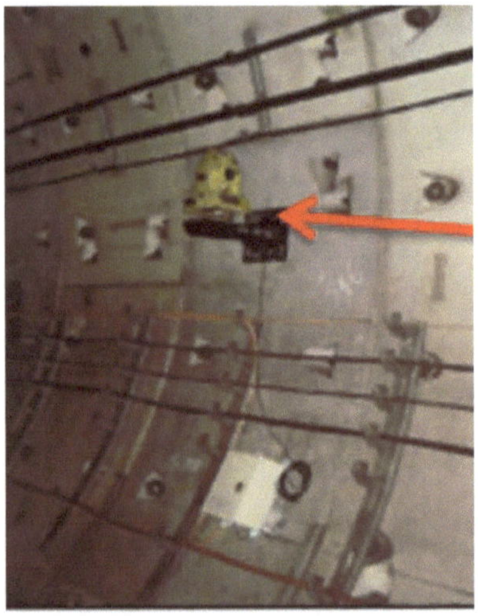

Automatic Total Station & Transmitter Inside Tunnel

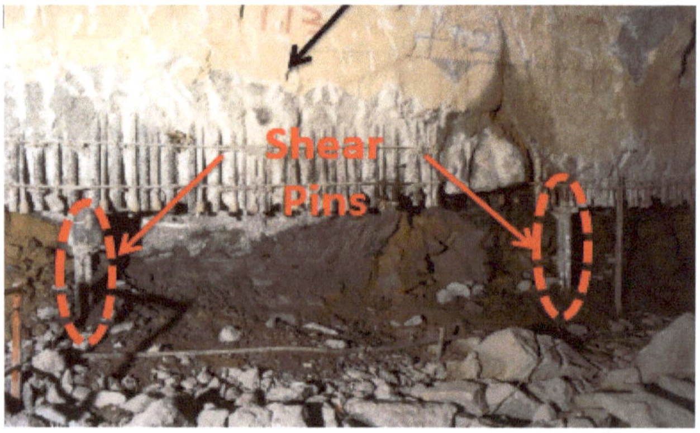

DW – Upper Section

Ground Settlement Meter

Load Cell Strain gauges

10.18.30 System

Systems adopted in line 3 are like those in line 1. MMRC awarded Alstom the overall contract for Line 3. The order includes the manufacture of 31 lightweight, fully furnished modern metro trains, each comprising 8 cars. Along with rolling stock, Alstom executed the power supply contract and equip Line 3 with Uralist 400, its latest generation of CBTC signaling technology. The scope of the signaling contract included unattended train operation (UTO), computer-based interlocking, centralised train supervision, an integrated telecom solution comprising CCTV, passenger information systems, and passenger announcements, a Gigabit network platform, screen doors, and the electrical and mechanical supervisory control and data acquisition system (E&M SCADA). The signalling and train control system on the line is planned for a design headway of 2.5 minutes to meet sustained train operation at 3-minute intervals during peak periods. 0.3.4 Telecommunication: The standards proposed to be adopted

The Mumbai Metro is being implemented with a standard gauge, and with the objective of uniformity, this corridor is also proposed to use the standard gauge (1435mm). Track on metro systems is subjected to intensive usage with very little time for maintenance. Therefore, it is essential that the track structure chosen for metro systems is durable, requires minimal or no maintenance, and ensures the highest levels of safety, reliability, and comfort, with minimal noise and vibrations. The ballast-less track structure has been proposed in line with this philosophy.

The ticketing system is simple, easy to use/operate, maintain, and is easy on accounting facilities. It can issue single/multiple journey tickets, is amenable to quick fare changes, and requires overall manpower. A computer-based automatic fare collection system is proposed

10.19 Metro Line 7

10.19.1 Background

Mumbai Metro Line 7 is an elevated corridor that covers approximately 16.5 km, connecting the western suburbs of Mumbai from Dahisar East to Gundavali. Along the way, the line serves 14 fully elevated stations. The DPR prepared by **DMRC,** DDC was **Ayesa-Spain** and LDC was **RUPL** (Dr BC Roy's team). References are also drawn from Wikipedia.

Andheri East and Dahisar East were proposed as Line 7 in the original Mumbai Metro master plan unveiled by the Mumbai Metropolitan Region Development Authority (MMRDA) in 2004. In June 2015, the MMRDA proposed building Line 7 as part of a plan to build 6 new metro lines. The MMRDA approved the detailed project report for Line 7 in August 2015. The report proposed a 16.5 km elevated line between Andheri East and Dahisar East.

The MMRDA invited bids for the project in December 2015. Tendering for the design and construction of the corridor was split into 3 packages. The first package included an elevated viaduct and five elevated stations – Andheri East, Shankarwadi, JVLR Junction, Mahanand and new Ashok Nagar.

The second package included an elevated viaduct and six elevated stations: Aarey, Dindoshi, Pathan Wadi, Pushpa Park, Bandongri, and Mahindra & Mahindra. It also included an elevated viaduct and five elevated stations: Magathane, Devipada, National Park, Ovaripada, and Dahisar East. A total of 16 companies expressed interest, and 9 bids were submitted. In April 2016, the MMRDA awarded contracts to Simplex Infrastructure, J Kumar Infraprojects, and Nagarjuna Construction Company for the first, second and third packages, respectively.

Fig.10.57: Mumbai Metro Corridor – during Operation (knocksense.com)

Construction work on the corridor began in August 2016.

The contractors were expected to complete construction of the corridor and all 14 stations within 30 months from the day of commencement of work. In February 2017, MMRDA officials stated that 15% of piling, pile caps and pier work on the corridor had been completed. By the end of April 2017. Line 7 stations were built on a single pier. The single-pier design was chosen to facilitate the construction of smaller stations and reduce the amount of land occupied on the Western Express Highway, which could otherwise obstruct traffic flow. **This project provides a much-needed transportation link to commuters in the region**.

The first U-girder, the concrete structure on which metro tracks are laid, was launched near Pathanwadi Junction in Malad in May 2017. In total, Metro 7 required 1,400 U-girders. Work on erecting the U-girders and disposing of and dumping muck generated from underground work was done only between 11:30 pm and 6:00 am. During this period, multi-axle trailers transported U-girders weighing 170 tonnes from the casting yard in Bandra to the location where it is to be erected. The trailer had to move slowly to avoid causing an imbalance to the heavy load, which can result in toppling, and took roughly two hours to complete the journey. Lifting the girder and attaching it to the pier caps was also a difficult process and was carried out at night. Traffic diversions were activated around the launching site to reduce the possibility of accidents. Officials erected an average of two girders every day.

Detailed design of Line 7 was carried out by **Ayesa, a Spanish Consultant and RUPL** (Dr BC Roy) carried out the LDC.

Fig.10.58: Metro Corridor- Line 7 (2).

Fig.10.59: Entry to Platform

Fig.10. 60: Entry from Unpaid to Paid Area

The MMRDA began a dynamic test and trial run on the 20 km stretch between Dhanukarwadi and Aarey (including a portion of Line 2) in June. The stretch between Dhanukarwadi and Aarey received provisional approval from the Research Designs and Standards Organisation (RDSO) in January 2022. The Commissioner of Metro Rail Safety (CMRS) began inspecting the line in February 2022.

A section of Line 7 from Dahisar East to Aarey (along with the section of Line 2 from Dhanukarwadi to Dahisar East) was opened in April 2022. The depth of the Pile for this tender was 12 m below the pile cut-off level, including the socket. The adjustments for extra/ less depth of Pile were made for an increase/ decrease in the cubic contents of Concrete

The Final location survey of the alignment and stations, as given in GAD, aims to accommodate statutory & other requirements, including adjustments to Vertical and horizontal alignments, without incurring any financial impact/cost.

The maximum cantilever pier cap, in general, was limited to 2m. In all other cases where the cantilever was more than 2 meters, a portal was provided. To protect the reinforcement bar from corrosion for the super-structure (components of Viaduct/Bridge/Stations above bearings) by "Bipolar Concrete Penetrating Corrosion Inhibiting Admixture" was used (RDSO specification No: M&C/PCN/126-2008). For the concreting of piles and pile caps, Portland Slag Cement (PSC) was used. To protect the reinforcement bar from corrosion for pile and pile caps, a coating of "Fusion Bonded Epoxy Coating" as per IS 13620:1993/Hot Dip Galvanising coating with a minimum thickness of 100 microns, was recommended for substructure, i.e. structure above pile cap and up to bearing, "Bipolar Concrete Penetrating Corrosion Inhibiting Admixture" used (RDSO specification No: M&C/PCN/126-2008).

Micro silica conforming to ASTM C1240 at 5% to 10% of cement volume as an admixture in concrete grade M-40 and above, where OPC was used, considering the environmental condition of Mumbai. No fly ash was allowed for elevated structures

Fig.10.61: Alignments of Line 7 and Line 2A

Fig.10.62: Internal view of a coach

10.19.2 Overview

The Mumbai Metro Line 7 (Dahisar East – Gundavali) metro corridor is 16.495 km long, fully elevated, and consists of 14 stations, spanning from Dahisar to Andheri. Metro Corridor 7 was initially planned to run between Dahisar and Andheri, but the state government later decided to extend the metro route. The Metro corridor at Dahisar has been extended up to Mira-Bhayander in the north. Andheri has been extended up to the T2 terminal international airport, Sahar, in the south. The extension for the metro corridor from Dahisar to Mira-Bhayander is known as Metro Line 9, and the extension from Andheri to T2 International Airport is referred to as Metro Line 7A. The official colour code for Metro lines 7, 7A, and 9 is Red, also recognised as the Mumbai Metro Red Line. Currently, Metro Lines 7 and 2A are fully operational. Depots are at Charkov (temporary) and Bhayander.

10.19.3 Salient Features

1. Gauge (Nominal): 1435 mm
2. Route Length: 16.475 Kms (Completely Elevated)
3. Number of Stations: 16(All Elevated)
4. Traffic: 528690 (2021) 667698 (2031)
5. PHPDT: 19156 (2021), 191156 (2031)
6. Train Composition: 6 coaches
7. Design Speed: 90 kmph
8. Scheduled Speed: 32 kmph
9. Traction System Voltage: 25KV AC
10. Current Collection System: Overhead Catenary
11. Rolling Stock: 3.20 m wide, air-conditioned stainless-steel body; Axle Load: 17 t
12. Signaling, Telecommunication, and Train Control

    a. Type of Signalling 'CATC' (Continuous Automatic Train Control System based on "CBTC"
    b. Train Control System ATP (Automatic Train Protection),

       ATO (Automatic Train Operation)

       ATS (Automatic Train Supervision)

    c. Telecommunication i. Integrated System with Optical Fibre cable, SCADA, Train Radio, PA system etc.
    d. Train information system, Control telephones and Centralised Clock System.

13. Fare Collection Automatic Fare collection system with POM

Civil/Structure

10.19.4 Viaduct Including Viaduct in Stations

The design of the viaduct was like that of other metros in Mumbai and consisted of the following.

i. Detailed survey of the alignment for viaduct, Geotechnical/soil investigation at every pier location, utility checking, fixing an optimised span configuration avoiding shifting of utilities as per contractor's design, subject to the obligatory requirements as shown in the GAD.
ii. Design & construction of pile foundations and sub-structure, including piers, pile caps, pedestals, pier caps, bearings and shear restrainers.
iii. Construction of superstructure with U-Girder for standard span (25m) for straight alignment

10.19.5 Superstructure

The choice of superstructure has been made with consideration for factors such as ease of construction, standardisation of formwork, and Optimal utilisation of formwork for wide spans. The segmental construction has been used since it is an efficient and economical method for structures with longer spans. Structures with sharp curves and variable super elevation can easily be accommodated.

i) Types of Superstructures for Elevated Section

Normally, in the metro elevated section, the following types of segmental superstructures are adopted.

(A) Pre-cast segmental box girder using external unbounded tendon.

(B) Pre-cast segmental U-Channel Superstructure with internal pre-stressing.

(C) Precast segment Twin U girders

ii) Pre-cast Segmental Box Girder using External Unbounded Tendons.

This consists of precast segmental construction with external prestressing, utilising proper jointing techniques. In this construction, the pre-stressing tendons are placed outside the structural concrete (inside the box section) and protected with high-density polyethene tubes, which are grouted with special wax or cement. The match-cast joints at the interface of two segments are provided with shear keys.

iii) Precast Segmental U-Channel Superstructure with Internal Pre-stressing.

The single 'U' type of viaduct structure is used for precast segmental construction with external pre-stressing and requires glueing and temporary pre-stressing of segments. Joints at the interface of two segments are provided with shear keys. The advantages of this type of structural configuration of superstructure are, i) a built-in sound barrier, ii) the Possibility to lower the longitudinal profile by approximately 1 m compared to the box girder. iii) built-in structural elements can maintain the trains on the bridge in case of derailment iv) built-in maintenance and evacuation path on either side of the track.

Fig.10.63: Metro Station – Line 7 (inkl.com)

10.19.9 Precast, pre-tensioned U-girder with Internal Pre-stressing

Girders of various spans (19 m, 22 m, 25m and 30 m) are cast in the casting yard, pre-stressed internally. These girders are transported to the site in trailers and launched into position by using double cranes of suitable capacity, one on either end. A significant advantage of these girders is that the launch is conducted at night, minimising disruptions to normal traffic. It will have better quality control because all the girders are shop-manufactured. Twin U Girders are generally more economical than segmental U girders and box girders.

10.19.10 Station Locations

Stations have been strategically located to serve major passenger destinations and facilitate convenient integration with other modes of transportation. The average spacing between stations is approximately one kilometre. All stations are two-level stations except Dahisar Terminal Station. The concourse includes passenger facilities and a station on the lower level, while the platforms are located on the upper level. Dahisar station is proposed to have two towers, one on either side.

10.19.11 Superstructure

A Box Girder having a soffit width of about 4 m (approx.) normally accommodates the two tracks situated at 4.2m centre to centre (c/c). The superstructure, featuring a box girder for all simply supported standard spans, was constructed using the 'precast pre-stressed segmental construction method' with epoxy-bonded joints.

The standard Spans with center-to-center piers of simply supported spans constructed by the precast segmental construction technique were proposed as 28.0m. The usual segments are 3.0m in length, except the Diaphragm segments, which are 2.0m each. The other spans (c/c of pier) comprise 31.0 m, 25.0 m, 22.0 m, 19.0 m & 16.0 m, which could be made by removing/adding usual segments of 3.0 m each from the centre of the span. The pier segment was finalised based on a simply supported span of 31.0m, and the same will also be kept for all standard spans of simply supported. For major crossings with spans greater than 31.0m, special continuous units, typically with a 3 m span, or steel girders are envisaged. All these continuous units (in case provided at obligatory location) will be constructed cast-in-situ by the 'balanced cantilever construction technique'.

Fig.10.64: Two U-Girders erected on Portal Pier (mumbaimetrolines.com)

Fig.10.65: Pier & Eccentric Pier Cap during Construction (mmrda / db-engineering.consulting.com)

10.19.12 Substructure

The superstructure of the viaduct was supported on a single cast-in-place RC pier. The shape of the pier follows the flow of forces. For the standard spans, the pier gradually widens at the top to support the bearing under the box webs. The design of the pier was found to be limited to a diameter of 1.8m to 2.0m, with a circular shape for most of its height, to occupy the minimum space at ground level, where the alignment often follows the central verge of existing roads. To prevent direct collision of a vehicle with the pier, a Jersey-shaped crash barrier 1.0 m in height above the existing road level has been provided all around the pier. A gap of 25 mm was also provided between the crash barrier and the outer face of the pier. The shape of the upper part of the pier has been so dimensioned that a required clearance of 5.5 m is always available on the roadside beyond the vertical plane drawn on the outer face of the crash barrier. In such a case, the minimum height of the rail above the existing road is 8.4 m. The longitudinal centre-to-centre spacing of the elastomeric/pot bearing over a pier would be about 1.8 m. The space between the elastomeric bearings will be used to place the lifting jack needed for replacing the elastomeric bearing. An outward slope of 1:200 will be provided at the pier top to facilitate drainage in the event of rainwater spillage.

10.19.13 Foundations

The foundations used were pile foundations, consisting of an in situ bored pile with a diameter of 1.2 m, supported by an RC pile cap that supported the viaduct piers.

10.19.14 Grade of Concrete

It is designed to carry out construction work with 'Design mix concrete' through computerised automatic batching plants, offering the following concrete grades for various members, considering design requirements and durability.

1. Piles - M -35
2. Pile cap and open foundation - M -35
3. Piers - M -40
4. All precast elements for viaduct and station - M -45
5. Cantilever piers and portals - M -45/M -60vi) Other miscellaneous structures - M -30

10.19.15 Reinforcement and pre-stressed Steel

It is proposed to use HYSD 500 or TMT steel as reinforcement bars. For pre-stressing work, low-relaxation, high-tensile steel strands with the configuration 12 K 15 and/or 19 K 15 are recommended (confirming to IS:14268).

10.19.16 Construction of Stations

At almost all locations, it is proposed to construct 'the elevated stations' with a concourse over the road to minimise land acquisition. To keep the rail level low, it is proposed not to take a viaduct through the stations. A separate structural configuration must be proposed, although this may necessitate a break in the launching operations at each station location. The substructure for the station portion will be like that of the viaduct and will be constructed in a similar manner. However, in the cross-section, there will be a single viaduct column in the station area, located on the median and supporting the concourse girders with a cantilever arm to eliminate columns in the right-of-way.

10.19.17 Traction System

Considering the ultimate traffic requirements, uniformity, standardisation, and other techno-economic considerations, a 25 KV AC traction system is the best alternative, and it has been proposed for this corridor. A 25 KV AC traction system offers economic advantages, including a minimal number of traction sub-stations and the potential to carry large traffic.

10.19.18 Permanent Way

i) Choice of Gauge

The issue of Broad Gauge vs. Standard Gauge for Metros in India has been widely debated, and the decision has been in favour of Standard Gauge.

ii)Track Structure:

Like other Metros in India, two types of Track Structure have been adopted. The ballastless track has been proposed for the viaduct, while the normal ballasted track has been proposed in the maintenance depot. The track will be completely welded, and even the turnouts will be incorporated into LWR/CWR. The rail section used will be UIC 60 (60 kg/m). The grade of rails on main lines will be 1080 Head Hardened as per IRS-T-12-96. As these rails are not currently manufactured in India, they will need to be imported. For the Depot lines, the grade of rails should be 880, which can be easily manufactured indigenously.

10.19.19 Signalling

The signaling system shall provide the means for efficient train control, ensuring safety in train movements. It assists in the optimisation of the metro infrastructure investment and the running of efficient train services on the network. Metro carries many passengers at a very close headway, requiring a very high level of safety enforcement and reliability. The requirements of the metro are planned to be achieved by adopting 'CATC' (Continuous Automatic Train Control System) based on "CBTC" (Communication-based Train Control System), which includes ATP (Automatic Train Protection), ATO (Automatic Train Operation) and ATS

10.19.20 Telecommunication

The telecommunications system acts as the communication backbone for Signaling systems and other systems such as SCADA, AFC, etc. and provides telecommunications services to meet a) Operational and administrative requirements of the metro. b) The telecommunications facilities proposed are helpful in meeting the requirements for, amongst others, c) Supplementing the Signalling system for efficient train operation. d) Crisis management during emergencies, e) Passenger information system, f) Maintenance Control, g) Emergency Control, h) Integrated Passenger Announcement System and Passenger Information Data Channels for Signalling, SCADA, Automatic Fare Collection etc.

(Automatic Train Supervision) sub-systems using radio communication between the trackside and the Train enable the commuters to travel.

Fig.10.66: View of Completed corridor including Station (scroller.com)

10.19.21 Automatic Fare Collection

The Mass Rapid Transit System carries many passengers. Ticket issues and fare collection play a vital role in the efficient and proper operation of the system. To achieve this objective, the ticketing system shall be simple, easy to use, and maintain, and minimise the burden on accounting facilities. It should be capable of issuing single/multiple journey tickets, amenable to quick fare changes, and require less overall manpower. In view of the above, a computer-based automatic fare collection system is proposed. The AFC system proves to be cheaper than the semi-automatic (Manual System). Seamless selection is now being considered for Mumbai. This system is recommended to be adopted.

10.19.22 Choice of Control Gates

Retractable flap-type or turnstile-type Control Gates are recommended.

10.19.23 Rolling Stock

The proposed rolling stock is the most advanced, featuring a sophisticated system with the latest state-of-the-art technology. The controlling criteria are reliability, light weight, low energy consumption, and high efficiency, leading to a lower annualised cost of service. The coach has a high rate of acceleration and deceleration. The Coaches are of stainless steel, approximately 22 m long, 3.2 m wide, and 3.9 m high. The overall length of a train comprising 6 coaches will be approximately 132 meters. The axle load will be 17 tonnes. The coaches will be equipped with asynchronous 3-phase AC squirrel-cage induction motors. Trains will have a regenerative braking system to save energy costs. Current will be drawn through the overhead catenary.

10.19.24 Power Supply

The high-voltage power supply network in Mumbai city was briefly studied. The city has 220, 110 and 66 kV networks to cater to various types of demand in the vicinity of the proposed corridors. Considering the reliability requirements, two Receiving Substations are proposed to be set up for the line. This is an economical solution that does not compromise reliability. It is proposed to avail power supply for traction as well as auxiliary services from the following grid sub-stations of TATA Power Company Limited at 110 kV voltage through cable feeders:

10.19.25 Traction system

There are three options available for the power supply system for MRTS

25 kV & 2X25 kV AC Overhead Catenary system
750 V DC third rail system
1500 V DC Overhead Catenary system.
Based on techno-economic considerations, a 25 kV AC traction system is recommended.

10.19.26 Standby Diesel Generator Set

In the unlikely event of simultaneous tripping of all input power sources or grid failure, the power supply to both stations and trains will be interrupted. It is, therefore, decided to provide a standby DG set of 180 kVA capacity at the elevated stations.

10.19.27 Environmental Impact Assessment

A detailed Environmental Impact Assessment study has been done along the full length of the proposed corridor. Baseline environmental data has been collected for all the relevant elements of the environment. Apart from the standard parameters of air pollution, water pollution, and noise pollution, a quantitative evaluation of 'Total Environmental Impact' has also been conducted. The total environmental impact score of this infrastructure project is obtained by aggregating all individual impacts. The impact assessment results indicate that the overall impact is relatively low for this type of infrastructure, making it acceptable even without remedial measures from an environmental impact perspective.

10.19.28 Issues

In March 2017, the Mumbai High Court temporarily stayed all construction activities at Metro 7's casting yard in Bandra Reclamation. The Court was hearing a PIL filed by locals who alleged that the site of the casting yard had been reserved by the Brihanmumbai Municipal Corporation (BMC) for other purposes, such as a cemetery.

Piling work in Kandivli resulted in a gas pipeline bursting in March 2017. The damage to the Mahanagar Gas pipeline resulted in temporary disruptions in the supply of CNG to Kandivli, Borivli, Dahisar, Mira Road and Bhayandar.

The BMC accused Nagarjuna Construction Company of operating an unlicensed concrete ready-mix plant illegally at its casting yard in April 2017.

10.20 Wadala – Ghatkopar – Mulund– Thane – Kasarvadavali Metro Rail Corridor

10.20.1 Background

The Wadala-Kasarvavali corridor is line 4 of the Mumbai Metro network's Master Plan in the Mumbai Metropolitan Region (MMR). MMR comprises seven municipal corporations, 13 municipal councils and 996 villages. The dominant feature of passenger movements in Mumbai is the overwhelming dependence on public transport modes and walking. In MMR, public transport systems are overcrowded, and the road network is congested, due to a large gap between demand and supply.

The MMRDA prepared the Master Plan of the Mumbai Metro network in 2003. The Master Plan totalled 146.5 km. The background, including the transport scenario and Recommendations are included in the beginning of the Section.

DMRC prepared the DPRs for Line-1: Versova – Andheri – Ghatkopar – 2005, Line-2: Colaba – Bandra – Charkop – 2008, Line – 3: Bandra – Kurla - Mankhurd – 2006.

Subsequently, corridors 2 & 3 were rearranged, and DMRC prepared another DPR for the corridor between Charkop, Bandra, and Mankhurd. The CES-led Consortia prepared the DPRs of Thane MRTS and Andheri – Ghatkopar, Line 1, in early 2000.

The Government of Maharashtra is keen to implement expeditiously the Master Plan Corridors recommended by DMRC on a fast-track mode and to complete them in the next 3-4 years. To start with, it is decided to take up the task of updating DPRs and preparing new DPRs.

One of the most impactful developments is the progress of Metro Line 4, connecting Wadala to Kasarvadavali. This elevated corridor is anticipated to reduce travel time between these points by 50% to 75%, bringing commute times down from 60-75 minutes to approximately 30-40 minutes. Such a significant reduction in travel time enhances Thane and Mulund's appeal for residential and commercial purposes. In addition, the upcoming Goregaon-Mulund Link Road (GMLR) is being transformed further to bolster connectivity between Mumbai's western and eastern suburbs. By providing a direct link, the GMLR will streamline traffic flow, reduce congestion, and make commutes more efficient.

Thane and Mulund stations are key railway junctions connecting the suburbs to Mumbai's vital areas. Enhancements in these stations are expected to improve the travel experience, making the region even more appealing to residents and investors. Being just 2.5 km from Mulund Station and 5 km from Thane Station, MTC offers easy access to Mumbai's central locations, positioning it as an ideal location for commuting professionals.

(**The Realty Today**)

A standout feature of Mumbai Metro Line 4 is its planned interconnectivity with several existing and upcoming transport systems. It will link with the Eastern Express Highway, Central Railway, Monorail, Metro Line 2B (D N Nagar to Mandale), Metro Line 5 (Thane to Kalyan), and Metro Line 6 (Swami Samarth Nagar to Vikhroli). This integration is poised to streamline commuting, offering seamless transfers across different modes of transport within the city.

Fig.10.67: Mulund-Thane Corridor (The Realty today)

Line 4 construction was carried out in two phases. Length between Mukund Naka and Kasarvadavali by the Consortium of Tata Projects Limited and China Harbour Engineering Company. The other phase was assigned to Reliance Infrastructure Ltd. And Astaldi. DB Engineering, Louis Berger and Hill International was engaged as GC. Funding for implementation of Line 4 was done by GoI, GoM & Asian Infrastructure Investment Bank (AIIB). The MMRDA appointed Systra MVA Consulting India Pvt. Ltd. as the detail design consultants for Line 4 in end 2017.

10.20.2 Alignment

First station on Wadala – Ghatkopar – Mulund – Thane - Kasarvadavali Corridor is named Bhakti Park Metro, and the last station is Kasarvadavali. Chainage of Bhakti Park station is taken as 0.0 for reference, and the end chainage of this station is (-) 450 m. The total length of the corridor from dead end to dead end is 32.322 km. The corridor proposed is elevated, with thirty-two stations on the corridor is approximately one kilometre long. The maximum and minimum The interstation distances are 1,632.7 m and 598.53 m, respectively. A depot For this corridor has been planned at Kasarvadavali. This corridor runs from South to North connects the eastern suburbs, marketplaces, Railway Stations, Mulund, Thane, and Kasarvadavali.

Fig.10.68: Route Map Line 4 (metrolinenewsindia.wordpress.com)

10.20.3 Salient Features

1. **GAUGE (NOMINAL):** 1435 mm
2. **ROUTE LENGTH:** 32.322 km (Entirely Elevated)
3. **NUMBER OF STATIONS:** 32 (All Elevated)

TRAFFIC PROJECTION:

YEAR	TOTAL TRIP	AVERAGE LEAD (KM)	MAXIMUM PHPDT
2021	870198	10.10	28107
2031	1213177	10.25	33417

TRAIN OPERATION:

Particulars	2021	2031
Cars/trains	6	6
Headway (Minutes)	3.75	3.00
Max. PHPDT Demand	28107	33417
PHPDT Capacity Available	28096* (35904**)	35120* (44880**)

* @ 6 persons per square meter of standee area

** @ 8 persons per square meter of standee area

Year	Headway (min)	No. of Rakes	No. of car per rake	No. of Coaches
2021	3.75	36	6	216
2031	3.00	44		264

4. Speed

i. Design speed 90 Kmph
ii. Maximum operating speed 80 Kmph
iii. Schedule (Booked) Speed 35 Kmph

5.Traction Power Supply:

a.	Traction system voltage	25 kV AC
b.	Current Collection	Over Head Catenary
c.	Receiving Sub Stations	One at Ghatkopar or Chembur, second at Thane and third at Kasarvadavali Maintenance Depot.

Power Demand Estimation (MVA)

Load	Year	
	2021	**2031**
Traction	27.04	33.94
Auxiliary	16.31	22.85
Total	43.35	56.79

6.Rolling Stock:

a.	3.20 m wide rolling stock with stainless steel body	
b.	Axle load	17 T
c.	Seating arrangement	Longitudinal
d.	Capacity of 3 coach units	
	With 6 standees/sqm.	900
e.	Class of accommodation	One (Air conditioned)

7.Maintenance Facilities

A Maintenance Depot has been proposed near Kasarvadavali Station (the main depot) and Ghatkopar (the stabling depot). The total land area proposed for maintenance facilities is 30.00 Ha.

8.Signaling, Telecommunication, and Train Control:

a)	Type of Signaling	„CATC" (Continuous Automatic Train Control System) is based on "CBTCS" (Communication-based Train Control System), which includes ATP (Automatic Train Protection), ATO (Automatic Train Operation) and ATS (Automatic Train Supervision) sub-systems using radio communication between Trackside and Train.
b)	Telecommunication	i. Integrated System with optical fibre cable, SCADA, Train Radio, PA system, etc. ii. Train information system, Control telephones and Centralised Clock System..

9.Fare Collection

An automatic fare collection system with POM, smart card, etc.

10.Structure

4. Viaduct: Precast twin „U" girders on a Single pier with pile / Open foundations up to a radius of 300m and flatter for sharper curves and location of Points & Crossings I-Girder.
5. Station structure on columns, independent of viaduct piers.

Civil Engineering

10.20.4 Alignment

First station on Wadala – Ghatkopar – Mulund – Thane - Kasarvadavali Corridor is named Bhakti Park Metro, with the last station being Kasarvadavali. An attempt has been made to locate stations about a kilometre apart. However, due to various considerations, including ridership, accessibility, land availability, and design considerations, a few stations could not be located within a 1 km radius. distance apart. The depot for this corridor has been planned at Kasarvadavali.

10.20.5 Geometric Design Norms

The geometrical design norms are based on international practices adopted for similar metro systems with standard gauges, assuming that the maximum permissible speed on the section is limited to 80kmph. The design parameters related to the Metro system described herein have been developed based on a detailed evaluation, experience, and internationally accepted practices. Various alternatives were considered for most of these parameters, but the best-suited ones have been adopted for the system.

Fig.10.69: Majivada Station- Line 4 (Indian Express)

The desirable minimum horizontal curve radius is 200 m, but it can be reduced to 120 m in extreme cases. The minimum curve radius at stations is specified as 1000 m. Vertical curves are proposed at every change of grade. The radius of the vertical curves is 2500 m desirable and 1500 m minimum. Per the IRC Code, the viaduct carrying the tracks will have a minimum vertical clearance of 5.5 m above road level. Usually, stations should be on a level stretch. In limited cases, stations may be on a grade of 0.1%. In this corridor, all stations are on a level gradient. Between stations, grades may typically not exceed 2.0%. However, where existing road gradients are steeper than 2%, gradients up to 4% (compensated) can be provided in short stretches. The maximum design speed has been proposed to be 90 mph, and the maximum sectional speed is 80 mph. The booked speed was taken as 35 mph.

10.20.6 Station and Terminals

The locations of the stations have been identified, taking into consideration the constraints in land acquisition and congestion issues. Stations are proposed to attract the maximum demand from the traffic nodal points.

Fig.10.70: Mumbai Metro Corridor 4 during Construction (83 Daily News)

Stations have been strategically located to serve major passenger destinations and facilitate convenient integration with other modes of transport, including Railway Stations and Bus Terminals. An effort has also been made to propose station locations so that inter-station distances are uniform. All stations are planned to be two-level stations. The concourse, comprising passenger and station facilities, will be located at the lower level, and the platforms are scheduled to be at the higher level. **Bhakti Park Metro Terminal** is located off the road parallel to Monorail Station, featuring a scissors crossover proposed at the rear end of the station. Kasarvadavali Terminal Station is proposed on the median of the road, before the depot location. Scissor crossovers, along with the depot entry, are proposed at the rear end of the station. It is suggested that a central depot be established at Kasarvadavali and a stable yard at Ghatkopar, utilising a 30 Ha plot of land.

10.20.7 Viaduct–Elevated Structure

The proposed Viaduct Structure uses four types of Superstructures to construct the Corridor. Structural forms proposed are (i) Segmental Box Girder, (ii) Segmental U Girder, (iii) I Girder and (iv) Double U Girder, depending upon the level of traffic congestion, available working space, etc. However, a double U-Girder is suggested in the superstructure for a radius of 300m. An I-Girder is indicated for a Radius of less than 300 m and at locations where points and crossings are to be provided.

10.20.8 Technical Investigations

The data available in the DPR prepared by **M/S RITES** for this metro rail corridor, from Wadala to Ghatkopar, Mulund, Thane, and Kasarvadavali, has been reproduced, as the corridor remains the same and the ground profile does not change for smaller deviations in the route. Initially, geological investigations were conducted from Wadala to Kasarvadavali as part of the preparation of the combined DPR for Wadala–Kasarvadavali.

Fig.10.71: Line 4 Corridor Approaching a Station

(mmrda.maharashtra.gov.in)

10.20.9 Foundation

In general, pile foundations are recommended considering the nature of the soil, the type of proposed structures, and expected loads on foundations, except at a few locations where the rock level is up to 6 m below GL open foundations have been recommended.

Considering the type of proposed structure and the subsoil strata, the length of the pile may be about 7.5 to 15 m and to be socketed in rock. Bored cast-in-situ piles, 1200 to 1500 mm in diameter (with a minimum diameter of 1000 mm), have been adopted. Piles transmit foundation loads through soil strata to deeper layers with higher bearing capacity. Piles carry loads as a combination of side friction and point-bearing resistance.

10.20.10 Utility Diversions

Several utilities, such as sewer lines, water pipelines, gas pipelines, power and communication cables, etc., are along and across the alignment. Some of these need to be diverted or kept at specific locations with temporary support, according to the final alignment and structural layouts.

10.20.11 Land

To minimise land acquisitions and ensure good accessibility in both directions, the metro alignments are primarily located along roads within the existing corridor. However, the existing roads, such as road turnings, may not be suitable for the metro rail alignment and structural arrangement at some locations. In such cases, besides Stations, Entry/Exit Structures/Substations, Temporary works, Depots, etc., require the necessary land parcels to be acquired. Temporary acquisition may also be necessary for construction activities.

Fig.10.72: Metro in Second Deck & Roadway on First Deck (sawarajyamag.com)

10.20.12 Safety & Security Systems

The standards and requirements for safety and security related to fires and unauthorised entry into premises are considered and installed to ensure the safety of passengers and premises in the Metro System. The safety design is to include

- Protection of passengers against fire in train services, stations, and other service buildings.
- The system shall protect vulnerable premises and include
- Fire alarm system.

- Fire Hydrant and Sprinkler System.
- Fire Extinguishers.
- Closed-circuit television with video analytics.
- Security Gates – Metal Detector.
- Baggage Scanner.

10.20.13 Depot cum Workshop

The requirement for a Maintenance Depot and a stable yard, along with its activities, has been included in the project preparation in Part 1. DPR has been prepared by **DMRC**.

This chapter includes the plan and design of the depots designed for the Mumbai Metro for ready reference. The primary activities of the depot are,

i. Major overhauls of all the trains.
ii. All minor schedules and repairs.
iii. Lifting for replacement of heavy equipment and testing thereafter.
iv. Repair of heavy equipment.

The planning is based on the following:

a. Sufficient space is a necessity for the establishment of a Depot- Cum- workshop. All inspection lines, workshop lines, and stable lines are designed to accommodate multiple train sets of 6 or 8 cars, in line with the system's design. And space is also to be earmarked for future provision.
b. Each stabling line must accommodate at least one train of 6/8- 8-car each. In general, all stabling lines are planned in the proposed depot-cum-workshop assuming adequate space availability. If space constraints If they exist, stabling facilities may need to be created at the terminal Stations or other locations to accommodate the required stability facilities.
c. The planning and design of a depot are based on the layout of the Stabling-yard, Inspection shed, minor repairs, heavy repair, overhaul workshop, and cleaning of Rolling stock.
d. Operational and functional safety requirements.
e. Ancillary buildings for other maintenance facilities.
f. Electrical & Mechanical Services, power supply and distribution system.
g. Water Supplies, Drainage & Sewerage

SEW–VSE JV was awarded the construction of Depot. The depot was designed to include:

* 64 stabling lines

* 20 maintenance lines

* Control centres

* Automated train wash systems

*Other heavy-duty maintenance facilities.

The depot for the line is located at Mogharpada in Thane. It will also serve as the depot for Line 10 and Line 11. The depot has a maximum capacity for 64 stabling lines, half of which are reserved for future use, in addition to 10 inspection bays lines and 10 workshop lines. The depot also houses the operational control centre, administrative building, maintenance and workshop building, auxiliary substation, and staff quarters.

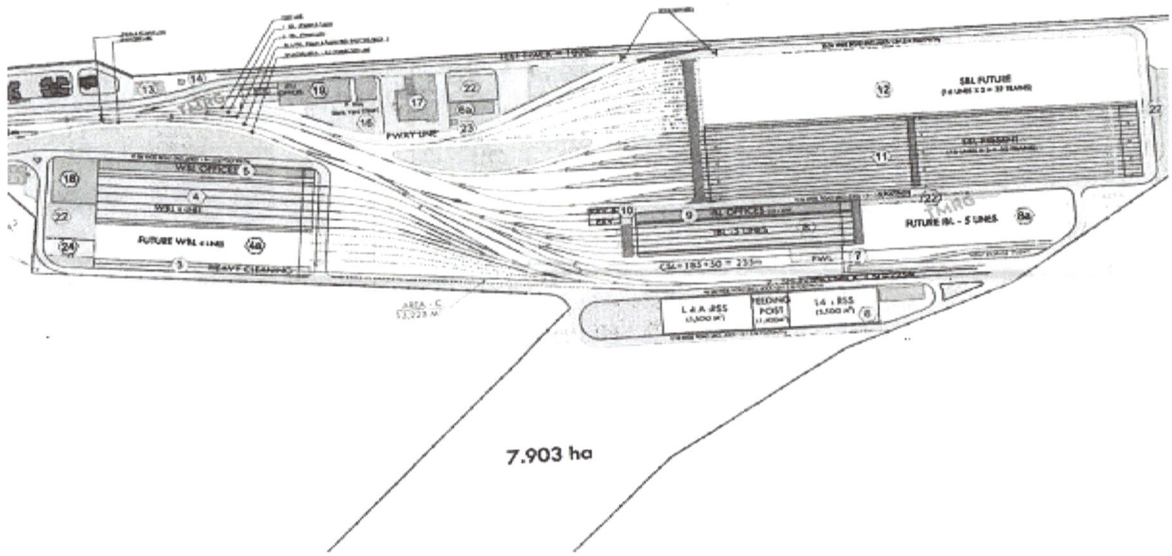

Fig.10.73: Plan of Mogharpada Depot (MMRDA)

SEW, between 2008-2014, had built 12 elevated stations and a cable-stayed bridge (in a JV with VSL) over the Western Express Highway for 11.2 km Line-1 between Versova – Andheri – Ghatkopar.

Mogharpada Depot, facility with double-decker sheds was originally planned as the largest metro depot in the city, slightly larger than the Yellow Line's Mandale Depot.

10.20.14 Permanent Way

10.20.14.1 Choice of Gauge

The issue of Broad Gauge vs. Standard Gauge for Metros in India has been widely debated, and the decision favours Standard Gauge. Even the Delhi Metro, which initially started with Broad Gauge, has switched to Standard Gauge. It is advantageous for many reasons and repeated below for ready reference.

- In general, alignment follows the road alignment, which has sharp curves. Standard Gauge permits the adoption of sharper curves.
- In Standard Gauge, 1 in 7 and 1 in 9 turnouts of lesser length can be used, while in Broad Gauge, 1 in 8 ½ and 1 in 12 turnouts are required.

- For Standard Gauge, optimised state-of-the-art rolling stock designs are available „off-the-shelf, " which is not the case for Broad Gauge.
- Standard gauge has been adopted for metros worldwide. Due to the large market, technology is constantly being upgraded, which is not available in Broad Gauge.
- For the same capacity, the gross weight of a metro coach is lower for Standard Gauge than for Broad Gauge. Thus, Standard Gauge rolling stock results in recurring savings in energy consumption during operation.

Once the Standard Gauge coach technology is absorbed and the manufacturing base for this setup is established in India, the coaches will have considerable export potential.

Despite the above, the implementation of the Mumbai metro remained very slow. Implementation was completed for one line between Versova—Andheri—Ghatkopar and is in PPT mode. An SPV was set up for the first corridor, comprising the Mumbai Metropolitan Region Development Authority (MMRDA) and a consortium formed by Reliance and a French Company, with MMRDA holding a planning and executive role. The third corridor, a 32 km-long underground metro line, Line 3, between Colaba and Seeptz via the International Airport, is in the advanced implementation stage.

10.20.14.2 Track Structure

Like other metros, regular ballasted track is used for the at-grade (surface) portion of the Depot, except within the Workshops, inspection, and washing plant lines. The ballast-less track is used for viaducts and underground tracks because regular cleaning and replacing ballast at such locations is difficult. Considering maintainability, riding comfort, and the reduction of vibrations and noise levels, the complete track is proposed to be jointless. For this purpose, even the turnouts preferred to be incorporated in LWR/CWR. Rail Section:

For rails, UIC-60 (60 kg. Rail sections are being adopted. For the Depot lines, the grade of rails is 880, which can be easily manufactured indigenously.

Systems

10.20.15 Rolling Stock

The process of procuring rolling stock was delayed due to a disagreement between the Central Government and JICA. The Centre wanted 75% of the total rolling stock procured for the line to be manufactured in India, as part of its 'Make in India ' program, even if the contract was awarded to a foreign company. However, JICA wanted the bid for rolling stock to have no restrictions. Under pressure from the NITI Aayog, the Department of Economic Affairs, and the Ministry of Urban Development, JICA agreed to accept the Union Government's domestic manufacturing requirement for rolling stock for all JICA-funded metro systems in India. As a result, the winning bidder may only manufacture a maximum of 25% of the metro cars required for Metro 3 outside India.

The MMRC received pre-qualification bids from seven firms - Alstom Transport India Ltd, India & Alstom Transport S.A., France (Consortium), CAF S.A., Spain, CRRC Nanjing, CRRC Changchun & CRRC International Corporation Ltd, China (Consortium), Hitachi Ltd, Japan, Kawasaki Heavy Industries, Japan & BHEL (Consortium), India, and Mitsubishi Corporation, Japan.

In September 2018, the MMRC awarded a contract to French transportation major Alstom and its Indian subsidiary to supply 336 coaches. The contractor would be responsible for the design, manufacture, supply, installation, testing, and commissioning of 42 train sets, each consisting of eight cars. The trains will have a carrying capacity of 3,000 passengers, with each coach featuring approximately 300 seats arranged in longitudinal seating. All coaches will be designed and developed at Alstom's engineering centre in Bangalore, and manufactured by the company's plant in Sri City, Andhra Pradesh.

The required transport demand forecast is the governing factor in choosing the Rolling Stock. The forecasted Peak Hour Peak Direction Traffic calls for a Medium Rail Transit System (MRTS). Initially, trains are proposed to have six coaches but can be expanded to 8 coaches if needed. All the platforms are proposed to be provided for 8 Coach Trains.

The following optimum size of the coach has been chosen for this corridor.

Table 10.7: Size of the coach

	Length*	**Width**	**Height**
Driving Motor Car (DMC)	**21.84 m**	**3.2 m**	**3.9 m**
Trailer car (TC)/Motor Car (MC)	**21.74 m**	**3.2 m**	**3.9 m**

***Maximum length of coach over couplers/buffers = 22.6 m**

A longitudinal seating arrangement is used in most metros in India to maximise passenger carrying capacity.

The train composition depends on the number of coaches in a rake and is specific to each line. For this metro, a rake with six coaches has the following composition: 6-car Train: DMC + TC + MC+ MC + MC + TC + DMC.

The carrying capacity is as given in Table No.10.8

Table 10.8: Carrying Capacity of Rail Vehicles

	Driving Motor car		**Trailer car/Motor car**		**6 Car Train**	
	Normal	**Crush**	**Normal**	**Crush**	**Normal**	**Crush**
Seated	42	42	50	50	284	284
Standing	120	240	124	248	736	1472
Total	162	282	174	298	1020	1756

NORMAL-3 Person/sqm of standee area CRUSH -6 Person/sqm of standee area

The recommended **performance parameters** are:

Traction Power Supply:	25Kv ac
Motoring capacity:	67%
Maximum Design Speed:	90 kmph
Maximum Operating Speed:	80 kmph Max.
Acceleration:	1 m/s^2+ 5%
Max. Deceleration:	1.0 m/s^2 (Normal brake)
	1.35m/s^2(Emergency Brake)

Rolling Stock is planned with an advanced, sophisticated system with state-of-the-art technology. The essential criteria for the selection of rolling stocks are:

i. Proven equipment with high reliability
ii. Passenger safety features
iii. Energy efficiency
iv. Lightweight equipment and coach body
v. Optimised scheduled speed
vi. Aesthetically pleasing Interior and Exterior
vii. Low life cycle cost
viii. Flexibility to meet the increase in traffic demand
ix. Anti-telescopic

The selection of coaches is based on reliability, low energy consumption, lightweight, and high efficiency, leading to a lower annualised service cost. The coach has a high rate of acceleration and deceleration. A low life cycle cost is achieved through reduced scheduled and unscheduled maintenance, as well as the high reliability of the subsystems. In Kolkata, high-tensile carbon steel, Carten Steel, was used for car bodies in the North-South metro. Almost all the coaches built by Indian Railways are of this type. Adopting stainless steel or aluminium for car bodies is now standard practice.

Aluminium car bodies have not been considered as they are yet to be manufactured in India. Stainless steel sections are available, and stainless-steel coaches have been specified. Stainless steel coaches save energy due to their lightweight. No corrosion repair is necessary on stainless steel cars during their service life, resulting in cost savings due to their ease of maintenance and repair.

Metro cars now universally adopt bolster-less lightweight fabricated bogies with rubber springs. These bogies require less maintenance, and the overhaul interval is prolonged. Using air springs at the secondary stage is considered to keep the car's floor level constant, irrespective of passenger loading. Perturbation from the track is also dampened inside the car body because of the secondary air spring and a suitable Vertical Hydraulic Damper. The primary suspension system enhances curve-running performance by reducing lateral forces through the application of a conical rubber spring. The provision of the above type of bogies ensures smooth.

The **braking system** consists of an electro-pneumatic (EP) service friction brake, a failsafe, and a pneumatic emergency brake with a spring-applied air-release parking brake. It also features an electric regenerative service brake, which provides a smooth and continuous blend of EP and regenerative braking.

10.20.16 Propulsion System

In Electric Rolling Stock, DC series traction motors have been widely used due to their ideal characteristics and good controllability for traction applications, though they require intensive maintenance.

Brushless 3-phase induction motors are used in traction applications, particularly in series motors. For the same power output, the induction motor is smaller, lighter, and ideally suited for rail-based transit applications. Variable Voltage regulates the motor tractive effort and speed, and frequency control can be programmed to suit the track profile and operating requirements. Another advantage of a 3-phase AC drive and VVVF control is that regenerative braking can be introduced by lowering the frequency and voltage to reverse the power flow and allow braking to very low speeds.

The AC catenary voltage is stepped down through a transformer and converted to DC voltage through a converter, which supplies the DC link. This DC link feeds the Inverter, operated with Pulse Width Modulation (PWM) control technology and utilising Insulated Gate Bipolar Transistors (IGBTs). Thus, the three-phase variable voltage variable frequency output drives the traction motors for propulsion.

Recently, an advanced IGBT has been developed for inverter units. The advanced IGBT comprises an Insulated Gate Bipolar Transistor (IGBT), a gate drive circuit, and protection features. It incorporates overcurrent protection, short-circuit protection, over-temperature protection, and low-power supply detection. The IGBT has internal protection from overcurrent, short circuits, overtemperature, and low control voltage.

The inverter unit uses an optical fibre cable to connect the control unit to the gate interface. This optical fibre cable transmits the gate signals to drive the advanced IGBT via the gate interface. It provides electrical isolation between the advanced IGBT and the control unit and is impervious to electrical interference. These are recommended for adoption in MRTS trains.

10.20.17 Passenger Doors

Four doors of adequate width, located on each side of the coach, have been considered for the swift evacuation of passengers in a short period. These doors shall be of such dimensions and location that all the passengers inside the train can evacuate within the least possible time without conflicting movement, and an automatic door closing mechanism is envisaged for consideration of passenger safety. Passenger doors are controlled electrically by a switch in the Driver's cab. An electrically controlled door operating mechanism has been preferred over a pneumatically operated door to prevent air leakage and sluggish door operation.

Interior View

Fig.10.74 & 10.75: Passenger Doors

10.20.18 Air conditioning

With heavy passenger loading of 6 (8) persons/sqm for the standee area and doors being closed for safety reasons, and with sealed windows to prevent noise transmission, the air conditioning of coaches has been deemed essential. Each coach shall be provided with two air conditioning units capable of cooling, heating and dehumidifying and thus automatically controlling interior temperature throughout the passenger area at 25°C with 65% RH all the time under varying ambient conditions up to full load. For emergencies, ventilation will be provided from the battery. Provisions should be made to shut off the fresh air intake and recirculate the internal air of the coach during emergency conditions, such as a fire outside the train, which would cause excessive heat and smoke to be drawn into the coach.

10.20.19 Cab Layout and Emergency Detrainment Door

The modern, stylish driver panel shall be FRP-molded, which provides the driver with maximum comfort, easy accessibility to various monitoring equipment, and clear visibility. A driver's seat is provided on the left side of the cabin. An emergency door for easy passenger deboarding on the track has been provided at the center of the front side of each cabin. It is easy to operate with a one-handle type master controller.

10.20.20 Noise and Vibration

The trains will pass through heavily populated urban areas. The noise and vibration of the metro railway become an essential criterion from a public acceptance viewpoint. The sources of noise are (i) rail-wheel interaction, (ii) noise generated from equipment like a Blower, Compressor, air conditioner, door, Inverter, etc., and (iii) traction motors in running trains.

Fig.10.76: Driving Cab

10.20.21 Signaling

The signalling system provides efficient train control, ensuring safety in train movements. It assists in optimising metro infrastructure investment and running efficient train services on the network.

The Metro operation requires high safety enforcement and reliability. The metro's requirements are planned to be achieved by adopting "CATC" (Continuous Automatic Train Control System) based on "CBTC" (Communication-Based Train Control System), which includes ATP (Automatic Train Protection), ATO (Automatic Train Operation), and ATS (Automatic Train Supervision) subsystems using radio communication between trackside and Train. The system provides.

- High level of safety with trains running at close headway, ensuring continuous safe train separation.
- Eliminates accidents due to drivers passing Signals in Danger by continuous speed monitoring and automatic brake application if the driver disregards the signal/warning.
- Temporary speed restrictions.
- Improve capacity with safer and smoother operations. The driver's cab will continuously display Target Speed and other information, enabling him to optimise the track section's speed potential. The system provides signal/speed status in the cab, even in adverse weather conditions.
- Increased productivity of rolling stock by increasing line capacity and train speeds and enabling the train to arrive at its destination sooner.
- Improve maintenance of Signalling and telecommunication equipment by monitoring the system status of trackside and train-borne equipment and enabling preventive maintenance.
- The sideline signaling and train Control system will be designed to meet the required headway during peak hours. The radio station for CBTC shall work in the license-free ISM band.

10.20.22 Telecommunication

Telecommunication facilities are required to supplement the Signalling system to ensure efficient train operation.

* Exchange of managerial information

* Crisis management during emergencies

* Passenger information system

The Telecom system is to cater for

- Train Traffic Control
- Assistance to Train Traffic Control
- Maintenance Control
- Emergency Control
- Station-to-station dedicated communication
- Telephone Exchange
- Integrated Passenger Announcement System and Passenger Information and Display System within and from Central Control to each station.
- Centralised Clock System
- Train Destination Indicator
- Instant online Radio Communication between Central Control and Moving Cars and maintenance personnel.
- Data Channels for Signalling, SCADA, Automatic Fare Collection, etc.
- Integrated Network Control System

10.20.23 Automatic Fare Collection

The ticketing system needs to be simple, easy to use/operate and maintain, easy on accounting facilities, capable of issuing single/multiple journey tickets, amenable to quick fare changes and require less manpower overall. Given the above, a computer-based automatic fare collection system is preferable to a Semi-automatic (Manual) system.

In the long run, an AFC system proves to be cheaper than a semi-automatic (Manual System) system due to reduced manpower costs, lower maintenance compared to paper ticket machines, the overall lower cost of recyclable tickets (Smart Cards/tokens) compared to paper tickets, and the prevention of revenue leakage. The relative advantages of an automatic fare collection system:

Seamless ticketing is being considered for all metros in Mumbai. This system will enable commuters to travel hassle-free using various modes of transport, including metro, suburban trains, buses, water transport (when introduced), and taxis, without needing to purchase separate tickets for each mode.

Automatic fare collection systems have the following advantages:

- A smaller number of staff.
- There is less possibility of revenue leakage due to 100% ticket checks by control gates.
- Recycling tickets fraudulently, avoided.
- Efficient and easy to operate.
- The system is amenable to quick fare changes.
- Management information report generation is easy.
- The system has multi-operator capabilities. The same Smart Card can also be used for other applications.

The AFC systems are like magnetic ticket-based systems, but contactless systems are more cost-effective due to reduced maintenance, less wear and tear, and lower susceptibility to dusty environments.

10.20.23.1 Choice of Control Gates

Retractable flap-type Control Gates, which offer high throughput, require less maintenance and are the latest in modern metros, are planned to be adopted.

10.20.23.2 Ticket Vending Machine (TVM)

Passenger-operated Ticket Vending Machines (TVMs) (Automatic Ticket Vending Machines) have been installed at all stations. TVMs enable passengers to bypass queues at ticket booths and offer them an international standard service.

10.20.23.3 Ticket Reader/Add Value Machine

These machines check the Card/Token balance and can be used as an add-value device if payment for Card top-up is made through an alternative Internet-based channel, such as net banking, Credit/ Debit, etc.

10.20.23.4 Recharge Card Terminal Machine (RCTM)

RCTM can be used to recharge a Card using a Credit Card, Debit card, prepaid card, or bank Note.

10.20.23.5 Integration of AFC with Other Lines and Modes of Transport

In the Indian Metropolis, the metro network includes several lines and operators. Given the importance of passenger convenience and operational efficiency, it is planned to integrate AFC for different metro lines and make innovative card-based fare products interoperable. The AFC system should consider implementing a revenue-sharing mechanism among different operators based on the number of journeys performed at each system. Single-ride tickets (tokens) may not be interoperable or limited to each operator's system. The AFC system may provide interfaces for other operators, such as Suburban Rail, Bus, Parking, and Toll, allowing these systems to be integrated with innovative card-based fare products.

10.20.24 Communication

The car-driving cabs are in continuous communication with the base Operational Control Center and station control, allowing easy monitoring of the individual train in all sections. A Public Address and Passenger Information Display System is provided in the car, allowing passengers to be continuously informed of the next stop, destination, interchange, emergencies, and other important messages. The rolling stock is equipped with Talk-Back Units inside the cars, which enable conversation between passengers and the drivers in the event of an emergency.

10.20.25 Maintenance - Rolling Stock

The maintenance schedule has been envisioned for the design of depots, assuming approximately 400 km of running per train per day and considering the maximum passenger load for intervals specified in the Maintenance Manual. The significant components of servicing are as follows:

a. Check the train's condition and function after every daily service. Vacuum the floor and walls at intervals.
b. Detailed inspection and testing of subsystems and underframe, as well as replacement/ topping up of oils and lubricants.
c. Check and test all subassemblies (Electrical + Mechanical). Overhaul of pneumatic valves and the compressor. Condition-based maintenance of subsystems to bring them to their original condition. Replacement of parts, rectification, and trial run.
d. Dismantling of all sub-assemblies, bogies suspension system, traction motor, gear, control equipment, air-conditioning units, etc. It is being overhauled to bring it to its original condition. Checking repair and replacement as necessary.
e. Inspection and changing heavy items such as bogies, traction motors, axles, gear cases, and axle boxes.
f. Washing of Rolling Stock

10.20.26 Inspection at Depot

Facilities required in the inspection bay for Systems / Equipment are

- Electronics; PA/PIS
- Mechanical components, couplers, etc.
- Batteries
- Air conditioner
- Brake modules
- Bogie
- Traction Motor

- Vehicle doors, windows and internal fittings
- Power system including converter, circuit breaker, etc.

10.20.27 Stabling lines at Depot

According to the dimensions of the Rolling Stock, the length of the car train depends on the number of cars (i.e., 4, 6, or 8 cars). For the design of the stabling lines in the depot and terminal stations, if required:

i. Pathway in the entry side=11m (or as per recommendation)
ii. Free length at outer ends (for cross pathway, Signal and Friction buffers) = 11m or as recommended.

The total length of the Stabling line for each will be the length of the car plus 2 x 11 m. The space between stable lines must include a 1 m wide pathway between tracks, (depending on the width of the coach recommended) to provide access to internal Train cleaning and undercarriage inspection. Each Stumbling line must have a water connection for local cleaning, and platforms must be located at suitable positions at each end of the stumbling lines to enable train operators to board or disembark conveniently.

a. Space shall be provided for HVAC, door, traction motor, etc., repairs. Distinct spaces shall be earmarked for dismantling/repairing/ assembling and testing each piece of equipment. Related machinery for overhauling, repairs, and testing activities of every piece of equipment is also to be housed in the earmarked space.
b. Washing and cleaning equipment should be on the workshop floor. A Bogey test stand must also be provided. Other heavy machinery shall also be suitably installed on the workshop floor. Each workshop column shall provide air circulators, lights, power supply points, and air supply lines.
c. Workshop lines must be interlinked through turn tables, each suitable for moving a train in unloaded condition and capable of rotating with a fully loaded bogie. Repair of heavy equipment, such as air conditioners, should be located to minimise its impact on the movement inside the workshop.
d. Bogie painting and battery maintenance cells must be in the workshop.
e. The workshops must be designed as service buildings with various closed spaces for activities like overhauling sections, offices, costly store items, locker rooms, toilets, etc. The buildings may have a column-and-beam structure enclosed with masonry work. Space for bogie/axle repair shops, including necessary infrastructure for disassembly, overhead, assembly, and testing of mechanical components of bogies/axles, must be included in space planning. This space should be equipped with appropriate ventilation, lighting, entry, and exits.

   The service building must include the following,

   1. Body furnishing
   2. Bogie
   3. Wheels

4. Traction Motors
5. Axle Box and Axle Bearing
6. Pantographs
7. Transformer, converter/inverter, circuit breaker
8. Battery
9. Air Compressor
10. Air-conditioner
11. Brake Equipment
12. Door actuators
13. Control and measuring equipment
14. Pneumatic equipment
15. Dampers and Springs
16. Couplers/Gangways
17. Coach Painting, where applicable.

10.20.28 Automatic Coach Washing Plant

Provisions should be made for the Coach exterior surfaces to be washed using a fully automated train washing system with a throughput capacity as envisaged in the operational Manual. Exterior surfaces on the incoming rakes should be cleaned before entry to the depot, and undesirable movement/shunting over ingress and egress routes within the depot should be avoided.

10.20.29 Car Entry / Delivery in Depot

One of the most critical aspects is the connectivity of operating lines and the depot. The rail connectivity between the Depot-cum- Workshop and metro lines and all trains due for scheduled/ unscheduled works should be such that trains can reach the depot-cum- Workshop safely. The provision of dedicated space must be included in the plan for the case of newly procured coaches transported by rail/road, which must reach the depot by trailers and be unloaded to the dedicated tracks. This area should have an insulated track about 40 m long embedded in the floor, facilitating the movement of a trailer. There should also be enough space available for heavy cranes to lift coaches.

10.20.30 Test Track

A test track of 1000 m. in length, covered and fenced, should be provided beside the depot workshop and equipped with a signalling system. Test tracks are essential for commissioning new trains, their trials, and testing trains after IOH and POH in compliance with safety norms.

10.20.31 Others

Auxiliary substations are part of the depot design to satisfy the power requirements for depot functioning. A standby power supply is required for adequate capacity, and an AMF panel is needed for a DG set. The capacity of the DG set will be sufficient to supply all essential loads without overloading. For **compressed air supply,** A silent type should be installed inside the depots at convenient locations to send compressed air to the workshop and for inspection. Each depot must have in-house facilities for water supply, Sewerage, and stormwater drainage to ensure efficient system functioning. Rainwater harvesting would be given due emphasis on charging the underground reserves. The Ancillary Workshop should have a line at the floor level with pits provided.

Arrangements for the repair of Shunters, Railroad Vehicles, and other ancillary vehicles are also to be provided. **Watch Towers:** an adequate number of watchtowers for the vigilance of the depot boundary. The **Administrative Building is generally** planned close to the main entrance**.** It can be suitably sized and features an architecturally designed access control system suitable for all staff working in the complex. For **Parking Facilities,** sufficient parking space should be provided.

10.20.32 Shed and Buildings

The depot typically provides (like other metro depots) sheds, buildings, sizes, and brief descriptions of their functions. Depending on land availability, these buildings can be designed in detail and grouped architecturally and functionally.

10.20.33 Plant and Machinery

a. A separate building is planned for housing the pit wheel lathe (PWL), approachable from the workshop, inspection bay, and stabling lines through rail and road for placing cars for re-profiling wheels within the depot, along with space for the depot of scrap.

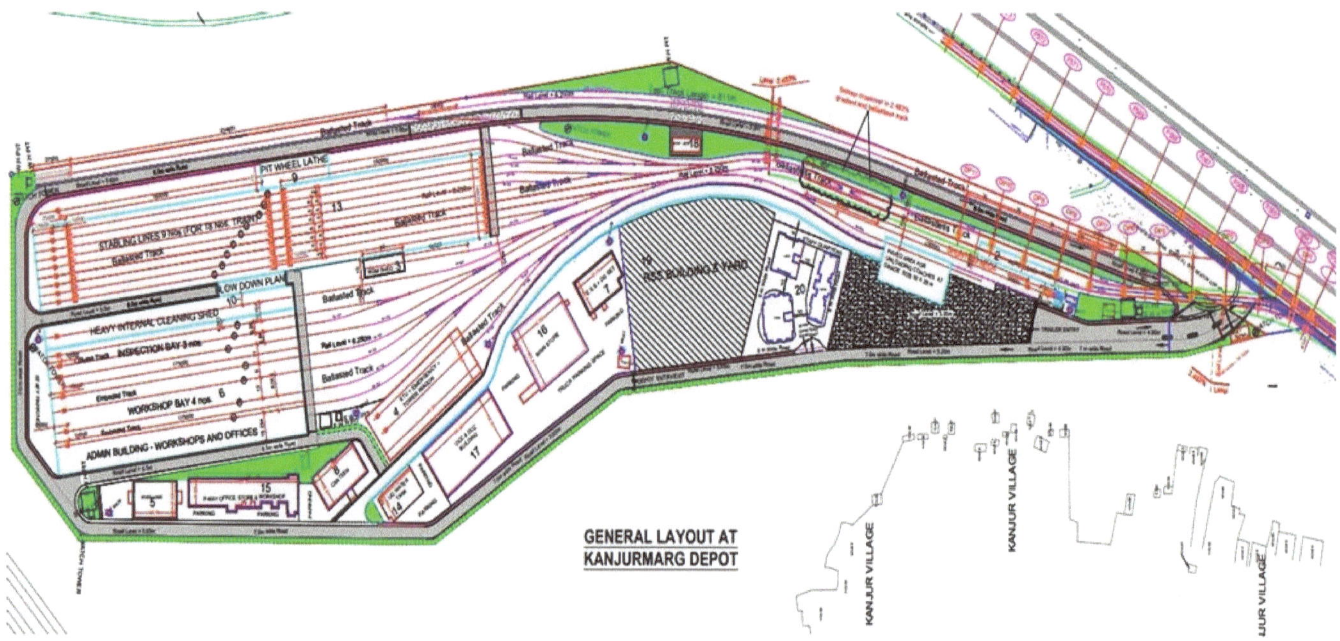

Fig.10.77: Plan of Kanjurmarg Depot (themetrorailguy.com)

10.20.34 Metro Rail in Mumbai Metropolitan Region- MMR

10.20.34.1 Mumbai Monorail

The Mumbai Monorail is part of a major expansion of public transport in the city and is operated by the Mumbai Metropolitan Region Development Authority (MMRDA). The monorail runs from Chembur in the city's eastern suburbs to Sant Gadge Maharaj Chowk at Mahalaxmi in south

Mumbai, effectively connecting the city's harbour line, the central line, and the western line. The monorail runs on a narrow, single track. A lightweight structure allows the trains to navigate sharp turns in congested urban areas. The Mumbai monorail is the first of its kind in India. The system began commercial operation in 2014 but has achieved only 10% of the estimated ridership. Perhaps it is due to being poorly connected to other modes of transportation.

Then, the Maharashtra Chief Minister laid the Foundation Stone for the monorail system, constructed in two phases. The first stretch linked Wadala, on the outskirts of the island city, with Chembur in the northeast, and the second connected Wadala with Jacob Circle in South Mumbai. A 108-meter test run was successfully conducted in January 2010. A one-kilometre test run from Wadala to the Bhakti Park monorail station was undertaken in February 2012. And the test run of the entire route was conducted in November 2012.

The only operational line of the Mumbai Monorail, Line 1, connects Sant Gadge Maharaj Chowk in South Mumbai with Chembur in eastern Mumbai. The 20.21 km line is fully elevated. Line 1 is owned and operated by the MMRDA. The monorail supplements the service of the Mumbai Suburban Railway in some heavily populated areas. The first phase consists of 7 stations from Chembur to Wadala Depot, and was opened to the public in February 2014. An extension for Line 1, with 11 stations, was built from Wadala Depot to Jacob Circle. It suffered delays due to a shortage of functional monorail rakes and finally opened in March 2019.

Fig.10.78: Mumbai Monorail (railway-technology.com)

10.20.34.2 Suspended Rail System (Sky Bus Metro)

Fig.10. 79: Suspended Car in Operation Outside India

Fig.10.80: 3D View of Skybus Metro – Planned

Fig.10.81: Prototype Sky Bus as Conceived by KMRCL

i. **Introduction**: A suspended driverless train, a system that runs on tracks primarily along the median of arterial roads. New technologies, such as Auto Driving Devices (ADD), Swing Arresters, Derailment Arresters, Collision Arresters, and Suspenders, enhance the performance and safety of the system. 750 V DC traction through third rail and using three-phase drive with Variable Voltage Variable Frequency (VVVF), Control is the power for propulsion

   System: The twin coach consists of a 20m-long vehicle with a 300-pax capacity suspended from a conventional standard-gauge bogie. The maximum speed is 160/100 kmph, and system acceleration/ deceleration is 1.3 m/s^2

   Capacity: The Proposed capacity of 18,000 pax per hour per direction, planned with a 60-second Headway.

   Supporting Structure: Bogies running on rails connected to the bottom slab of a 3-cell box beam integrated monolithically with piers spaced every 20/30m, as per site configuration

ii. **Assessment of System**

   Aim

   Study the structural adequacy and stability of the suspended rail system to ascertain passenger comfort

Steps for Assessment

i. System configuration
ii. Load combination
iii. Structural model
iv. Mathematical simulation of the system as a whole
v. Adequacy – Section-wise
vi. Static & dynamic behaviour

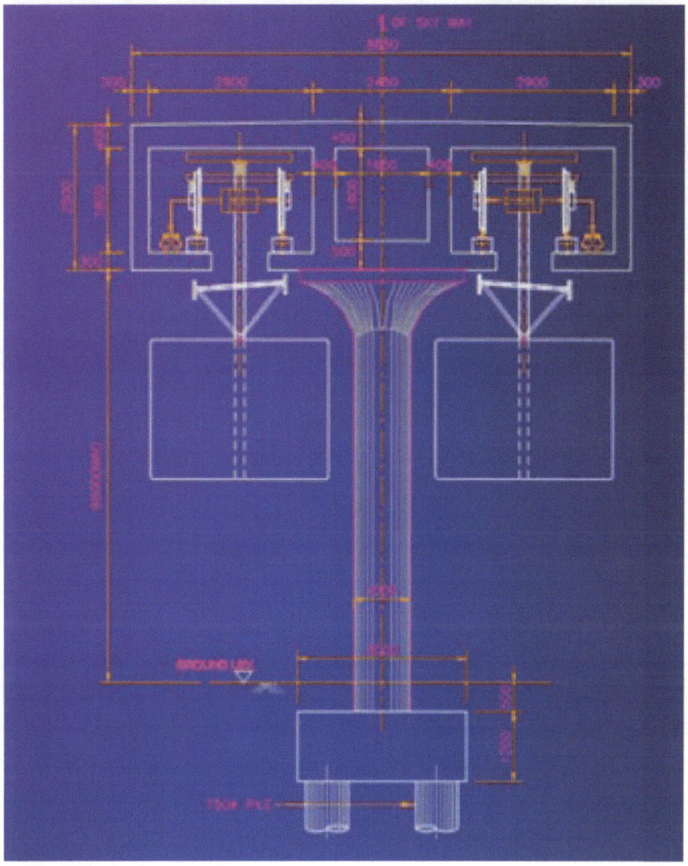

Fig.10.82: Schematic Sectional View

iii. **Types of Analyses**
   i. Static analysis for supporting structure without carriage, but including loads due to carriage and loads due to wind & seismic
   ii. Dynamic analysis without carriage
   iii. Static & Dynamic analysis of carriage

iv. **Dynamic analysis of the total system**

Boundary Conditions

- **Base of all piers** – fixed
- **Rotation** of the end piers was restrained to take the effect of continuity

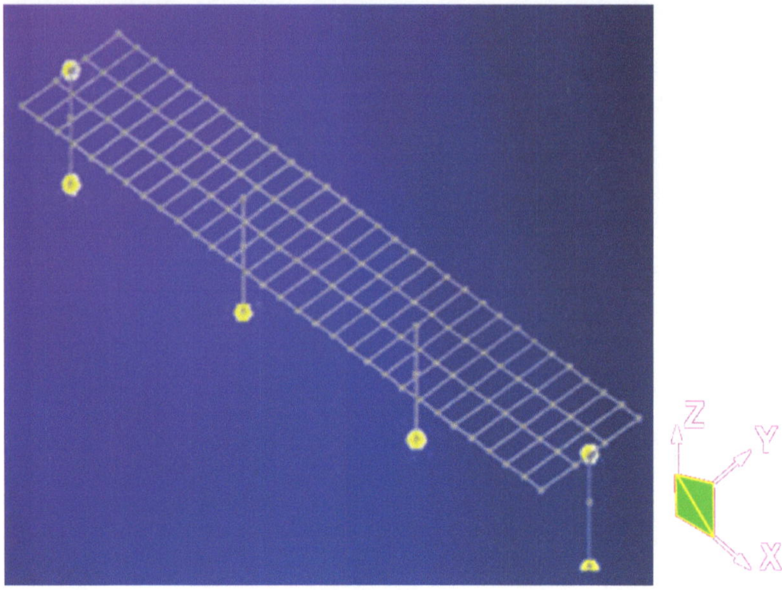

Fig.10.83: Grillage Model of Skybus (Beam elements = 226)

v. **Loads and Load Cases**

Solid – Dynamic

Besides forces that act on the system from/due to long welded rail, racking, derailment, braking & traction, erection, differential settlement, thermal rail, load cases adopted:

Structure

Three-Span Continuous Structure

Three-cell split box structure supported on piers at 20m intervals

Size of Coach - 9.25 m x 3.15 m x 2.4 m

Loads

Self-weight of box structures transferred through pier heads

Self-weight of carriage through axles on rail supported on box slab

Live load through carriage

Wind/Earthquake load

Load Cases

Static

(Self-weight + Wind load)-structure

(Self-weight + Wind load)-structure + (Self weight + Live load + Wind load)-carriage

Dynamic

Moving load of carriage + (Live load + Wind load)

Seismic Load

Carriage Alone

Static – Self Weight Live & Wind Load

Dynamic – Support Acceleration +(Wind + Live) Load on carriage

Modeling and Idealization

Model- i) Grillage using Beam Elements

ii) FEM using Solid Elements

Aim- i) Modelling for Preliminary Check

ii) Elaborate Modelling

Analysis- Static and Dynamic for both Grillage & FEM

vi. **Grillage – Static**

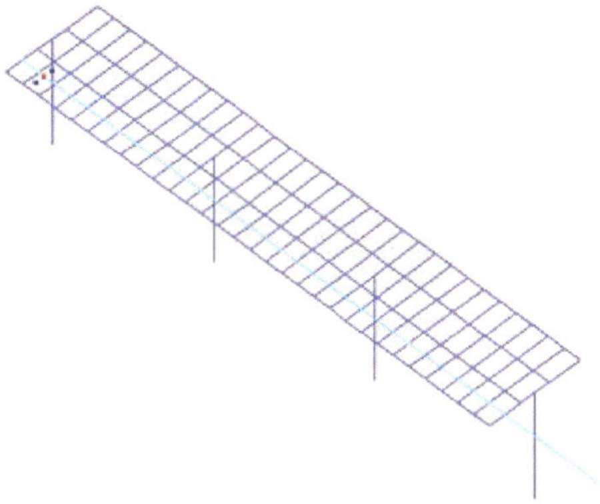

Fig.10.84 Grillage – Moving Loads – Animated Isometric View (Exaggerated Scale)

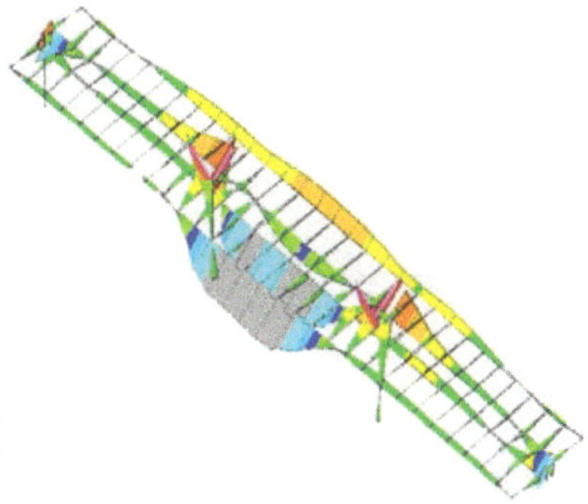

Fig.10.85 Bending Movement Diagram for one position of the Moving Load

Grillage Dynamic

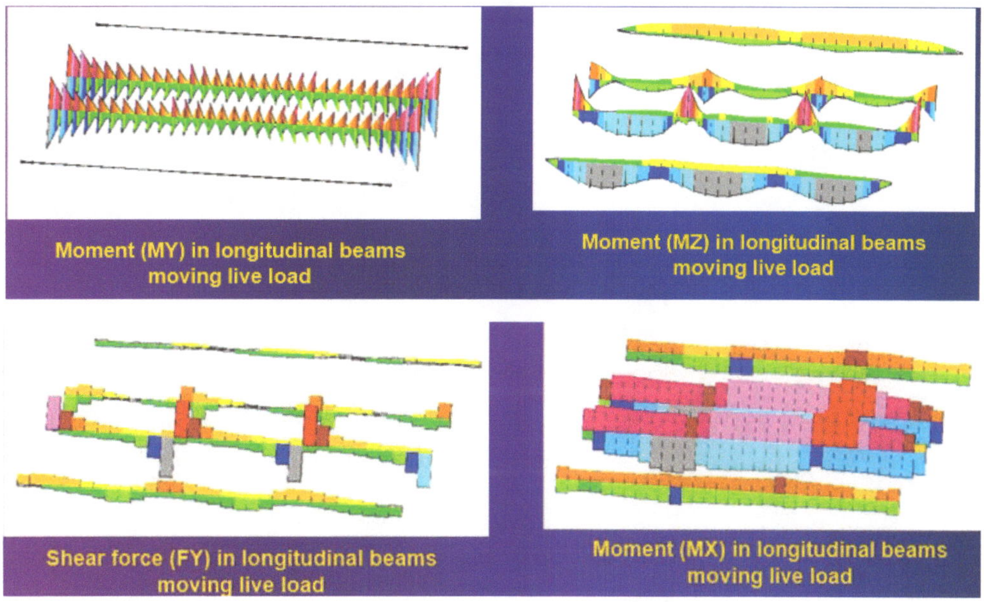

Fig.10. 86, 10.87, 10.88 & 10.89

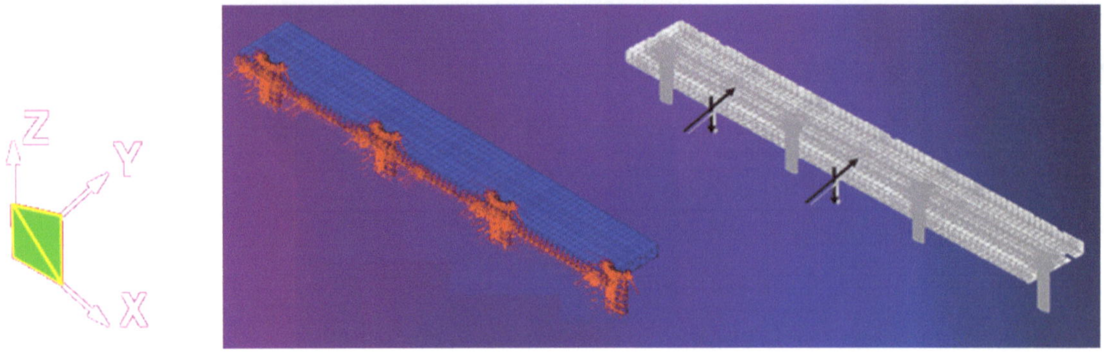

Fig.10.90: 3D Solid Model

3D FEM with solid model for true system performance and to avoid approximation with center line dimensions

More realistic stiffness & mass distribution

Structure – complex with many re-entrant corners & geometric discontinuities

Static and dynamic analyses have been carried out for the same load cases as considered for the grillage model

Von Mises stress – DL+ Wind Deformation Plot DL+ Wind

Von Mises stress – DL+ Wind

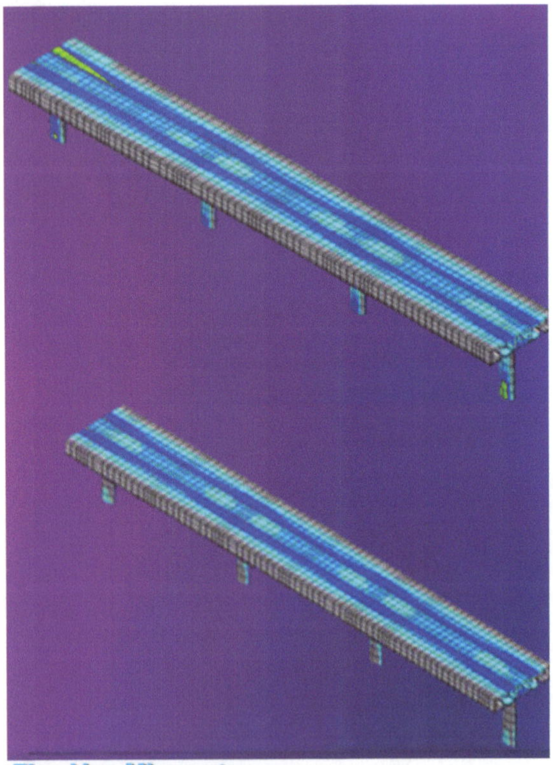

Fig. Von Mises stress – DL + wind + LL (one track

Fig.10.91: Displacement Contours – Isometric View

Deformation Plot DL+ Wind

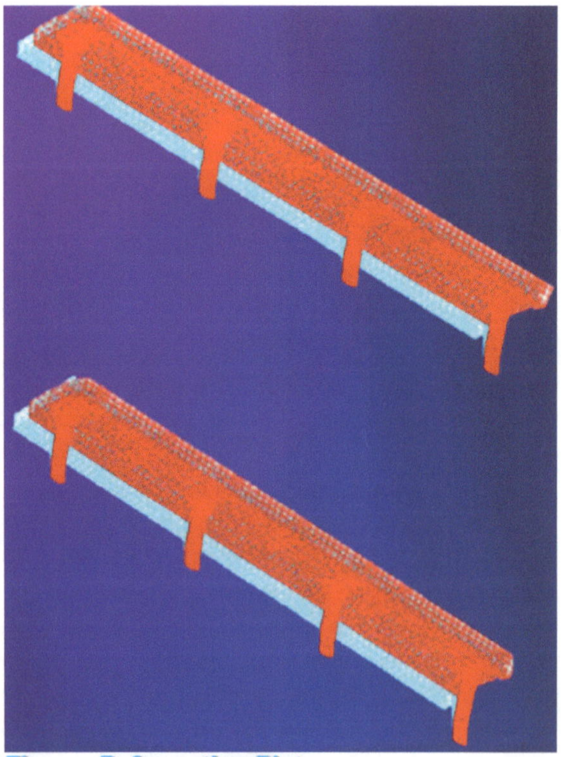

Fig. Deformation Plot DL + wind + LL (one track)

Fig.10.92: Stress Contours – Isometric View

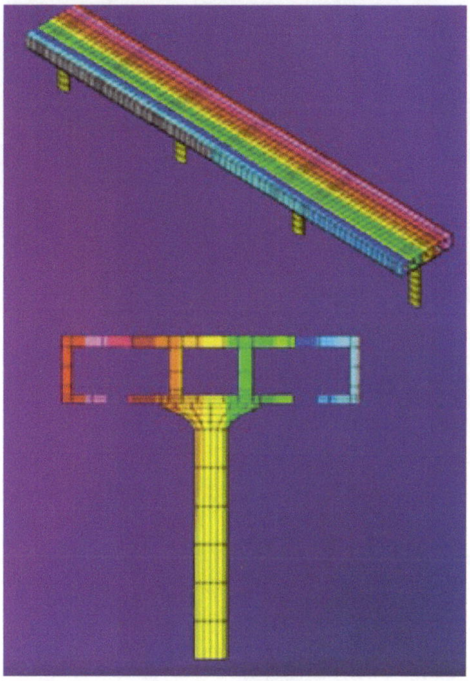

Fig.10.93: Displacement Contour

Fig.10.94: Stress Contour

vii) Carriage

i. Each carriage is modelled as suspended from the bogie through two suspenders
ii. Carriage was analysed as an isolated component for static and dynamic response
iii. Load consideration

    Self-weight + Live load from carriage – 4 axles of 16 MT each

    Wind load – 150 kg/m^2

iv. For static analysis, the self-weight for acceleration due to gravity is considered.

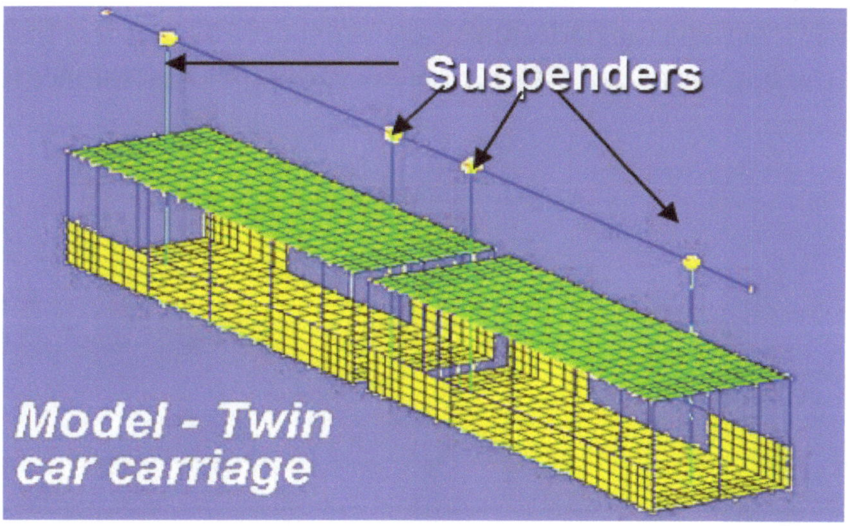

Fig.10.95: Elements for Twin Carriage

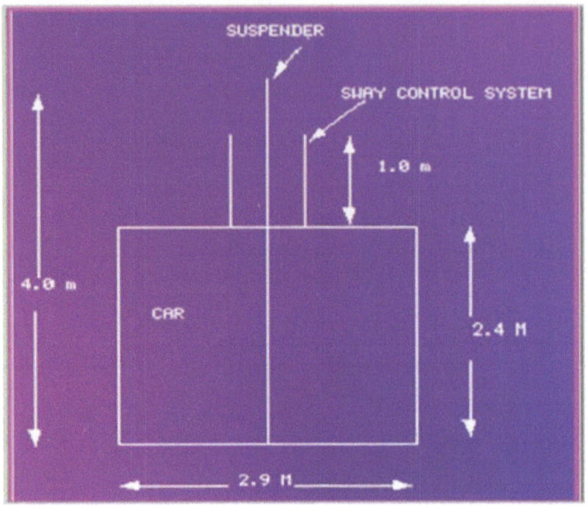

Fig.10.96: Line Diagram of Carriage Cross Section

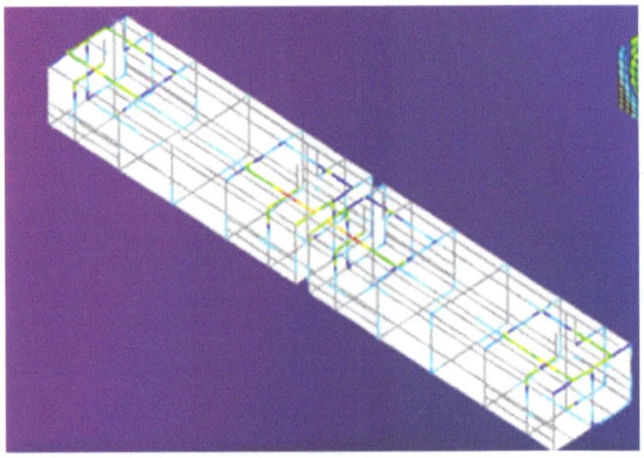

Fig.10.97: Carriage - Stress in beam elements

Fig.10.98: Carriage - von Mises stress in shell elements

viii) Carriage – Swing

i. Swing arrester to limit the swing of the coach, preventing it from hitting the support
ii. Mounted on a coach and provided with hard rubber wheels at both ends.
iii. Total swing depends upon the design of the suspenders, bogie, frame, and suspension spring
iv. Computer simulation and dynamic analysis show that the lateral sway of the carriage under wind load is 1.5 °, resulting in a horizontal displacement of 105 mm, less than the gap proposed of 600 mm at the bottom level of the coach. However, minimum clearance is required from the supporting column.

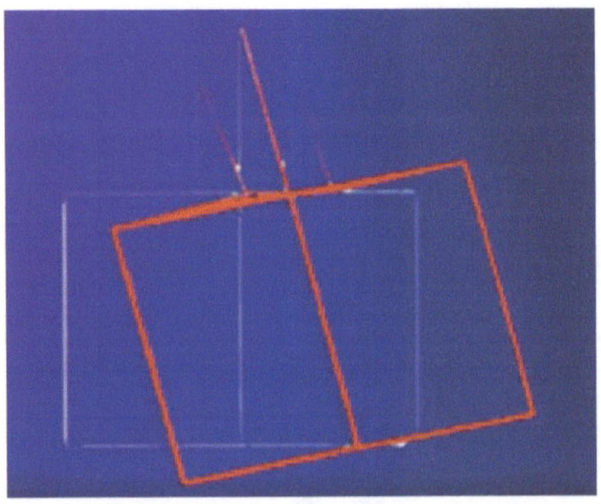

Fig.10.99 Total System– Deflection (Swing)

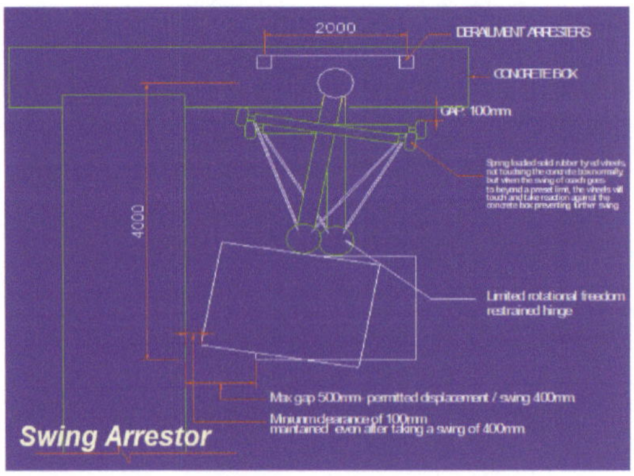

Fig.10.100: Deflection Using Swing Arrestor

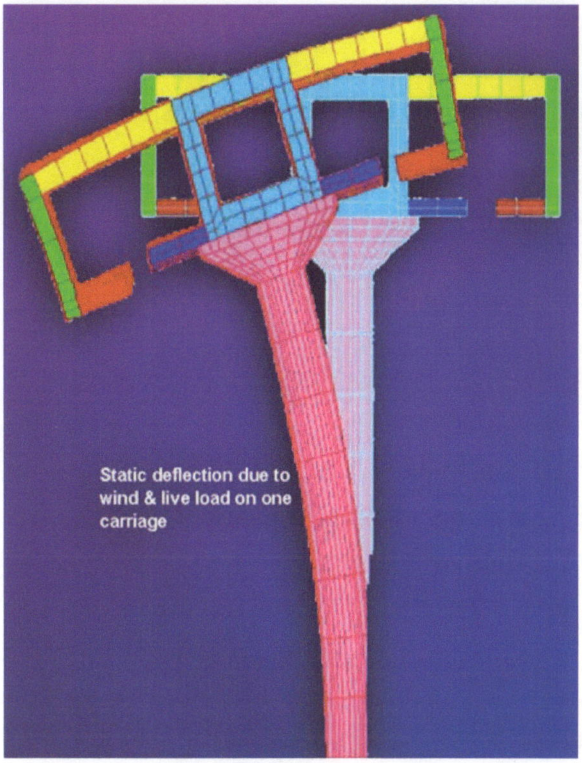

Fig.10.101: Static Deflection due to LL & Wind

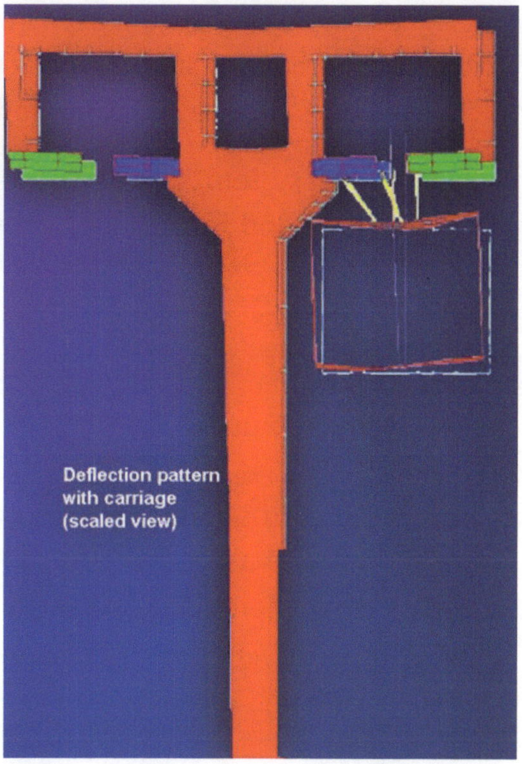

Fig.10.102: Deflection Pattern with Carriage

ix) Summary

i. Stress contour plot for seismic excitation in the vertical and transverse direction for different combinations studied in adopting the system (Not included)
ii. Model for suspended-rail system evolved, design checked and analysed for different load combinations
iii. Grillage & solid FE models are analysed for system performance
iv. Carriage alone is modelled separately & analysed under dynamic loads due to support accelerations.
v. System stability under dynamic loads is found to be okay in general, for the materials and section used for review.
vi. Swing Characteristics under dynamic loads have been established for the gap planned between the two tracks
vii. A suspended rail system with its associated features depends upon its overall system.
viii. The clearances of the bottom level of the platform have to satisfy the Indian Code of Provision. It means the suspended system has to be more than others transportation structures.
ix. It is noted that two tracks have been planned with a gap of 600mm, and the headway of the system is considered as 30 seconds.
x. Which appears to be absurd when no provision for automatic transfer even with a driver.
xi. It is said that transfer would be done mechanically, the system not checked as in the absence of a plan.

xii. It is said that the coach turning system used in the Indian Railway may be used. It means that there have to be two turntables at both ends.

xiii. Turn turntable used in the Railway is for at-grade and supported on tracks, not suspended.

The conclusion is that the full system, stability, passenger comfort, operation and test run have to undergo for prototype carriage and operation system and satisfy the requirements of the concerned Department.

10.22 Accidents

An accident occurred at the construction site of the Thane-Bhiwandi-Kalyan Metro Line 5 when a long iron rod fell from the metro bridge. The rod directly pierced the forehead of a passenger travelling in an autorickshaw on the road below. The passenger sustained serious injuries. The Mumbai Metropolitan Region Development Authority (MMRDA) imposed a penalty on two companies involved in the construction of Metro Line 5. The GC was also penalised.

Thane Viwandi-Kalyan Corridor (ToI)

Chembur (Free Press Journal)

The Indian Express

The Indian Express

A major accident took place at Suman Nagar Society in Chembur when a 20-foot concrete structure, part of the ongoing Metro Line 4 construction towards Wadala, collapsed onto the residential area.

The incident, which occurred around 9:30 pm on Thursday, caused panic among residents. The mishap occurred on the Sion-Trombay Road, a major traffic artery in Mumbai. The collapse happened during the removal of temporary shuttering. A turnbuckle fastener, which was attached to the concrete block, reportedly broke after being struck by an unknown object. This caused the ropes securing the reinforcement cage to loosen, leading the structure to topple onto the security cabin and damage its metal sheet.

The Mumbai Metropolitan Region Development Authority (MMRDA), which is constructing Metro 2A connecting DN Nagar to Dahisar, has termed the incident an "unfortunate accident" caused by a technical failure. It has established a three-member committee to investigate the accident. November 01, 2020, Indian Express

Sep 14, 2019 · Following the death of a construction worker on Metro Line 3, the Mumbai Metro Rail Corporation (MMRC) has initiated an inquiry into the incident,

www.msn.com › en-in › newsMMRCL piling rig falls during Metro 3 work at Girgaon, no one ...

Fortunately, no injuries or casualties have been reported. The MMRCL said that a crane has been dispatched to the site to restore the toppled machine. The MMRC assured that safety measures are

Sep 15, 2019 · MMRCL has stopped work at the tunnel site of Metro 3 between Powai and SEEPZ, where a worker was killed and another injured after a rock fell late Friday night.

Following the death of a construction worker on Metro Line 3, the Mumbai Metro Rail Corporation (MMRC) has initiated an inquiry into the incident, while the Mumbai Police have...

timesofindia.indiatimes.com › city › mumbai

Tragic Accident in Mumbai: Transit Mixer Operator Dies ...

Dec 5, 2024 · A transit mixer operator was killed in a tragic accident near Mumbai when the mixer toppled into an excavation. MMRDA imposes a fine of Rs 30 lakh on the contractor, J Kumar.

On Tuesday, fire engines, rescue vans and ambulances were rushed to the spot near Hotel Leela at Andheri, where several were feared injured as a slab of the under-construction Metro rail line that links Andheri to Ghatkopar came crashing down.

There are previous instances of accidents on the metro line.

2008: A Pile rig collapses on an autorickshaw, killing one.

In May 2008, a person was killed and another was seriously injured after a pile rig at a Mumbai Metro Rail construction site in Andheri (West) collapsed on an autorickshaw.

The Indian Express report stated. An inquiry into the mishap at the Metro construction site at Saki Naka, where scaffolding had collapsed, found the civil contractor infrastructure, at fault. The committee set up by Mumbai Metropolitan Region Development Authority, the nodal agency for Mumbai Metro Rail Project, too had similar observations, according to an Indian Express report.

2009: Four injured after a steel reinforcement cage collapsed

In 2009, the near-collapse of a steel reinforcement cage and temporary scaffolding for a concrete pillar in Andheri (East) left four workers injured, according to a Times of India report. A case of negligence had been registered against the site engineer and the contractor working.

2012: Crane crashes into hospital, no one injured

In April 2025, part of a 400-tonne crane being used for work on the metro line veered off the truck carrier and crashed into a portion of the Sarvodaya Hospital in Ghatkopar, the Mumbai Mirror reported. Luckily, the building was empty at the time of the incident, and no injuries were reported.

CHAPTER 11

Pune Metro

11.1 Introduction

Pune has witnessed enormous industrial growth since the 1990s. The traffic in the city is rising at an alarming rate. The roads in the city accommodate a variety of vehicles simultaneously. Being a densely populated area, Pune's traffic needs cannot be met only by a road-based system. Pune Mahanagar Parivahan Mahamandal Ltd. (PMPML), the public transport provider that operates buses and BRT services in Pune, is inadequate to meet the transport demand. According to reports published in April 2018, the number of vehicles registered in the city stands at 3.62 million, surpassing the city's population. The need for a robust public transportation system, such as a rail-based mass transit system, has been discussed, and in August 2008, the **DMRC** was appointed to prepare a Detailed Project Report (DPR). Hon'ble Prime Minister Narendra Modi laid the foundation stone of the Pune Metro Project (Phase 1) in 2016.

Fig.11.1: The Hon'ble Prime Minister of India laying the Foundation for Purple and Aqua Lines of the Pune Metro

The Pune Metro system comprises three lines with a combined length of 66.27 km, of which 32.97 km on two lines are operational as of November 2024. The extended metro line from Ruby Hall Clinic to Ramwadi, including the Bund Garden, Kalyani Nagar, and Ramwadi metro stations, covers 6 km. The 16.59 km Purple Line from PCMC Bhavan to Swargate runs on an elevated viaduct between PCMC Bhavan to Range Hills, followed by an underground section.. The Aqua Line runs from Vanaz to Ramwadi, covering a distance of 14.66 km at an elevation. The 23.33 km elevated Line 3 Puneri Metro runs from the Rajiv Gandhi Infotech Park in Hinjawadi via Balewadi. Sections of the purple and aqua lines were inaugurated in March 2022. Additional sections on both lines became operational in August 2023. Line 3 of the Pune Metro is being implemented by the Pune Metropolitan Region Development Authority (PMRDA) and is a joint venture of Tata Projects and Siemens, Pune IT City Metro Rail Limited, on a public–private partnership basis.

11.2 Phases of Pune Metro

The Pune Metro network was divided into three phases: Phase I, Phase II, and Phase III, for execution.

i) Phase 1

The Delhi Metro Rail Corporation (DMRC) submitted the DPR for Pune Metro Phase 1 in July 2009 and received approval from the government in June 2012. However, the project immediately encountered problems with various organisations demanding alignment changes and environmental issues. Considering the various issues and cost revisions, the DPR was updated in January 2013, August 2014, and again in November 2015 to reflect current prices. The project received final approval from the Union Government's Cabinet in December 2016, and in January 2017, Nagpur Metro Rail Corporation Ltd. (NMRCL) was renamed to Maharashtra Metro Rail Corporation Ltd. (Maha-Metro) to execute Pune Metro's Phase 1 consists of two lines: the purple line, running from PCMC Bhavan to Swargate, and the Aqua line, running from Vanaz to Ramwadi. The two lines comprise a total of 42.84 km, including extensions.

ii) Phase 2

The Pune Metro has begun preparing for the second phase of the metro project. The routes from Khadakwasla to Kharadi via Swargate and Hadapsar metro rail route, 25.862 km; Paudphata to Manikbaug via Warje route, 6.118 km; Vanaz to Chandni Chowk route, 1.112 km; and Ramwadi to Wagholi route, 11.633 km in length, have been approved by the Pune Municipal Corporation.

11.3 Routes / Lines

11.3.1 Line 1 (Purple)

Fig.11.2: Underground station at Civil Court metro station.

The Purple Line runs from PCMC Bhavan in Pimpri to Swargate. The 16.59 km line is elevated until Range Hills, with 9 stations, and goes underground up to Swargate, with 5 stations, via Nashik Phata, Khadki, and Shivajinagar. The maintenance depot for this line is planned near the Range Hills station on the land acquired from the College of Agriculture. The 5.09 km long underground section of the Purple Line from Agriculture College to Swargate is being carried out in two packages: 1. Agriculture College to Budhwar Peth; 2. Budhwar Peth to Swargate. Maha Metro is involved in the design and construction of the stations and associated tunnels. The underground station at Swargate is planned to be part of a multimodal transit hub integrating intra- and inter-city bus services run by PMPML and MSRTC, as well as autorickshaws, and a parking facility for private vehicles. MahaMetro is undertaking the project on a PPP basis. In April 2018, Maha Metro began construction on the elevated stretch in the Kharalwadi and Morwadi areas. The PCMC conducted trials on the newly built BRTS route in April 2018, near the metro lines, and intended to launch services; however, it ordered Maha Metro to stop construction. Maha Metro revised the alignment in May 2018 and started the construction one year later.

In Pune's metro rail development, PMRDA and Maha Metro are two separate entities with different roles and responsibilities, which is why they operate separately. The Pune Metropolitan Region Development Authority (PMRDA) had proposed taking over the Metro project, but the PMC and PCMC declined it. The opposing representatives stated that the Metro rail will be implemented by a Special Purpose Vehicle. Instead, the civic bodies suggested including PMRDA in the SPV to increase the reach of the Metro rail. PMRDA will execute only Line 3, i.e., the Shivajinagar to Hinjawadi (Phases I, II, and III) line. They are following the PPP model for the project. For the implementation of the initial two lines, the former Nagpur Metro Rail Corporation was reconstituted to form Maharashtra

Metro Rail Corporation Limited (MahaMetro). Maha Metro operates in core urban areas (within PMC and PCMC), while PMRDA covers wider metropolitan areas.

Fig.11.3: Elevated station at PCMC Bhavan metro station. **Fig.11.4: Pune Metro Coach**

i) Line 1A (Extension)

Ever since the Centre gave its approval for the first two metro corridors, there has been a demand for extending the Purple line from PCMC Bhavan to Bhakti Shakti Chowk, Nigdi. Considering these demands, PCMC decided to prepare DPRs for the extensions in their respective jurisdictions and bear the associated costs. In October 2018, Maha Metro submitted the DPR for the 4.5 km-long extension to the PCMC. The state approved the DPR in February 2021, and in October 2023, the centre approved the extension of the route from PCMC to Bhakti Shakti Chowk in Nigdi. Maha Metro had started the soil testing.

ii) Line 1B (Extension)

There have also been demands for extending the line from Swargate to Katraj. The PMC approved the DPR prepared by Maha Metro for the 5.4 km-long extension in September 2021. The extension is planned to be entirely underground with 3 stations. In August 2024, the extension from Swargate to Katraj was approved by the Government of India. The official ground-breaking ceremony took place in September 2024. In December 2024, the Pune Municipal Corporation and Maha Metro mutually decided to construct a fourth metro station at Balajinagar.

11.3.2 Line 2 (Aqua)

Aqua line runs from Ramwadi to Vanaz via Mangalwar Peth and Gymkhana. The elevated line, which is 14.66 km in length, has 16 stations. It connects with the Purple line and Line 3 at the Civil Court interchange station. The maintenance depot for the Aqua line is located near the Vanaz station on the former site of the garbage depot. The Aqua line currently terminates at Vanaz in the west; it was decided to extend the line to Chandani Chowk. The decision came in February 2018, when a longstanding dispute over the erstwhile garbage depot land in Kothrud was resolved by allotting it to Maha Metro for the development of a metro depot. The station at Chandani Chowk will bear the name of the Shivsrushti

11.3.3 Line 3 (Pink)

Puneri Metro is under construction by Pune IT City Metro Rail Limited and will run from Civil Court, Pune, to Megapolis Pune (including Maan and Bhoirwadi) in Hinjawadi. The 23.3 km line will be completely elevated, featuring 23 stations, and will align with the Maha Metro lines at the Civil Court interchange station. The construction will be taken up in three phases. The section between Hinjawadi and Balewadi is expected to be taken up first, followed by the section between Balewadi and the Civil Court in Shivajinagar. A metro car shed will be built in Maan, Hinjawadi. The MIDC will provide 22 ha of land in Hinjawadi for setting up a Metro rail depot. The line will also connect to the multimodal transit hub planned along the National Highway 48 in Balewadi, which will integrate the inter- and intra-city bus services and BRTS operated by the MSRTC and PMPML. The hub will be constructed on a PPP basis under the Smart City mission. The hub will comprise a built-up space of 176,516 m², including 162,580 m² of commercial office space, parking for 80 buses, 1,942 cars, and 3,884 two-wheelers. As of October 2018, offers from international and national real estate developers have been invited for the hub, and the PMC is in the process of transferring the land to the executing agency, Pune Smart City Development Corporation Ltd. (PSCDCL). The PMRDA has announced that the line will be extended from the Civil Court intersection to Phursungi IT Park via Hadapsar. **DMRC** is preparing the DPR for the extension.

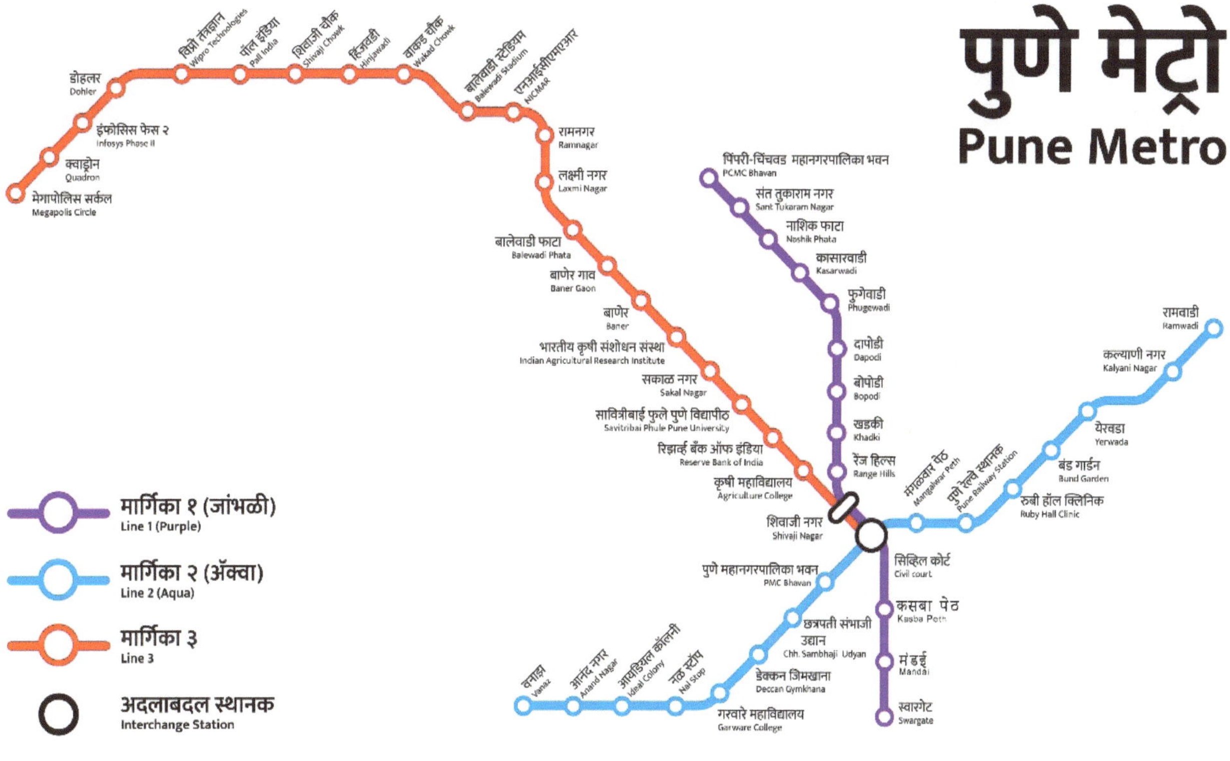

Fig.11.5: Pune Metro Final Network Plan (DMRC Final DPR)

Fig.11.6: Completed Station During Operation

11.4 Important Features

Gauge (Standard) ... 1435 mm

Route Length

Corridor 1: PCMC – SWARGATE - Underground- 5.019 Km

Elevated -11.570 Km

Total -16.589 Km

Corridor-2: VANAZ – RAMVADI - Underground- NIL

Elevated- 14.665 Km

Total- 14.665km

Number of Stations

Corridor-1: PCMC – SWARGAT Underground- 6, Elevated- 9

Corridor-2: VANAZ– RAMVADI Underground- NIL, Elevated-16

Speed - Designed Speed -80 kmph, Scheduled speed - 33 kmph for Corridor -1 & 31 kmph for Corridor -2,

Traction Power Supply - a) Voltage ... 25 KV ac

b) Current Collection

Overhead **Power Demand (**MVA): Year 2031

Corridor 1 34.2

Corridor 2 15.9

Traffic Forecast (Average in Lakhs): – Corridor I 382577 (2018)

- Corridor-2 189307 (2018)

Headway (2018): - Corridor – 1 - 4 min.

- Corridor – 2 - 8 min

Train Composition: 4 Cars/Train (for both corridors)

Power: SCADA system

Rolling Stock: 2.90 m wide rolling stock with stainless steel body. Axle load - 16 T Seating arrangement - Longitudinal Capacity of 4 coach unit - 1034

11.5 Geometric Design

The design parameters were worked out in line with the client's requirements, the city's demands, and based on national and internationally accepted practices. Various alternative alignments were considered to finalise the best-suited ones.

Alignment was selected through densely populated areas, touching important locations depending on ridership, while minimising or avoiding land acquisition. The Alignment was also planned to minimise the underground corridor, reducing costs and avoiding existing utilities as far as practicable.

The geometric design parameters considered are.

Horizontal curves:

Underground Section – Desirable Minimum- 300 m Absolute Minimum - 200 m

Elevated Section – Desirable Minimum – 200 m Absolute Minimum - 120 m

Minimum curve radius at stations1000m (both underground & Elevated)

Maximum permissible cant (Ca) (for both)	– 125 mm
Maximum cant deficiency (Cd) (for both)	– 100 mm
Vertical clearance for Viaduct (above road level) minimum	**–** 5.5 m
Rail level for Elev. Corridor with PSC Box Section (Pune Metro)	– 9.8 m
At the central median, with a concourse (depending upon station location)	12.5 m
Rail level for the Underground midsection in tunnelling min.	– 12.0 m (**below the ground)**

At station location (Cut & Cover) – 12.5 m

(Using TBM) – 15,05 m

Track Center (Elevated): – 4.m

(Underground): – 4.5 m

11.6 Viaduct and Elevated Stations

The viaduct structure for the Pune Metro is a precast segmental box girder, carrying two tracks supported on a single pier located on the median of the road. Road clearance of 5.5 m is ensured below the viaduct structure. The foundations are pile foundations at most locations. Open foundations are located at specific isolated sites.

The 140 m-long stations are generally planned on the road median and designed in two levels. The 70 m-long concourse is in the mid-section, with staircases leading from either side of the road. The maximum width of the concourse is 35 m and reduced to 16.5 m on both sides. Passenger facilities, including ticketing and information services, as well as operational areas, are available at the concourse. The concourse is divided into public and non-public areas.

The non-public area includes the Station Control Room, Station Master's Office, Waiting Room, Meeting Room, UPS and Battery Room, Signaling Room, Train Crew Room & Supervisor's Office, Security Room, Station Storeroom, Staff Toilets, etc. The public area is further divided into paid and unpaid areas. An auxiliary service station is provided on the ground nearby, subject to land availability.

Since the station is in the middle of the road, a minimum vertical clearance of 5.5 m has been provided under the concourse. Platforms are about 12.5 m from the road. Structures with maximum transparency and a light appearance have been designed.

A slim and modern concrete form would look compatible with better environments and low-rise developments, as and where required. Platform roofs have been designed with steel frames and aluminium cladding, resulting in a light look. Platforms would be protected from the elements by providing an overhang of the roof, and sidewalls would be avoided, thereby enhancing the transparent character of the station building. To allow unhindered traffic movement below the stations, portals have been proposed across the road in the concourse area, and a single column at the central verge supports the rest of the station.

11.7 Underground sections

The rail level at the midsection in the tunnelling portion was kept at least 12.0 m below ground level, ensuring a cover of 6m is available over the tunnels. At stations, the desirable depth of rail below ground level is 12.5 m, and the track center in the underground section to be constructed by Tunnel Boring Machine (TBM) is 15.05 m to accommodate a 12 m wide island platform. The track center in the underground section planned with the cut-and-cover method is 4.5m. For the underground section to be built using a TBM, each track is accommodated in a single tunnel with a diameter of 5.5m, resulting in two parallel tunnels with a gap of one tunnel diameter. The typical underground station features a two-level design, with platforms located at the lower level and a concourse on the upper level. Concourses are provided at the two ends only to minimise the total height of the underground station, and hence the cost. The upper level has, in addition to the concourse, all the passenger amenities, ECS plant rooms, electrical and S&T equipment rooms, station operation areas such as Station Control Room, Station Master's Office, Waiting Room, Meeting Room, UPS & Battery Room, Signaling & Train Crew Room, Train Crew Supervisor's Office, Security & Station Storeroom, Staff Toilets, etc. The lower-level features platforms, tracks, a seepage sump, a pump room, and similar ancillary spaces located beyond the platforms on either side.

Ventilation shafts, equipment hatches, entrances and chiller plants for the ECS plant are aboveground structures associated with the underground station and are on the open spaces by the roadside. Four entrances have been provided to the station, two at each end. Two emergency staircases are also planned in the event of traffic congestion on the islands. The reinforced concrete box, 20 m wide, 225 m long and 14.6 m high, has been designed for the UG Station structure.

11.8 Stations

Station locations were strategically placed to serve major passenger destinations and facilitate convenient integration with other modes of transportation. The plan has been to propose station locations such that inter-station distances are uniform and the average spacing of stations is as close to one km as possible.

Stations have been divided into two distinct areas: public and non-public. The public area is further subdivided into an unpaid and a paid area. The paid area includes provisions for escalators at all stations from the start. The provision in structural plans at stations has been kept providing lifts for disabled passengers in the future.

Integration facilities at Metro stations include approach roads, circulation facilities, pedestrian routes, and circulation areas for various modes likely to use these important stations, including feeder buses. Parking for private vehicles has not been proposed, given the scarcity of land along the alignment

11.8.1 Platforms

All the underground stations have island platforms. In the elevated section, stations have been planned with side platforms to prevent the viaduct structure from flaring in and out at stations, which may affect the road traffic below. Care has been taken to locate stations on a straight alignment. However, in some stations, site constraints have become the deciding criteria, and a curve of a 1000 m radius has been introduced.

11.8.2 Utilities

The Metro alignment, primarily along major arterial roads of the city's road network, serves institutional, commercial and residential areas. Several surface and sub-surface utility services, viz. sewers, water mains, stormwater drains, telephone cables, electric poles, traffic signals, etc., exist along the designed alignment. Details of the existing utility services have been collected from the concerned authorities. The design and construction of metro corridors are carried out with minimal conflict and diversion, wherever possible. Utilities that cannot be diverted have been maintained at their existing locations, with appropriate supporting arrangements in place.

Fig.11.7: Pune Metro during Operation

11.9 Ticketing

Ticket Counters and Ticket Issuing Machines (TIMs)

It is planned to deploy manual ticket issuing at the beginning of the operation, and at a later stage, automatic TIMS will be used, for which space provision has been made in the concourse. future. The capacity of manual ticket vending counters is 10 passengers per minute, and it is assumed that 40% of the commuters would purchase tickets at the stations.

The rest are expected to buy prepaid tickets, cards, etc.

11.10 Rolling Stock

Rolling stock for Pune Metro has been selected based on the following criteria:

- Proven equipment with high reliability
- Passenger safety feature
- Energy efficiency

- Lightweight equipment and coach body
- Optimised scheduled speed
- Low Life cycle cost

The purple and aqua line of the Pune Metro uses standard gauge rolling stock manufactured by Titagarh Firema.

The trains consist of 3 coaches each, have a total capacity of 970 passengers and can reach a maximum speed of 95 km/h.

The aluminium-bodied coaches are 21.34–21.64 m long, 2.90 m wide, and 3.90 m high.

Each coach has a capacity of 320 passengers with longitudinal seating for 44 passengers.

The trains are fully air-conditioned and feature CCTV, panic buttons, emergency doors, public address systems and audio-visual systems for indicating the opening and closing of doors. One coach in each train is reserved for women, and there are dedicated spaces for passengers in wheelchairs.

i) Train Operation Plan

The underlying operation philosophy is to make the MRT System more attractive and economical. The main features are:

- Selecting the optimum frequency of Train services to meet sectional capacity requirements during peak hours on most sections.
- Economical & optimum train service frequency not only during peak period, but also during off-peak period.
- A short train consists of 4 coaches with high-frequency service, which can be increased to 6 Coaches to meet future requirements.
- Multi-tasking of train operation and maintenance staff.

Salient Features of the proposed train operation plan are: -

- Running of services for 19 hours of the day (5 AM to Midnight) with a station dwell time of 30 seconds,
- Make up time of 5-10% with 8-12% coasting.
- Scheduled speed for these corridors has been taken as 33 Kmph for Corridor-1 & 31 Kmph for Corridor-2. Requirements of coaches are calculated based on the following assumptions:

The basic unit of a 4-car train comprising DTC-MC-MC- DTC configuration has been selected for the Pune Metro.

ii) Requirements of coaches

Calculated based on the following-

i. Train Composition planned as 4 Car Train Composition: DTC-MC-MC-DTC
ii. Train Carrying Capacity of 4 Cars
iii. Train Capacity: 1034 passengers
iv. The coach requirement based on the headway during peak hours.
v. Traffic reserve - one train per section to cater to the failure of a train on the line and to make up for operational time lost.
vi. (vi) Repair and maintenance reserve has been estimated as 8 % of the total requirement (Bare +Traffic Reserve).
vii. (vii) The calculated number of rakes in a fraction is rounded off to the next higher number.

11.11 Track Structure

Track on Metro Systems is subjected to intensive usage with very little time for day-to-day maintenance. The track structure must be long-lasting and should require minimal maintenance. The design ensures the highest levels of safety, reliability, and comfort, with minimal noise and vibrations. Ballastless track with continuously welded head-hardened rails has been designed for elevated and underground sections. Ballasted track structures were planned for the at-grade section and at the depot.

11.12 Traction System

Based on the ultimate traffic requirements, standardisation, and other techno-economic considerations, a 25 kV AC Overhead Electrification (OHE) traction system is designed for the Pune Metro System.

11.13 Signaling and Communication

As rolling stock necessitates optimisation of its capacity to provide the best services, Pune Metro opts for ATP (Automatic Train Protection) and ATS (Automatic Train Supervision) signalling systems to achieve this. Automatic Train Operation (ATO) is planned to be added in the future. The telecommunication system serves as the backbone for signalling and other systems, such as SCADA and AFC, and provides services to meet operational and administrative requirements.

The telecommunication facilities are to cater for

*Supplementing the Signaling system for efficient train operation.

* Exchange of managerial information

*Crisis management during emergencies

*Passenger information system, besides the following.

- Train Traffic Control
- Maintenance Control
- Emergency Control
- Station-to-station dedicated communication
- Telephone Exchange
- Passenger Announcement System and Passenger Information and Display System
- Centralised Clock System
- Train Destination Indicator
- Data Channels for Signaling, SCADA, Automatic Fare Collection, and more.

11.14 Automatic Fare Collection

Ticket issues and fare collection play a vital role in the efficient and proper operation of the system. And so, the ticketing system was made simple, easy to use/operate and maintain, easy on accounting facilities, capable of issuing single/multiple journey tickets, amenable to quick fare changes and requiring overall lesser manpower, resulting use of computer-based Automatic fare collection system. The AFC system proves to be cheaper than the semi-automatic (manual system) due to reduced manpower cost for ticketing staff, reduced maintenance in comparison to paper ticket machines, and overall, lower cost of recyclable tickets (Smart Card/Token) and help prevent revenue leakage.

11.15 Pune Metro Corridor 1

The Detailed Design Consultancy Services for 9 (Nine) Elevated Stations from PCMC to Range Hill, Excluding the Viaduct in the Station Portion on Corridor-1 of the Pune Metro Rail Project. The project was awarded to **M/s Ayesa Ingenieria y Arquitectura S.A.U**. by Maharashtra Metro Rail Corporation Limited (MAHA-METRO)

Detailed Design Consultancy (DDC) services for Corridor-1 (Purple Line) of the Pune Metro Rail Project, extending from PCMC to Swargate, a stretch of approximately 16.59 km with both elevated and underground sections. Specifically, the consultancy work involved designing nine elevated metro stations between PCMC and Range Hill, excluding the viaduct portion within the station areas. The stations under this package included PCMC, Sant Tukaram Nagar, Bhosari (Nashik Phata), Kasarwadi, Phugewadi, Dapodi, Bopodi, Khadki, and Range Hill.

As part of the assignment, the DDC was responsible for preparing architectural designs to facilitate efficient passenger movement, circulation, and aesthetics, along with detailed structural designs for station buildings, concourses, platforms, and ancillary facilities. The scope also included MEP (Mechanical, Electrical, and Plumbing) designs covering HVAC, fire detection and firefighting systems, and water supply, as well as the integration of E&M systems such as signalling, telecom, traction power, and Automatic Fare Collection (AFC). Preparation of tender drawings, Good for Construction (GFC) drawings, and detailed design reports formed a key part of the work, alongside coordination with viaduct, track, and system contractors to ensure seamless interface and integration. The designs were required to comply with NBC guidelines, fire safety norms, seismic standards, and universal accessibility requirements.

The nine elevated stations designed under this package form the northern entry stretch of the Pune Metro, connecting the Pimpri-Chinchwad Municipal Corporation (PCMC) area with central Pune. The designs were prepared with a focus on integration with feeder transport systems such as PMPML buses, adequate parking facilities, and safe pedestrian access, while also considering passenger comfort, safety, and provisions for future capacity augmentation. This consultancy work contributed to the timely completion of Corridor-1, enabling efficient connectivity from PCMC to Range Hill and beyond.

The project has significantly enhanced mobility for commuters in densely populated areas of PCMC and Pune, improved accessibility to workplaces, educational hubs, and residential areas along the corridor, and promoted sustainable urban development by reducing vehicular traffic and pollution.

11.16 Brief on Line 3

- * **Route:** Hinjawadi – Civil Court
- * **Length**: 23.33 km
- * **Type**: Elevated
- * **Depot**: Maan Village (20 hectares)
- * **Number of Stations**: 23

Depot: Mann Village (For Line 3)

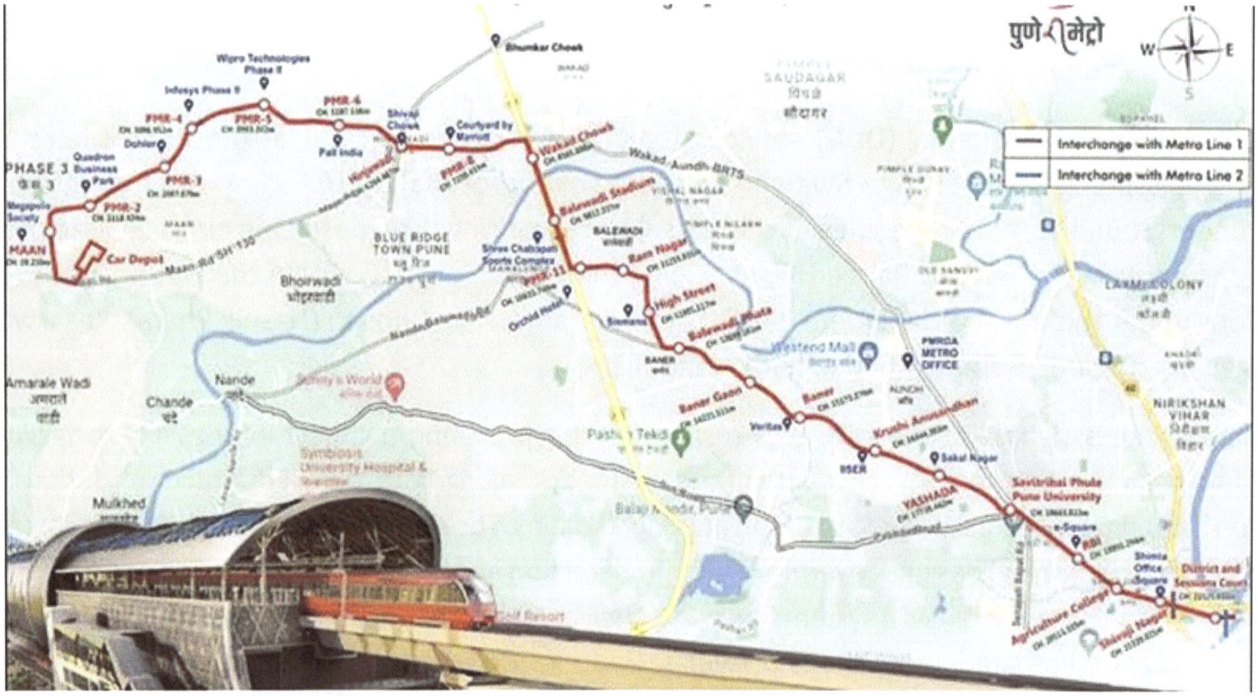

Fig.11.8: Route map for Line 3

23.3 km long, Line 3 is an elevated metro that connects Hinjewadi to Shivaji Nagar in Pune. This line was approved by Maharashtra's state government in February 2018 for implementation on the public-private partnership (PPP) model on a design, build, finance, operate, and transfer (DBFOT) basis. The project has a 39-month duration, and the contract for the design and construction of the viaduct and stations, including finishings, MEP, track, and Depot.

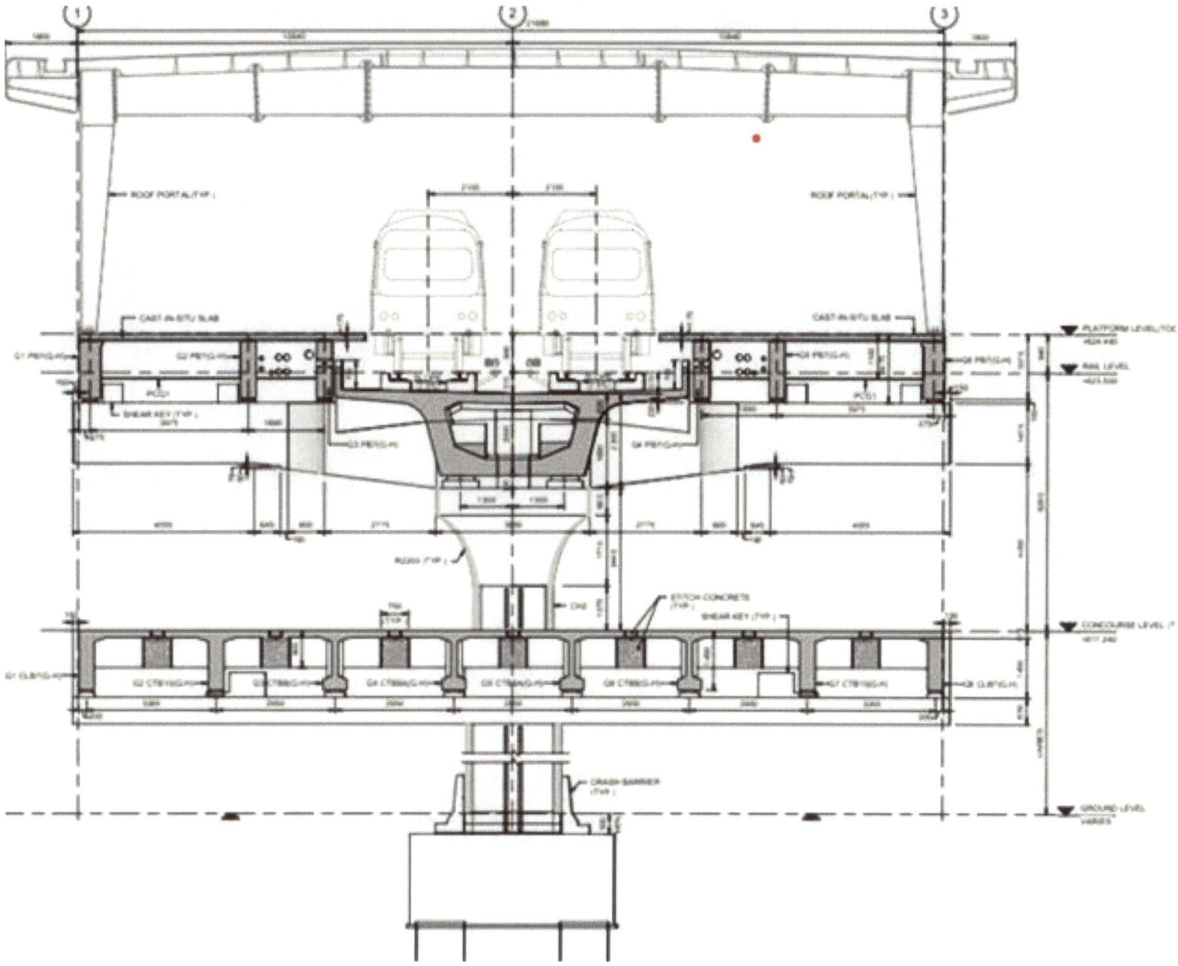

Fig.11.9: Cross-section through Station Building

Signalling and traction to the single agency to minimise interface issues and expedite construction.

11.17 Important Features

- Elevated viaduct and stations.
- Viaducts are with pile foundation, and a Segmental superstructure
- Standard gauge
- Deck width 8.8 m
- Third rail power system
- Major stakeholder is the Government
- Stations planned for precast construction

11.18 System

- **Top Speed**: 80 kmph
- **Average Speed**: 34 kmph
- **Track Gauge**: Standard Gauge – 1435 mm
- **Electrification**: 25 KV, OHE
- **Traction**: 750V DC Third Rail
- **Signalling**: Communications-based Train Control (CBTC)
- **Rolling Stock** by Alstom Transportation

Fig.11.10: Photographs taken During Construction

11.19 Casting Yard

All precast segments were cast in the Casting Yard. The casting of elements, such as girders and box segments, was done in parallel with site execution works. The casting yard brings factory-controlled production techniques, efficiency, quality control, and time savings to metro construction. Fabricating elevated segments in a separate area removes casting operations from the critical construction path, thereby reducing overall construction time.

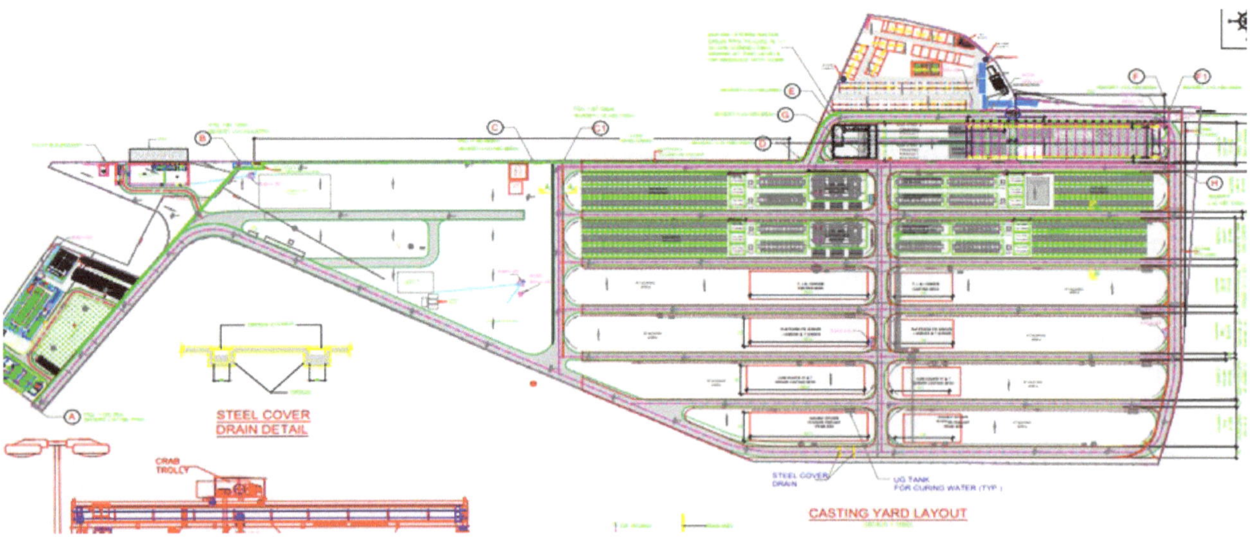

Fig.11.11: Casting Yard Plan

Fig.11.12: Aerial View of Casting Yard as Implemented

Casting Layout Planning has been carried out considering the steel yard, QC laboratory, stacking area, and gantry to optimise use and logistics. Additionally, entry and exit gates, stores, and offices have been varied through a Systematic Layout Planning (SLP) relationship chart. A rainwater drainage system and water recycling have been implemented.

Construction initiative and smart planning on the viaduct concerning time, cost and safety

1. Planned a casting yard to accommodate all establishments along the segment and station elements
2. Robust drainage arrangement to work on full swing in the monsoon
3. Recycling of water in curing
4. Good facilities of a labour colony for 1000 workers
5. Electro-Hydraulic Storm Wedge Brake mechanism
6. Storm and wedge locking
7. Automatic bar cutting machine in the centralised yard

Fig.11.13: Photos During Construction

11.20 Construction Initiative Concerning time, Cost and Safety: Viaduct and Stations Construction:

- Designed precast station and planned to run LG through stations
- Segmental span designed over the station viaduct
- Prefabricated reinforcement cage designed for viaduct and station pier cap
- Frictional collar designed instead of staging for pier cap casting in situ
- Special circular reaction frame to pile load test to accommodate within the barricading area
- Precast portals with bearing and an integrated portal with bar continuity
- Used a coupler to simplify and decongest the steel fixing
- Used twin LG to launch up to a 200 m radius segmental span
- 12 m pier casting in a single lift
- Crash barrier with a metal barrier instead of an MS board for more stability
- Stations are designed using cantilever portions on both sides to reduce challenges in doing the foundation, and for less road blockage
- 30% saving in foundation and substructure, converting a three-legged foundation to a one-legged foundation.
- Prefabricated cage designed for the station pier cap
- Used special launching girder for station element erection
- Derrick is used for arm erection instead of a crane
- 70% concrete in precast form to reduce time and cost.
- Superstructure: Platform construction in stations is planned with 70% pre-cast concrete for faster, higher quality and safety
- Bed gantry used for station girders used in PD stations as an acceleration measure

Green initiation and water wastage reduction: Grid power usage, curing compound, transplantation instead of tree removal, with 95%.

11.21 Photos During Different Stages of Construction

Fig.11.14: Overview of Different Levels of Construction

Fig.11.15: Elevated Corridor

Fig.11.16: Shows the Structural Arrangement of Station Roof **Fig.11.17: Approach Viaduct to Station During Construction**

Fig.11.18: More Photos during Construction

11.22 Innovative & technical initiatives adopted on PUNE METRO LINE-3

Pune IT City, Metro Rail Limited, has taken a special technical initiative and is implementing innovative approaches, which are in the advanced stage of construction. Innovations are in Environment, Health, and Safety (EHS), also known as SHE (Safety, Health & Environment).

On PML3, the major EHS innovations implemented are:

1. Initiatives of Transportation and Segment Erection
2. Learning Center
3. Communication of Green Initiatives through Quarterly Environment Newsletter "Green Updates"
4. Eco-friendly steps

11.22.1 Safety in Transit, Awareness on the Move

"The installation of hazard markers on trailer beds is a unique safety measure designed to alert commuters and mitigate the high risks associated with transporting precast elements, thereby enhancing public safety on the road."

The construction of metro projects involves significant high-risk activities, and the transportation of precast elements from the casting yard to the project location is a major safety challenge that requires the intervention of public and road users, particularly during the night shift.

The major safety challenges for transportation precast elements include embedded weight, segment size, insecure loading, poor road conditions, vehicle selection, and inadequate visual clearance to road users, which pose a significant risk of traffic-related accidents to road commuters. The general practice adopted across the metro is to use two utility vehicles, along with traffic marshals deployed to pilot the long-bed trailers. These logistical arrangements, in certain cases and locations, lead to encroachment of commuters during transportation.

The Pune Metro Line-3 project has successfully implemented an innovative solution to enhance traffic-related safety during the transportation of precast elements. Hazard markers have been installed on all three sides of Over Dimension Carrier (ODC) trailers, significantly improving the visibility of hazards associated with transporting these elements from casting yards to work locations and along commuter routes. This solution has improved the transportation process, contributed to a safer working environment, and reduced the risk of accidents.

Key benefits achieved: Enhanced Visibility for Commuters, Improved Road Safety, Proactive Risk Mitigation, and Compliance with Safety Standards.

Fig.11.19: Transportation of Heavy Segment to Site

11.22.2 EHS Learning Centre

"Empowering workforce and stakeholders with essential Environment, Health & Safety knowledge for safer project execution", as part of its commitment to fostering a culture of safety and sustainability, the Pune Metro Line-3 project has developed an **Environment, Health & Safety (EHS) Learning Centre** at the Casting Yard. **Key Features** are Personal Protective Equipment (PPE**), Fire Safety Awareness, Critical Activity Guidelines, and Environmental Protection Practices:**

a. **Recycling & Reuse Initiatives**.
b. **Tool & Tackle Safety Awareness**

Fig.11.20: Transportation of Other types of Heavy Segments

11.23.3 Communication of Green Initiatives

"Green Updates – Showcasing Commitment to the Environment" as part of its commitment to Environmental stewardship, the Pune Metro Line-3 project has launched a quarterly environmental newsletter titled **"Green Updates."**

Key Benefits of "Green Updates" are Promoting Environmental Awareness, Showcases Sustainability Efforts, Encourages Stakeholder Engagement, Supports Regulatory Compliance, Inspires Continuous Improvement, and Strengthens Project Image

The Pune Metro Line 3 project exemplifies innovation, responsibility, and leadership in environmental management. Through its zero-waste-to-landfill practices, resource conservation, use of green materials, and community-driven initiatives, the project has set new benchmarks in sustainable infrastructure.

11.23.4 Eco-friendly steps taken on the project

Sustainability works in Project- Water and Energy conservation till March '25

Tree Conservation: PITCMRL successfully saved 838 trees by adjusting the alignment & modifying the work methodology.

Water Conservation: Approx. 68,588 kiloliters of water were saved through the use of admixtures during concrete production and curing compounds.

Water Reutilisation: 27.8% –Water reused for Sprinkling, board washing, wheel washing

Energy Conservation: 95% of electricity consumption is sourced from the grid, resulting in a savings of 39,84,421 litres of diesel.

Waste to Wellness

i. "No waste is waste until we waste", replenishing natural resources and minimising the environmental impact of activities.
ii. This involves using recycled products as raw materials, implementing waste minimisation, and promoting biodiversity through initiatives such as developing an Ayurvedic garden using waste materials.

CHAPTER 12

Hyderabad Metro

12.1 Introduction

With increasing industrial development and growth in the IT sector, particularly in Hyderabad, there is a tremendous population growth, resulting in increased traffic. The government of Andhra Pradesh (AP) implemented a rail-based metro system to address its transport needs. The government awarded the well-planned Metro project to L&T to implement through a Special Purpose Vehicle (SPV), **"L&T Metro Rail (Hyderabad) Limited",** with the goal of completing the project within five years from the date of appointment. The project was to be executed on a design, build, finance, operate and transfer (DBFOT) basis with a concession period of 35 years. The length of the Metro network was 71.16 km, including 66 stations.

1. Corridor - I Miyapur - LB Nagar (28.87 km - 27 stations)
2. Corridor - II JBS – Falaknuma (14.78 km - 16 stations)
3. Corridor - III Nagole – Shilparamam (27.51 km - 23 stations)

The metro corridor was planned with elevated tracks connecting significant offices, retail, and residential areas. The up-and-down tracks pass through the city's arterial roads. The metro lines connect major bus stations: Miyapur, MGBS, Koti, Dilsukhnagar, Charminar, and Jubilee Bus Station. It integrates with existing rail terminals at Secunderabad, Begumpet, and Nampally. It links to MMTS services at Bharatnagar, Begumpet, Khairatabad, Malakpet & Falaknuma. GOAP is planning a feeder bus service for stations from different city areas.

This section outlines the standards for geometric design, general features, specifications, and requirements for accommodating structures within the Rail system. The basis of the horizontal and vertical alignments was derived from the Detailed Project Report. Detailed engineering for the Civil Structural, System, Stabling Yard, and Depot were carried out by the Concessionaire, including the maintenance equipment, to satisfy the following requirements.

i. The rail system met the requirements specified.
ii. Safe, healthy and efficient operation plans.
iii. Passenger flow to ensure a safe and comfortable manner.
iv. Structures to ensure all functions for operation.

All other buildings/structures required for operation and maintenance

12.2 Performance requirement

All permanent and temporary civil works are constructed to satisfy existing and future traffic flow requirements.

a. Minimum interference to existing traffic. Care was taken during the construction of the foundations.
b. Most of the superstructure was built with precast concrete segments. The scaffolding system was used for foundations. Where scaffolding was used for superstructures, the scaffold system provided the required horizontal and vertical clearances for the moving traffic.
c. **Minimum carriageway width**

   During the construction period, the minimum carriageway width on the existing roadway at ground level was maintained at 8.75 m in each direction, except for stretches where the existing ROW was insufficient. The traffic management/diversion plan for these sections was developed through discussions with the relevant authorities.

d. **Environmental**

   The requirements for environmental quality had been met as per regulations

12.3 Elevated Corridor

The superstructure was chosen considering its ease of construction, fast track, and maximum standardisation. Different types of superstructures, as given below, were considered, and the most suitable form was selected.

i. Pre-cast segmental box girder using external unbonded tendons.
ii. Pre-cast segmental U-Channel superstructure with internal pre-stressing; and
iii. Steel/Concrete composite girders.

Any other superstructure found more suitable and appropriate for the site condition was also considered. When necessary, in situ construction was considered in specific locations.

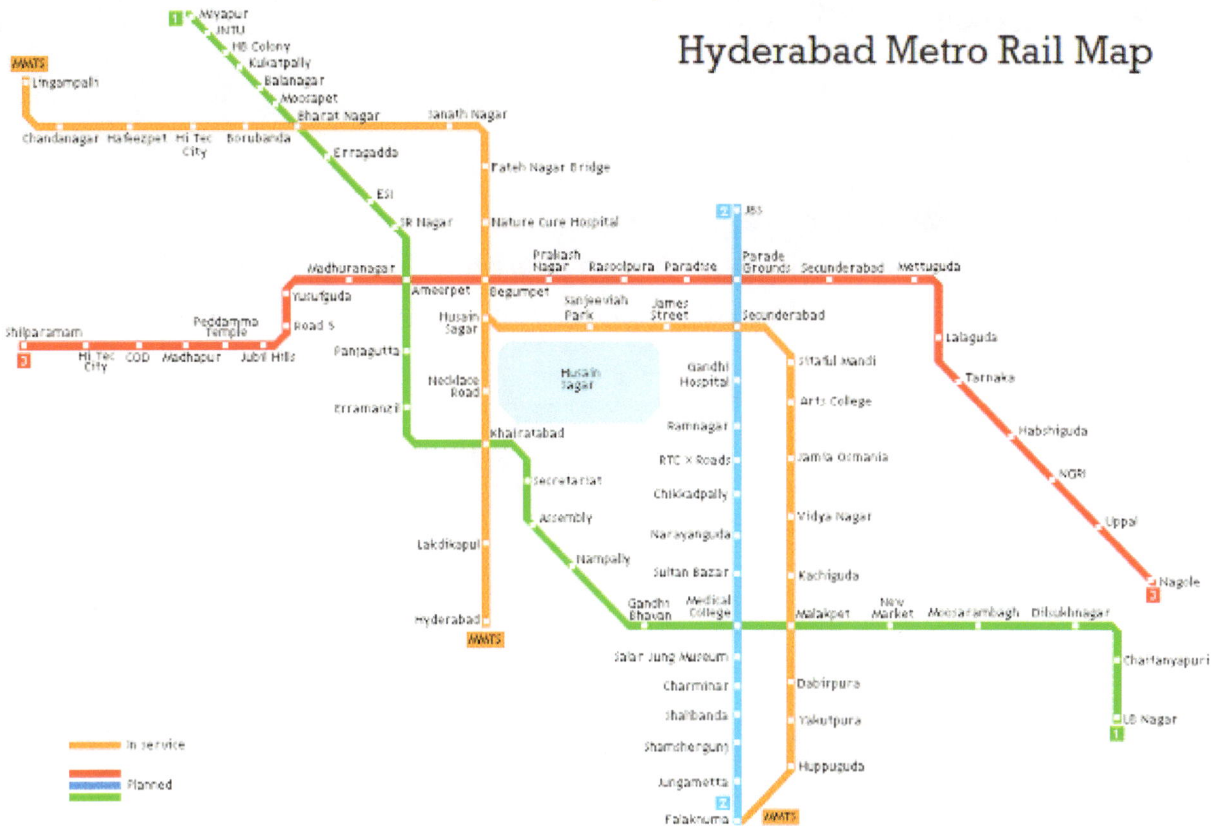

Fig.12.1: Hyderabad Metro Network Plan (hyderabadmetrorailworks.blogspot.com)

12.4 Salient Features

Salient features of the metro corridor include

1. Green & Eco-friendly mode of travel – reduces carbon emission, fuel consumption, and pollution.
2. Faster, safer, and comfortable air-conditioned with reduced travel time.
3. Modern coaches.
4. High frequency of trains- reducing waiting time.
5. User-friendly stations with lifts, staircases and facilities for the disabled.
6. Parking facilities at 25 strategic locations along the route.
7. Automatic ticket vending machines – reducing waiting time in queues and counters.
8. The Automatic Fare Collection system allows free entry and exits from the stations.
9. Essential Facilities at Stations- toilets, public address and information system, telephones.

12.5 Geotechnical

Investigations were conducted along the corridors in both soft and hard rock to a depth of 10 m. The top layer of soil is generally reddish, silty sand with clay, and below it is a layer of soft rock, followed by a layer of hard rock. Soil and rock samples were collected and tested in a laboratory.

For the elevated section, the foundation is shallow, either on soft rock or on a pile foundation socketed in rock.

12.6 Utility

Services such as sewers, water mains, stormwater drains, telephone cables, electric poles, and traffic signals exist along the proposed alignment. Details of existing utility services were collected from the relevant authorities, including the Hyderabad Municipal Corporation, Hyderabad Water Supply, and others.

Fig.12.2: Hyderabad Metro During Operation (thenewsminute.com)

Fig.12.3: Metro Corridor during Construction (newindiaexpress.com)

Fig.12.4: Ameerpet Metro Station (magicbricks.com)

Affected sections were identified, and temporary diversion & relocation were carried out. Regarding Land, efforts were made to minimise the requirements and the acquisition of private property.

12.7 Structural System-Superstructure

The superstructure is supported on a single row of RCC piers erected mainly on the central median. The spacing of piers is mostly 25 m. Longer/continuous spans, pier locations, and superstructures were adjusted to suit the site's requirements. RCC portal frames or other suitable support systems were necessary when the alignment deviated from the central median. The decision regarding the erection system for the precast segments was made by the Concessionaire, which may include the following.

i. Span-by-span erection with launching gantry.
ii. Balanced cantilevering erection with launching gantry.
iii. Full span pre-casting and erection; and
iv. Incremental launching.

Static and dynamic rail live load responses at essential movement joint locations complied with the design requirements. In accordance with international practice, a rail-structural interaction analysis was conducted using a continuous welded rail with direct fixation.

12.8 Substructure

The superstructure is designed to be supported on a single cast-in-place RC pier. The pier gradually widens at the top to accommodate the standard spans that support the bearing under the box webs. Piers are circular for most of their height to minimise ground-level space, allowing it to be accommodated within the existing central verge. A New Jersey-shaped crash barrier of 1.0m above the existing road level was provided to prevent direct collisions between vehicles and the pier. A gap of 25mm has also been provided between the crash barrier and the outer face of the pier. The longitudinal centre-to-centre spacing of the elastomeric/pot bearing is 1.8m. The space between the elastomeric bearings is placed in the lifting jack required to replace the elastomeric bearing. An outward slope of 1:200 was provided at the pier top for drainage., The transverse spacing between bearings would be 3.2m.

12.9 Foundations

Open and pile foundations were adopted depending on the suitability of the soil strata encountered. The pile cap/open foundation was kept at least 500 mm below the road level to facilitate necessary drainage from the viaduct and utility crossing. Circular/Rectangular/Square piers, as per design, including pier caps, were cast in a single lift to give a good finish without joints, except for very tall piers. To protect the piers against collision from moving vehicles on the road, concrete crash barriers were provided up to a height of 1.0 m above road level around the piers.

12.10 Deck Drainage and Waterproofing

RCC surface drains at the deck level were constructed along the viaduct. Runoff shall be collected through surface drains, which shall lead to the down. All metallic components of the drainage system were welded and then coated with two layers of bituminous paint before placement.

12.11 Roadway width during construction

Most of the construction took place in the middle of the road; two central lanes, including the median, were reserved for construction activities. During piling and/or open foundation work, a maximum width of 8 m was needed to facilitate construction. It was essential to maintain two lanes for traffic on either side by widening the roads as necessary. In cases of insufficient space at a particular site, one-way traffic was proposed in discussions with the relevant authorities. Additionally, a traffic management plan was prepared in consultation with the Relevant Authorities. Safety issues were given the highest importance during construction. They are primarily required for Depots, entry-exit to Stations, Traffic Integration, and Substations.

12.12 Codes and Standards

The following codes shall be applicable:

IRS, IRC, IS, AASHTO, and BS.

IRS: includes Bridge Rules for loading (Min. of Railways), a Code of practice for steel bridges, a Code of practice for plain, reinforced, and pre-stressed concrete for general Bridge construction, the latest revision, and a Code of practice for the design of substructures and foundations of bridges.

IRC: 18 – Design Criteria for Pre-stressed Concrete Road Bridges (Post-Tensioned Concrete)

21 – Standard Specification and Code of Practice for Road Bridges, Section III – Cement Concrete (Plain and Reinforced)

22 – Composite Construction for Road Bridges

78 – Foundation and Substructure

83 (Part I) – Metallic Bearings

83 (Part-II) – Elastomeric Bearings

83 (Part-III) – POT cum PTFE Bearings

SP – 47 Guidelines on Quality Systems for Road Bridges (Plain, Reinforced, Pre-Stressed and Composite concrete)

12.13 Materials Used

The following concrete grades were used

Table.12.1: Concrete strengths

Structural Element	Grade (fc') Cylinder Strength	Grade (fcu) Cube Strength	E (short term) Modulus of Elasticity
Superstructures	48 N/mm^2	60 N/mm^2	36 kN/mm^2
Precast - 3 Span	48 N/mm^2	60 N/mm^2	36 kN/mm^2
Precast 44/44 straight	56 N/mm^2	70 N/mm^2	38 kN/mm^2
Precast 44/44 curved	48 N/mm^2	60 N/mm^2	36 kN/mm^2
Precast 36m curved	48 N/mm^2	60 N/mm^2	36 kN/mm^2
Precast Type 1 station deck	40 N/mm^2	50 N/mm^2	34 kN/mm^2
Precast – Other			
Insitu continuity stitches:			
Precast 44/44 straight	48 N/mm^2	60 N/mm^2	36 kN/mm^2
Precast 44/44 curved	56 N/mm^2	70 N/mm^2	38 kN/mm^2
Type 1 station deck	48 N/mm^2	60 N/mm^2	36 kN/mm^2
Insitu Structures	40 N/mm^2	50 N/mm^2	34 kN/mm^2
Piercaps and Bearing Plinths	40 N/mm^2	50 N/mm^2	34 kN/mm^2
Pier Columns (3-span internal piers)	40 N/mm^2	50 N/mm^2	34 kN/mm^2
Other Pier Columns	32 N/mm^2	40 N/mm^2	31 kN/mm^2
Abutment walls, Bases and Pile, Caps, Piles	32 N/mm^{2*}	40 N/mm^{2*}	31 kN/mm^{2*}

Allowances were made for the loss of strength due to the placement of the concrete under drilling fluid. The design strength, fcu of 50 and 40 N/mm2, was the characteristic strength before and after placement. Where appropriate, the long-term modulus of elasticity was taken as half the short-term modulus.

12.13.1 Steel Reinforcement:

Hot-rolled reinforcement to BS 4449: 1997 will be specified with the following properties:

Type	Designation	Characteristic Strength	Elastic Modulus
Mild Steel	R	250 N/mm2	200 kN/mm2
High Yield Deformed Type 2	T	460 N/mm2	200 kN/mm2

12.13.2 Grade of Concrete

The concrete used was prepared using computerised automatic Batching Plants, and the grades of concrete used are as follows.

i. Pile Cap and Open Foundations: M -30
ii. Piers: M -40
iii. Precast Elements: M 50
iv. Cantilever Piers and Portals: M -45, M -60
v. Other miscellaneous structure: M - 30

12.13.3 Reinforce and Prestressed Steel

HYSD-415 or TMT steel was used as the reinforced steel, and for pre-stressing work, low-relaxation, high-tensile steel strands with the configurations 12T13 and/or 19K15 were used (IS:14268).

12.13.4 Space During Construction

As most of the construction was carried out on the existing roads, two central lanes, including the median, were used during construction. A minimum width of about 8m was required during piling and open foundation, and the area was barricaded. However, during the construction of the elevated corridor, two lanes were provided for existing traffic by widening existing roads as necessary. In a few cases, one-way traffic was resorted to. All these actions required a minimum period of 6 months. During this period, the following works were taken up:

i. Preliminary action for diversion of utility and preparation of estimates thereof.
ii. Reservation of land along the corridor, identification and survey for acquisition.

12.14 General Features

12.14.1 General

This section outlines the standards for geometric design, general features, specifications, and requirements for accommodating structures within the Rail system.

The basis of the horizontal and vertical alignments was from the Detailed Project Report, and the detailed engineering for the Civil Structural, System, Stabling Yard and the depot was carried out by the Concessionaire, including the maintenance equipment, to satisfy the following.

i. The rail system met the requirements specified.
ii. Safe, healthy and efficient operation plans.
iii. Passenger flow to ensure a safe and comfortable manner.
iv. Structures to ensure all functions for operation.
v. All other buildings/structures required for operation and maintenance.

12.14.2 Performance requirement

All permanent and temporary civil works are constructed to satisfy existing and future traffic flow requirements.

a. Minimum interference to existing traffic. Care was taken during the construction of the foundations.
b. Most of the superstructure was built with precast concrete segments. The scaffolding system was used for foundations. Where scaffolding was used for superstructures, the scaffold system provided the required horizontal and vertical clearances for the moving traffic.
c. **Minimum carriageway width**

   During the construction period, the minimum carriageway width on the existing roadway at ground level was maintained at 8.75 m in each direction, except for stretches where the existing ROW was insufficient. The traffic management/diversion plan for these sections was developed through discussions with the relevant authorities.

d. **Environmental** The requirements for environmental quality have been met in accordance with regulations.

12.15 Superstructure

The superstructure was chosen considering its ease of construction, fast track, and maximum standardisation. Different types of superstructures, as given below, were considered, and the most suitable form was selected.

i. Pre-cast segmental box girder using external unbonded tendons.
ii. Pre-cast segmental U-Channel superstructure with internal pre-stressing; and
iii. Steel/Concrete composite girders.

Any other superstructure found more suitable and appropriate for the site condition was also considered. When necessary, in situ construction was considered in specific locations.

12.16 Sub-structure and Foundations

Open and pile foundations were adopted depending on the suitability of the soil strata encountered. The pile cap/open foundation was kept at least 500 mm below the road level to facilitate necessary drainage from the viaduct and utility crossing. Circular piers, as per design, including pier caps, were cast in a single lift to give a good finish without joints, except for very tall piers. To protect the piers against collision from moving vehicles on the road, concrete crash barriers were provided up to a height of 1.0 m above road level around the piers.

12.17 Design Criteria for Elevated Structure

The Concessionaire carried out detailed topographic and geotechnical investigations along the corridor handed over to finalise the alignment and design of the superstructure, substructure, and foundations.

The elevated corridor was designed in line with the following:

a. The elevated structure was designed for an Axle Load configuration based on Coach design and Train configuration.
b. The ballast-less track structure used continuously welded rails. The resultant forces from rail temperature variation were considered.
c. The Indian Railway's Concrete Bridge Code was followed to design RCC sub-structure and foundations and prestressed concrete girders with internal and external prestressing as required.
d. Box Section, 'U' Section or any other efficient structural form was recommended for the main spans. The design was to provide a suitably designed sound barrier all along the corridor.
e. The superstructure WAS supported on single columns erected chiefly along the road's median. Portal frame support was considered at a few locations where single-column support was not feasible.
f. The overall width of the superstructure was 9.6 m, with the track centre at 3.7 m - 4.0 m.
g. The minimum clearance above the road level was 5.5 m. In the case of minor roads/streets, a lower level was adopted with the specific approval of the authority owning and/or maintaining the road/street. The vertical clearance above the rail level, where the alignment crosses the existing railway track, was provided by the Schedule of Dimensions of Indian Railways, to be not less than 5.8 m.
h. A provision for the passage of a width not less than 700 mm on both sides of the guideway along the corridor was maintained for emergency evacuation.
i. Provision for ducts for laying signals, communication, and illumination was kept.
j. Adequate drainage provision was kept.
k. Prefabricated elements were used to the maximum extent due to the need for fast-track work and the corridor's heavy congestion.
l. The structural form had to be aesthetically pleasing.
m. Approach slabs between the viaduct and the grade section were provided.

12.18 Elevated Structure

The Hyderabad Metro proposed an elevated corridor. The structure was chosen considering its ease of constructability, the use of standard elements, and the typical formwork for various structural elements. The following types of superstructures were considered.

i. Precast segmental box girder using external unbonded tendon.
ii. Precast segmental U-Channel superstructure with internal pre-stressing.

Segmental construction was chosen due to the following,

- Segmental construction is an efficient and economical method for an extensive range of span lengths and types of structures. Sharp curves and variable superelevation could be accommodated.
- Segmental construction permits reduced construction time as segments may be manufactured while substructure work proceeds and assembled rapidly afterwards.
- Segmental construction reduces environmental hazards. Segments are easy to stack and have more than one layer, saving space.

  Segments with moderate sizes are transportable by road trailers on city roads. The superstructure was chosen considering the ease of constructability and the standardisation of formwork for a wide range. Segmental construction was chosen mainly due to the following advantages.

- Segmental construction is an efficient and economical method for an extensive range of span lengths and types of structures. It accommodates structures with sharp curves and variable superelevation.
- Segmental construction permits reduced construction time as segments may be manufactured while substructure work proceeds and assembled rapidly afterwards.

Fig.12.5: Nagole Station

Fig.12.6: Uppal Station

Fig.12.7: Uppal-Depot: OCC-Building

*Segmental construction reduces the environmental hazard. The superstructure is manufactured away from busy areas and placed using a system erected from piers at various heights.

- Segments are easy to stack in the yard in more than one layer, thereby saving in the requirement of space.
- Segments are transportable from the casting yard to the site in Trailer Trucks, as the size and shape can be accommodated in available Trucks/Trailer Trucks.
- It can incorporate last-minute changes in span configuration if needed.
- Interference to existing traffic during construction is significantly reduced.
- Segmental construction contributes to an aesthetically pleasing appearance.
- The overall labour requirement is less than that for conventional methods.
- Better quality control and safety than conventional construction methods.

The viaduct superstructure is supported on a single cast-in-place RCC pier. The profile and size of the piers were designed to match the superstructure profile. The piers were circular, 1.6m in diameter, and flared at the top. A Jersey crash barrier of 1.0m height above the existing road level was provided all around the pier to prevent the direct collision of vehicles. A gap of 25mm was kept between the crash barrier and the outer face of the pier. The shape of the upper part of the pier is planned to provide a required clearance of 5.5m. The centre-to-centre spacing of the elastomeric/pot bearing is 1.8m.

12.19 Design Criteria for Elevated Structure

The Concessionaire carried out detailed topographic and geotechnical investigations along the corridor, handing over the results to finalise the alignment and design of the superstructure, substructure, and foundations.

The elevated corridor was designed in line with the following:

a. The elevated structure was designed for an Axle Load configuration based on the Coach design and Train configuration.
b. The ballast-less track structure used continuously welded rails. The resultant forces from rail temperature variation were considered.
c. The Indian Railway's Concrete Bridge Code was followed to design RCC sub-structure and foundations, and prestressed concrete girders with internal and external prestressing as required.
d. Box Section, 'U' Section or any other efficient structural form was recommended for the main spans. The design was to provide a suitably designed sound barrier all along the corridor.
e. The superstructure was supported on single columns erected chiefly along the road's median. Portal frame support was considered at a few locations where single-column support was not feasible.
f. The overall width of the superstructure was 9.6 m, with the track centre at 3.7 m - 4.0 m.

g. The minimum clearance above the road level was 5.5 m. In the case of minor roads/streets, a lower level was adopted with the specific approval of the authority owning and/or maintaining the road/street. The vertical clearance above the rail level, where the alignment crosses the existing railway track, was provided by the Schedule of Dimensions of Indian Railways, to be not less than 5.8 m.
h. A provision for the passage of a width not less than 700 mm on both sides of the guideway along the corridor was maintained for emergency evacuation.
i. Provision for ducts for laying signals, communication, and illumination was kept.
j. Adequate drainage provision was kept.
k. Prefabricated elements were used to the maximum extent due to the need for fast-track work and the corridor's heavy congestion.
l. The structural form had to be aesthetically pleasing.
m. Approach slabs between the viaduct and the grade section were provided.

Fig.12.8: Stations

12.19.1 Superstructure

The Hyderabad Metro proposed an elevated corridor. The structure was chosen considering its ease of constructability, the use of standard elements, and the typical formwork for various structural elements. The following types of superstructures were considered.

i. Precast segmental box girder using external unbounded tendon.
ii. Precast segmental U-Channel superstructure with internal pre-stressing.

Segmental construction was chosen due to the following,

* Segmental construction is an efficient and economical method for an extensive range of span lengths and types of structures. Sharp curves and variable superelevation could be accommodated.

Fig.12.9: Hyderabad Depot Internal- Views

Fig.12.10: Depot Workshop

Fig.12.11: Elevated Corridor

Fig.12.12: Maintenance Bay -- Depot

* Segmental construction permits reduced construction time, as segments may be manufactured while substructure work proceeds and assembled rapidly afterwards.

- Segmental construction reduces environmental hazards. Segments are easy to stack and have more than one layer, saving space.

*Segments of moderate size are transportable by road trailers on city roads. The superstructurewas chosen considering its ease of constructability and the standardisation of formwork for a wide range.

Segmental construction was chosen mainly due to the following advantages.

- Segmental construction is an efficient and economical method for an extensive range of span lengths and types of structures. It accommodates structures with sharp curves and variable superelevation.
- Segmental construction permits reduced construction time as segments may be manufactured while substructure work proceeds and assembled rapidly afterwards.
- Segmental construction reduces the environmental hazard. The superstructure is manufactured away from busy areas and placed using a system erected from piers at heights.
- Segments are easy to stack in the yard in more than one layer, thereby saving in the requirement of space.
- Segments are transportable from the casting yard to the site in Trailer Trucks as the size and shape can be accommodated in available Trucks/Trailer Trucks.

- It can incorporate last-minute changes in span configuration if needed.
- Interference to existing traffic during construction is significantly reduced.
- Segmental construction contributes to an aesthetically pleasing appearance.
- The overall labour requirement is less than that for conventional methods.
- Better quality control and safety than conventional construction methods.

The viaduct superstructure is supported on a single cast-in-place RCC pier. The profile and size of the piers were designed to match the superstructure profile. The piers were circular, 1.6m in diameter, and flared at the top. A Jersey crash barrier of 1.0m height above the existing road level was provided all around the pier to prevent the direct collision of vehicles. A gap of 25mm was kept between the crash barrier and the outer face of the pier. The shape of the upper part of the pier is planned to provide a required clearance of 5.5m. The centre-to-centre spacing of the elastomeric/pot bearing is 1.8m.

12.19.2 Structural System-Superstructure

The superstructure is supported on a single row of RCC piers erected mainly on the central median. The spacing of piers is mostly 25 m. Longer/continuous spans, pier locations, and superstructures were adjusted to suit the site's requirements. RCC portal frames or other suitable support systems were necessary when the alignment deviated from the central median

Fig.12.13: Track Abutting Existing Road

The decision regarding the erection system for the precast segments was made by the Concessionaire, which may include the following.

i. Span-by-span erection with launching gantry.
ii. Balanced cantilevering erection with launching gantry.
iii. Full span pre-casting and erection; and
iv. Incremental launching.

Static and dynamic rail live load responses at essential movement joint locations complied with the design requirements. In accordance with international practice, a rail-structural interaction analysis was conducted using a continuous welded rail with direct fixation to the structure.

12.19.3 Substructure

The superstructure is designed to be supported on a single cast-in-place RC pier. The pier gradually widens at the top to accommodate the standard spans that support the bearing under the box webs. Piers are circular for most of their height to minimise ground-level space, allowing it to be accommodated within the existing central verge. A New Jersey-shaped crash barrier of 1.0m above the existing road level was provided to prevent direct collisions between vehicles and the pier. A gap of 25mm has also been provided between the crash barrier and the outer face of the pier. The longitudinal centre-to-centre spacing of the elastomeric/pot bearing is 1.8m. The space between the elastomeric bearings is placed in the lifting jack required to replace the elastomeric bearing. An outward slope of 1:200 was provided at the pier top for drainage., The transverse spacing between bearings would be 3.2m.

12.19.4 Photographs Showing Activities During Construction

Fig.12.14: Excavation for Metro rail footing

Fig.12.15: Laying of Lean Concrete

The proclaimer excavated foundations 10m x 8m x 4.5m.

After excavation, the ground at the bottom was levelled, followed by laying a layer of low-strength concrete. A mixture of fly ash, stone dust, and cement was laid and cured for at least seven days.

Fig.12.16: Laying of Reinforcement in Foundation

Fig.12.17: Reinforcement details of Pier

Figures 12.16 & 12.17 show the laying of the reinforcement for the foundation, of size 6m x 5.5m x 0.9 m, and Figure.12.18 includes a pedestal and pier. Fe500 grade TMT bars of 20mm dia. with 10mm dia. stirrups were used for the reinforcement. The ready-mix concrete used was M40.

Fig.12.18: Footing after curing and reinforcement

Fig.12.19: Concrete Pedestal & Pier Reinforcement

Pedestal size was 3.6m x 3.6m x 0.6 m.

Fig.12.20: Pier Shuttering and Staging

Fig.12.21: Pier after concreting

Shutters were provided for pier construction up to the top curvature, and curvature shuttering was also used. After the shutters were provided, a boom-pressure concrete Pouring Vehicle poured the concrete. Concrete was poured into a 1.8m x 1.8m pier, up to 1m in the first step, and then the remaining 1m or 2m for the pier height.

Fig.12.22: Concreting of Pier using Pump Concrete

12.19.5 Structural System

a. The guideway superstructure shall be supported by a single row of RCC piers erected mainly on the central median. The piers' spacing shall be less than 25 m unless otherwise justified. Where field restraints dictate the use of longer/continuous spans, pier locations and superstructure may be adjusted to suit the proposed span lengths.
b. Where the alignment deviates from the central median, RCC portal frames or another suitable support system may be necessary at some locations. The size of piers and crash barriers at ground level shall be within the median width.
c. The span arrangement for the viaduct, as indicated in the DPR, is for general guidance only. The Concessionaire shall determine the final span arrangement and structural system, considering the ground realities and a feasible, fast-track erection system. For a pre-cast segmental viaduct, a decision for the erection system shall be taken by the Concessionaire, which may include the following methods adopted internationally:

    i. Span-by-span erection with launching gantry.
    ii. Balanced cantilevering erection with launching gantry.
    iii. Entire span pre-casting and erection; and
    iv. Incremental launching.

    The Concessionaire shall provide, by suitable choice of span lengths, a sufficiently stiff deck and supporting substructure to resist loading as defined in the above-mentioned paragraphs.

Static and dynamic rail live load responses at essential movement joint locations shall comply with the design requirements.

Rail/Structural interaction analysis due to continuous welded rail with direct fixation on the structure shall be performed in accordance with proven international practice.

12.20 Design Loads

12.20.1 General

The railway loading applied to structures on the Project shall be guided by the selected Rolling Stock parameters and generally be in accordance with BD 37/88 (revised version of BS 5400: Part 2) or IRS Bridge Rules to the extent applicable below.

Nominal Loads: For computing stresses and deformations, the following loads, Accommodating Structures and consequential effects shall be considered as applicable:

i. Dead loads.
ii. Superimposed dead load.
iii. Live loads.
iv. Dynamic effects.
v. Forces due to curvature or eccentricity of track.
vi. Temperature effects.
vii. Frictional resistance of expansion bearings.
viii. Longitudinal forces.
ix. Long welded rail forces.
x. Racking forces.
xi. Forces on parapets.
xii. Wind pressure effect.
xiii. Forces and effects due to an earthquake.
xiv. Erection forces and effects (Construction Stage Safety check);
xv. Buoyancy; and
xvi. Differential settlement.

12.20.2 Loading Combinations

The various combinations of loads and forces to which components of the structures can be subjected of IRS CBC. Each structure element shall be designed/checked for all applicable combinations of these loads and forces. They shall resist the effect of the worst combination.

The loading combinations indicated are not exhaustive. The Concessionaire shall analyse the effects of any other combination as deemed appropriate.

12.20.3 Loading effects

The design shall include the effects of

a. **Static Loading:** These shall consist of loads due to:

   i. Track bed: RCC blocks or concrete pour or pre-cast slabs in RCC with inserts and fittings in case of ballastless track or PSC sleepers over ballast for ballasted track; and

   ii. other loads: As per Indian Railway Standards (IRS) and Bureau of Indian Standards (BIS).

b. **Fatigue Loading:**

   The nominal loading for the design of members, as per BS 5400: Part 10, shall comprise Trains with eight individual cars each.

c. **Dynamic Loading:**

   The static and fatigue loading given above shall be multiplied by an appropriate dynamic factor. Dynamic loading shall not be applied to piles, pile caps, centrifugal loads or braking/traction loads.

d. **Longitudinal Loads:**

   i. Longitudinal loads from braking and traction shall be appropriate percentages of live load per track as per the IRS Bridge Rules.
   ii. When a structure carries two tracks, both tracks shall be considered occupied simultaneously. Traction forces shall act on one track, and braking forces act on the other, with both acting in the same direction to produce the worst loading condition.
   iii. Longitudinal forces acting on the track shall be dispersed through the track before being transmitted to the substructure. This shall be calculated based on IRS Bridge Rules, IS Codes and relevant BS Codes.
   iv. Provision shall be made for the effect of horizontal and longitudinal forces in the rail, especially in the girders with ballastless deck.
   v. Forces shall be calculated for continuous welded rail with structure interaction resulting from the temperature differential of rail and concrete, and
   vi. Longitudinal forces shall consider the effects on stability and safety arising from a broken rail on ballast-less track.

e. **Derailment Load:**

   Check for derailment loads shall be made in accordance with the IRS Bridge Rules.

Wind Loading:

The viaduct structure shall be designed for wind loading in accordance with IS 875.

Temperature Loading:

Overall temperature and differential temperature effects shall be determined as per the provisions of IRS or IRC Codes.

Seismic Loading:

Seismic effects shall be considered on all structures, except culverts, consistent with a horizontal acceleration. Seismic forces will be considered to act in any horizontal direction and in the vertical direction. It is also required to check the structures for seismic forces as per IS: 1893-2001.

Erection Forces and Effects:

The weight of all permanent and temporary materials, together with all other forces and effects that can operate on any part of the structure during erection, shall be considered. Allowance shall be made in the design for stresses caused in any member during erection.

Shrinkage and Creep:

Provision shall be made for the effects of shrinkage and creep of the concrete in the structure.

Differential Settlement:

The forces resulting from differential settlement shall be considered where the nature of the chosen foundation system and the ground conditions indicate that such a condition may arise but not more than 10 mm long-term settlement and 6 mm short-term settlement.

Design Criteria for Steel / Concrete composite girder

Guiding of Height structure	BS: 5400 (Part 3 & 5) Code Varying. To be decided as per the site requirement
Grade of Concrete Deck Slab M45 Grade of Reinforcement	Fe415 or Fe500 HYSD bars conforming to BIS standards
Grade of structural Steel for Bottom Flange, Top Flange, Webs and Stiffeners	Grade Fe540B
Grade of Structural Steel for Deck Plate and Rolled Sections	Grade Fe410B
Permanent Bolts for Site Splice of Box Girder Grade 8.8	
Permanent Bolts for Connection to Bracings and Cross Frames Grade Grade 4.6	
Grade of Structural Steel for Headed Stud Connector	Grade 540B
Construction stage loading	1.5 kN / m2

The Concessionaire shall propose any additional significant design parameters in consultation with the IE.

Parapets

Parapets shall be provided along the entire length of the structure on both sides of all viaducts. They shall be designed as the support structure to the railway cabling and junction boxes as appropriate. Parapets shall be designed to resist a horizontal and a vertical force, each of 150 kg/m applied simultaneously to the top of the railing or parapet. Parapets shall be provided for all transition structures to protect the guideway from intrusion by trespassers, vandals and road vehicles. Parapets may be designed to function as Noise Containment Barriers. Parapets shall be designed to cater to the forces of Signal masts (if any).

Vertical profile

Profile grade

The superstructure shall be so designed that, when subject to dead load only, the rail level would be above the theoretical vertical profile of the system by an amount equal to the permissible LL deflection for the structure. Provision for super-elevation shall be made as part of the track structure over the deck.

Camber

The superstructure deck, including the soffit of any overhead structure above the deck, shall be cambered to compensate for the combined effect of vertical curvature, if any, and dead load deflection.

Span/Depth Ratios

The span-to-depth ratio should, as far as possible, be restricted as indicated below:

Reinforced concrete member – 10

Pre-stressed concrete member – 14

Composite members – 16

For Box girders, these ratios shall be further subject to stipulations regarding internal dimensions required for inspection and future prestressing.

Design surface crack width

For the serviceability limit state of cracking, the design surface crack width of reinforced concrete and pre-stressed concrete viaduct structures shall not exceed the values given in BS 5400.

12.20.4 Temperature Effects

Where applicable, temperature effects shall be considered by the requirements of BS 5400 Part 2.

Minimum thickness of members

Desirable minimum thickness of any concrete member

Deck – 200 mm

Web of T-beam – 250 mm

Web of pre-stressed girders – 150 mm + d

If there are 2 cables at any level – 150 mm + 3d

(Where d is the outside diameter of the cable duct)

Box Girders: minimum thickness of member:

Deck slab – 200 mm

Bottom flange – 300 mm

Web – 250 mm

Or as required by IRS Concrete Bridge Code, whichever is greater.

Vibration and Deflection Limitations

The amplitude and frequency of vibrations of the viaduct structure shall be limited to international standards. The overall deflection for an elevated structure will be limited, considering the effect of vibration, as well as other factors.

12.20.5 Bearings

Bearings for the viaducts were designed to accommodate the span (between center to center of bearing in the longitudinal direction) of the different loading combinations. Suitable provisions were to be kept for bearing replacement and any repairs during service.

Bearings were designed in accordance with the requirements of BS 5400.

12.20.6 Stray Current Corrosion Control

Precautions per EN 50122-2 or equivalent to minimise stray current corrosion caused by DC traction power return through rails. A continuous electrical path shall be ensured by providing a complete and reliable electrical connection throughout the structure. The electrically continuous path shall be provided through the steel reinforcement, either by constant welding of structural reinforcement or by providing additional welded mesh reinforcement. Where a weld is used to form a grid, welded cross connections shall be at a minimum spacing of:

i. For longitudinal bars, 600 mm was measured in the transverse direction, and
ii. For transverse bars, 6 m was measured in the longitudinal direction.

The Concessionaire shall demonstrate during construction that the electrical continuity between all metallic structures has been achieved.

12.21 Proof checking

All the general arrangement designs prepared by the Concessionaire's design consultant were submitted to the Authority or his representative for review and approval. The Concessionaire was supposed to take up all the detailed design/construction drawings, which would be proof checked by a reputed consultant appointed by the Concessionaire with the authority's approval. It was the sole responsibility of the Concessionaire to arrange all proof-checked "Good for Construction Drawings."

12.22 Casting of Segments

Segments were cast in a single pour. All segments were marked on the inside with a unique identification.

Tolerances for Finished Segments of pre-cast box girder as shown below:

Length of match-cast segment – 25 mm

Length of totally assembled span – ±2.5 mm

Web Thickness – ±10 mm

Depth of Bottom Slab – ±10 mm

Depth of Top flange – ±6 mm

Overall Top Flange Width – ±25 mm

Diaphragm Thickness – ±12 mm

Tendon hole Location – ±3 mm

Position of Shear Keys – ±6 mm

Handling of Segments

The Concessionaire handled, lifted, stored, transported, and erected all segments properly without damage. Segments were always maintained in an upright position and stored, lifted, and/or moved to prevent torsion and other undue stress. After only 14 days, the segments were transferred to the store at the casting yard, and the concrete reached the required strength.

Segments were moved on the road at night. Enough Mobile cranes of adequate capacity and boom length were mobilised. Low-bedded trailers transported and moved the segments at ground level. Launching girders (LG) involved lifting and assembling the segments. An adequate number of LGs for erection was mobilised to satisfy the target milestone requirements.

A full-scale load test of the Launching Girders, lifting, and temporary holding hardware was performed to demonstrate the launcher's adequacy before beginning the erection of the segments.

12.23 Safety During Erection and Data Sheets

The Concessionaire enforced a strict Safety System with all necessary precautions and instructions, including safety tools, nets, railings, personal protective equipment, and workers, as well as first aid. Survey Control during Pre-casting operations and erection

The data sheet for each segment was recorded and included essentially:

i. Theoretical basic data supplied by the design office, allowing the preparation of the horizontal alignment and the two parallel bolt lines.
ii. Corrected values defined graphically/with computer aid.
iii. Survey control readings.
iv. Linear measurements on the segments.
v. Schematic representation of the segment to rapidly verify the relative positions of the segment axes.
vi. Level checks to pick up any gross error in level readings on the same segment, comments on the casting operations, and
vii. Survey control during construction.

12.24 Design Checks During Construction

In addition to the design calculations for permanent structures, the Proof Checking Consultants received design calculations for falsework, erection devices, formwork, and other temporary construction elements.

i. Construction Stage Stress Check in the superstructure during all stages of erection using Launching Girders as per the Concessionaire's proposed erection scheme.
ii. Calculations for pre-camber correction required in pre-casting moulds/beds, taking care of all post-tensioning forces, temperature, creep, shrinkage, etc., computations of deflections and required camber due to dead loads, post-tensioning forces, creep and shrinkage.
iii. Computation of jacking forces is required at pre-cast glued joints during temporary post-tensioning, resulting in compressive stress at the interface of segments.

12.25 Load Tests

The Authority might have instructed the Concessioner to conduct load tests, covering UPV, cover meter, rebound hammer, etc., during the work progress or maintenance period.

a. The site-made concrete test cubes failed to attain the specified strength.
b. The shuttering for concrete works was removed prematurely.
c. Overloading during construction of the Works or part thereof.
d. Concrete improperly cured.
e. If any portion of the identified work was carried out without prior approval in writing of the Engineer (Authority) or his representative.

f. Any other circumstances attributed to alleged negligence on the part of the Contractor which, in the opinion of the Authority, result in the works or any part thereof being of less than the expected strength.
g. Any reason other than the foregoing when there was a fear of safety being jeopardised.

12.26 Stations

Fig.12.23: View of a Completed Station

The Stations are designed site-specific, reflecting local culture and flavour, functional aesthetics, user-friendliness, energy efficiency, and environmental friendliness. The design features significant architectural elements, including stations covered by an overlapping roofing system that is open on the sides, which also covers the platform, as well as staircases and escalators designed with functional aspects in mind.

The structural system for balanced cantilever (Spine & Wing) stations is unique. Station piers are located at the central median and are cantilevered on a single column. The structure consists of a concourse floor at the first level and a platform at the top level. Both floors cantilever out from the central columns, providing clear space for road traffic at ground level.

The stations' layout is commensurate with track geometry, operational requirements, predicted passenger flows, ROW availability, and electrical and mechanical requirements. They are divided into public and non-public areas. The layout follows state-of-the-art Railway engineering equipment installations at the station concourse level for smooth Metro Rail operations and Passenger Safety.

Ticketing Counters/Booths and ticket vending machines have been provided at all Stations, making the ticketing process simpler and faster. Pedestrian facilities have been provided from the road to the concourse/car park and/or to the ticket counters to enable the unhindered and smooth flow of pedestrians. Facilities for people with disabilities are provided at all stations to ensure their safe movement between parking areas and platforms. Lifts and escalators are also offered.

Public access telephones are provided at all Stations.

Signages are designed to provide instant message transfer to passengers, from street-level guiding at the concourse level to the platform level.

Each station features advanced equipment for commuter security, including baggage screening devices, handheld devices for CCTV monitoring, and an access control system. Station planning is based on normal passenger flow and emergency evacuation scenarios. Required lifesaving facilities such as fire protection, emergency lighting and signage, PA and PIDS, auto-opening of ticketing gates, sizing of the staircases, etc.

12.27 Permanent Way

The permanent way consists of rails of section UIC 60, grade 1080, Head Hardened for the complete corridor and rails of section UIC 60, grade 880, for depots, which are 18/13 meters long. Restraining rails of section 33 C1, grade 880, 13 m long, have been used for all curves of radius less than 190 m.

Rails for the corridor are continuously welded by Flash Butt Welding/Aluminothermic Welding processes and laid on plinth-type ballastless track, with rails discretely supported by fastenings on Concrete plinths of M35 grade

Fig.12.24: Pier & Pier cap for Top-Down Constructioin

The ballastless rail fastenings are spaced at 650 mm c/c on the viaduct and ramp, with a 1 in 20 rail inclination. The track in the depots shall be laid on ballasted track with Standard Gauge Monoblock PSC Sleepers, with a density of 1540 sleepers per Km using an ERC MK III fastening system, GFN Liners of Indigenous make, and 65 mm Stone ballast with a minimum cushion of 250 mm under the sleepers. The cushion for the test track shall be 300 mm. A portion of the depot track should be laid as an embedded track with special base plates. All the turnouts have a 1 in 20 rail inclination. 1 in 9 switches of radius 300 and 190 for the corridor and 1 in 7 switches of radius 190 in the depot with Zu 60 profile (thick web) have been used.

12.28 Rolling Stock

The rakes procured are standard-gauge rolling stock, air-conditioned, and made of lightweight stainless steel/aluminium-bodied three-car formations, with a trailer car between two motored driving cars. Internal wide gangways provide ease of passenger movement and assist load distribution. LCD screen type route map indicators are installed over all the doors, and LCD TVs are placed at suitable locations inside the cars to provide infotainment. The safety of commuters and operations is paramount, and the trains have foolproof safety features and onboard fire & smoke detection.

The trains have CCTV cameras in and outside the cars, mobile and laptop charging sockets inside the vehicle, better humidity control, microprocessor-controlled brakes, secondary air suspension for better ride comfort, and will be capable of maintaining an average speed of 33 km/h and a maximum operating speed of 80 km/h. The trains with regenerative electric braking convert the momentum into electrical energy and feed it back to the power supply system while braking, resulting in a contribution towards the CDM and reducing the energy requirement of the grid.

The maximum capacity per three-car train is approximately 965, including longitudinal seating, resulting in a high proportion of the floor area dedicated to standing passengers. Each car will have longitudinal seats for approximately 40 persons.

A total of 171 cars were planned to be procured for 57 trains as part of the project's first phase. Although with many automated functions, the ATO-enabled trains will be under driver control. All salient features are planned to add to the comfort and safety of the commuter. Aesthetics combined with comfort and safety will win the metro users. Hyundai Rotem (Korea) was chosen not only for its latest technology but also for its cost-effective maintenance, service, and availability of spare parts.

12.28.1 Salient features of Rolling Stock

Rolling Stock for Hyderabad Metro

* Train Set3 Cars train set (DMC-TC-DMC), extendable to 6 Cars train set (DMC-TC-MC-MC-TC-DMC).
* Gauge Standard Gauge (1435 mm).
* Traction25 kV AC Overhead Catenary, Single phase and 50 Hz frequency.
* Maximum Design Speed90 Kmph, Maximum Operating Speed80 Kmph, Average Speed 33 kmph.
* Headway is 3 minutes during peak and 15 minutes during lean hours. Dwell Time 20 seconds at stations. Maximum Axle Load 17 Ton with 8 Passenger/Square metre.
* Automatic Train Protection (ATP)/ Automatic Train Operation (ATO)Automatic.

Train Supervision Passenger Emergency Alarm Passenger Addressing System

* CCTV in cars, Saloon Door opening and closing, Alarm Fail Safe Pneumatic
* Smoke and fire detectors in the driver's cab and saloon car:
* Fire extinguishers inside vehicles and the driver's cab.
* Emergency brake
* Platform screen door.
* Rolling Stock Car Body Light Weight Stainless Steel/Aluminium.
* Energy regeneration during electro-dynamic braking.
* Dedicated space to accommodate wheelchairs (for people with reduced mobility).
* Longitudinal Seats, Grab Poles and Rails for standing passengers
* Air-conditioned with Humidity Control
* Mobile and Laptop Charging Points, LCD TVs for Entertainment, information and advertisement.
* Dynamics Route Display.
* Exterior Modern and Aesthetic, unpainted, covered with vinyl sheet for advertisement
* Current Collection System Overhead Pantograph Electro-pneumatic friction brake system (EP)
* Electric-regenerative brake system and others

12.29 Signaling and Communication

The Hyderabad Metro has adopted the advanced signalling and train Control technology, Communication-Based Train Control (CBTC), to control the trains. It will be the first in India to claim train control by CBTC technology. The trains running on three corridors will be controlled and monitored from Uppal's state-of-the-art Operation Control Centre (OCC).

The trains can run with a headway of 90 Sec to tackle commuters during office peak hours in the morning and evening. The overview display panel facilitates all three corridors and indicates the current locations of all trains running on these corridors. The OCC is the primary control centre, and a backup control centre at Uppal depot is in place to take over the operation of the OCC in the event of an evacuation due to unforeseen circumstances.

The Fall-Back system is also adopted as a secondary train detection system in all three corridors in the unlikely event of CBTC operation failure.

The Uppal and Miyapur depots have facilities to can be controlled and monitored from independent control centres at Uppal and Miyapur, respectively. Test tracks at the Uppal depot facilitates the testing of trains to ensure safe and reliable train operation.

The normal mode of operation of trains is automatic train operation (ATO). The Automatic Train Protection (ATP) system continuously monitors safe train operation and initiates necessary action if a train doesn't perform as required. Station equipment, such as Computer-Based Interlocking (CBI) and Wayside ATP, is vital signalling equipment redundantly arranged to ensure safe and uninterrupted train operation.

12.30 Power Supply, Traction and SCADA

The Hyderabad Metro consists of three elevated corridors with a standard-gauge system. Electrification is achieved through 25 kV, single-phase, 50 Hz overhead equipment, which transmits electrical energy. A Booster Transformer is provided throughout the corridors at specified locations to minimise interference with communication lines.

Receiving Substations (RSS) are located at Uppal, Miyapur, Yosufguda, and MGBS, which will receive the power supply from the proposed APTRANSCO at a 132 kV voltage level.

This 132 kV supply will be divided into 3 phases: 33 kV for lighting & general purposes, and a single phase of 25 kV for traction purposes.

All RSSs are AIS (Air-Insulated Substations), and TSSs are GIS (Gas-Insulated Substations). ASS (Auxiliary Substations) is located at each Metro Station, Depot, and OCC to provide LV power supply with 33 kV/415 V for S&TC, Communications, MEP, etc.

For maintenance purposes, OHE (overhead equipment) is to have sectioning, sub-sectioning, and paralleling posts at various locations.

12.31 SCADA System

Supervisory control and data acquisition systems provide the highest monitoring, controlling, and command levels. A centralised OCC (Operation Control Centre) shall be located at Uppal Depot to control the entire Power Supply and Traction system Equipment through SCADA.

Fig.12.25: Top View of Depot

12.32 Depot

The Hyderabad Metro system was planned with three depots, one for each line, for stabilisation and maintenance. These maintenance depots are adequately manned and equipped with all facilities required for efficient and effective maintenance.

The depots planned in the 72 km network are located at:

- Miyapur for Corridor 1
- Falaknuma for Corridor 2
- Uppal for Corridor 3

The Uppal depot is the main depot cum workshop where the overhaul and corrective maintenance facilities for the whole fleet are planned. The other two depots have basic facilities for scheduled preventive maintenance and minor corrective maintenance.

- Automatic train wash plant
- Inspection and workshop bays
- Offices
- Depot control center
- Covered & open Stabling lines
- Infrastructure maintenance facilities
- Electrical & Mechanical workshops
- Electronic repair shops
- Open and outdoor storage facilities
- Wheel profiling lathe
- M&P for repair & overhaul
- Training, conference rooms, Cafeteria etc.

To allow trains to be in service, they are cleaned, inspected, checked, and certified 'fit-for-service'. This requires a thorough analysis of data downloaded from the train, a visual inspection by a team of engineers and technicians, and the successful completion of onboard tests of vital safety systems and other comfort functions of the train.

The facility of the depot includes,

- Automatic train wash plant
- Inspection and workshop bays
- Offices
- Depot control center
- Covered & open Stabling lines
- Infrastructure maintenance facilities
- Electrical & Mechanical workshops
- Electronic repair shops

- Open and outdoor storage facilities
- Wheel profiling lathe
- M&P for repair & overhaul
- Training, conference rooms, Cafeteria etc.

To allow trains to be in service, they are cleaned, inspected, checked, and certified 'fit-for-service'. This requires a thorough analysis of data downloaded from the train, a visual inspection by a team of engineers and technicians, and the successful completion of onboard tests of vital safety systems and other comfort functions of the train.

12.33 Automatic Fare Collection System

The ticketing solution proposed is a state-of-the-art Automated Fare Collection System based on Contactless Fare Media Technology. The fare collection system divides the Metro Stations into two areas: 'paid' and 'unpaid'. Automatic entry/exit gates, along with fencing, separate the paid and unpaid areas. This ensures that only individuals with valid tickets are allowed to enter the system. The system is designed as a closed one, with ticket checking at both entry and exit, thereby ensuring 100% ticket check-in within the system.

12.34 AFC System Key Features

- State-of-the-art Automatic Fare Collection System
- Technologies best suited for MRTS and recognised worldwide
- Web application to top-up the contactless smart card through the internet
- System accepts banknotes, coins, bank cards (credit/debit) and smart cards as a payment mechanism
- TVMs facilitated with Coin recyclers and Bank Note recyclers
- Integrated Parking Solution for future implementation

12.35 The proposed system comprises the following key sub-systems

- Automatic Gates
- Ticket Vending Machines
- Add Value Machines
- Ticket Office Machines
- Ticket Readers
- Portable Ticket Analysers

12.36 Fare Media & Fare Products

There are two types of tickets: Contactless Smart Card (CSC) and Contactless Smart Token (CST). Contactless Tokens are pre-paid fare media for journeys between fixed stations. They can be purchased for a single or return trip and are a more suitable fare media option for non-frequent users.

Contactless Smart Cards are stored-value cards. They are available in either monetary (e-Purse) or non-monetary (Pass) form or a combination of both, with various options to attract regular and frequent users. The value equal to the fare between the two stations is deducted from the stored value at the exit Gate.

12.37 A few of the proposed fare products are listed below

- Electronic/Stored Value Purse (e-Purse)
- Tourist Pass
- Trip Pass (with the options like 10/30/50, etc.)
- Daily Pass
- Weekly Pass
- Monthly Pass
- Holiday Pass

The fare media can be procured from the Ticket Offices or Ticket Vending Machines.

12.38 Telecommunication System Key Features

- A state-of-the-art communication system facilitates continuous communication between central control, train operators, and station controllers.
- Modular, Reliable, Redundant and Proven Systems
- Distributed Architecture—If the link between OCC and the Station fails, all Station Communication equipment shall continue to operate independently in Local mode.
- Gigabit Ethernet-based IP Backbone System—All Communication Subsystems shall support IP and be integrated through the GE (IP) Network.
- Broadcast & display of train-related information in different languages
- Emergency Call Point / Help-point phone at stations to assist passengers if any help is required.
- PoE-based complete IP-based CCTV surveillance of all passenger areas for all stations.

12.39 Conclusion

Green & Eco-friendly mode of travel – reduces carbon emission, fuel consumption and pollution. Faster, safer and more comfortable air-conditioned travel with reduced travel time. Ultra-modern coaches. High frequency of trains- reducing waiting time. User-friendly stations with lifts, staircases, and facilities for people with disabilities. Parking facility at 25 strategic locations along the route. Automatic ticket vending machines reduce waiting time in queues and counters. The automatic fare collection system provides free entry and exit from the stations. Station facilities include toilets, public addresses, information systems, and telephones.

CHAPTER 13

Chennai Metro

13.1 Introduction

This section is an extension of the brief included in Part 1 (Vol 1) of the book.

Chennai, the fourth-largest metropolis in India, is a hub of economic and social development, as well as cultural activity. It is growing at a rapid pace. To achieve the objectives of making this metropolis more liveable and of international standard, a coordinated and integrated approach among the various agencies involved in urban development and service provision is needed, including a participatory process in planning and implementation.

The first Master Plan laid down policies and programmes for overall development and was approved in 1976. It addressed the distribution of future population, policies for economic growth, the location of economic activities, physical development, circulation patterns, plans for traffic & transportation, and land use plans. The population of CMA, as per the 2001 census, was 70.41 lakhs, which has increased multi-fold in the last two decades. Regarding Chennai city, a detailed development plan for 52 planning units has been prepared.

The study area for the Assignment is the administrative boundary of the Chennai Metropolitan Area (CMA), which encompasses Chennai city and parts of Tiruvallur and Kanchipuram districts, covering an area of 1,189 sq. km. CMA comprises the City of Chennai, 16 Municipalities, 20 Town Panchayats and 214 Village Panchayats in 10 Panchayat Unions. According to the 2011 Census, the population of the CMA and the City area is 89.2 Lakh and 46.8 Lakh, respectively. CMA and City areas have registered annual growth rates of 2.4% and 0.8%, respectively, over the decade from 2001 to 2011. Large-scale urbanisation in IT/ITES and industrialisation, accompanied by the rapid growth of the vehicular population, has placed severe stress on the urban transport system in the city. The city has approximately 48 lakh vehicles, according to Tamil Nadu government vehicle statistics. The usage of private modes is increasing unabated, mainly due to inadequate public transport facilities. Phase I of the Chennai Metro covers 54 km in two corridors: Washermenpet to Airport (23.085 km) and Chennai Central to St. Thomas Mount (21.96 km), with an extension from Washermanpet to Wimco Nagar (9 km) in Thiruvottriyur. A stretch of 10.7 km from Koyambedu to Alandur in Corridor 2 became operational in June 2015. Additionally, Little Mount to Airport (7.7 km) & Alandur to St. Thomas Mount (1.3 km) in Corridor 1 opened in September/October 2016. Furthermore, sections from Thirumangalam to Nehru Park (8.0 km) in May 2017, as well as Little Mount Station to AGDMS

Station and Egmore to Chennai Central Station (7.3 km) in May 2018, are also operational. With a view to developing an effective and efficient mass transit system in addition to the existing public transportation and Phase-I Metro rail system, a Feasibility Report for Phase II Corridors with a total length of 88.9 km was prepared by CMRL in 2015. The Detailed Project Report for Chennai Metro Rail Phase-II Corridors, covering 107.55 km across 3 corridors (C3, C4, and C5), was prepared in March 2017. A Detailed Project Report for the Extended C4 corridor (total length 26 km) was also prepared in October 2018. Govt. of Tamil Nadu, through CMRL.

The Chennai Mass Rapid Transit System (MRTS) between Chennai Beach and Chepauk opened in 1995 as India's first elevated rail line. The MRTS was extended to Thirumayilai in 1997, Thiruvanmiyur in 2004, and Velachery in 2007. In 2006, a modern metro rail system for Chennai was planned, to be modelled after the Delhi Metro.

Fig.13.1: Chennai Metro Phase II (currentindia.com)

Fig.13.2: Chennai Metro Route Plan (housing.com)

The **Chennai Metro** is operated by Chennai Metro Rail Limited (CMRL) – a joint venture between the Government of India and the State Government of Tamil Nadu. The network consists of two lines and spans 54.1 km. The system uses standard gauge and has 41 underground and elevated stations. Planning for the Metro started in 2007-08, with construction commencing in February

2009. After testing in 2014, the first segment between the Alandur and Koyambedu stations on the Green Line began operations in June 2015. In September 2016, operations on the Blue Line commenced between Chennai Airport and Little Mount. The underground stretch from AG-DMS to Washermanpet of the Blue Line opened in February 2019, completing phase 1 of the Metro. Three more lines covering 118.9 km are under construction as part of the second phase. A light rail has been proposed to connect Tambaram station of the Chennai suburban rail network with Velachery on the MRTS. Further extensions are also planned for the three lines under construction. The phase-wise details are included in the following paragraphs.

Fig.13.3: Aerial View of Chennai Metro (CMRL)

13.2 Phases

The Chennai Metro Rail Project has been implemented in different phases as detailed below.

Phase I

Phase I of the Chennai Metro Rail project spans a 45.046 km network and comprises two corridors: Corridor-I, which extends from Washermenpet to the Airport (23.085 km), and Corridor-2, which runs from Chennai Central to St. Thomas Mount (21.961 km). About 55% of the Corridors in Phase I are Underground, and the remaining are elevated. Out of 32 stations, 13 are Elevated Stations and the remaining 19 are Underground Stations. Both Corridor interchange Stations are available in Alandur (Elevated) and Chennai Central (Underground) respectively.

Phase I was commissioned in a stage-wise wise. The first segment of the section between Alandur to Koyambedu was commissioned in June 2015. The first Underground line between Thirumangalam Metro Station to Nehru Park Metro Station was opened in May 2017. The Underground stretches from Nehru Park Metro Station to Chennai Central Metro Station and from Saidapet Metro Station to AG-DMS Metro Station and was opened a year later in May 2018. In February 2019, the Underground stretch from AG-DMS to Washermenpet of the Blue Line was opened, completing a 45 km network under Phase I of the Metro Rail Project.

Phase I Extension covers a length of 9.05 Km from Washermenpet to Wimco Nagar. It consists of 9 stations and an Elevated depot. In this, 7 stations are Elevated, and the remaining 2 stations are Underground. The estimated base cost of this project is about 3770/- Cr (Rupees Three thousand seven hundred and seventy crores only). Out of this, 57% of the loan amount was granted by the Japan International Cooperation Agency (JICA) and the remaining 43% was contributed by the Central and State Governments. General Consultants who were awarded for Phase I of the project were assigned with the consultancy.

The Phase I Extension was commissioned and opened in February 2021, except for Thiruvottriyur Theradi Station and Wimco Nagar Depot Station. The remaining two stations were commissioned and opened for the Public in March 2022.

Phase II

Phase II expansion has been planned for a 118.9 km network with 128 stations. It consists of three corridors: Corridor-3, spanning from Madhavaram to SIPCOT (45.8 km); Corridor-4, from Lighthouse to Poonamalle Bypass (26.1 km); and Corridor-5, from Madhavaram to Sholinganallur (47.0 km). The estimated cost of the project is Rs. 63,246 Cr (Rupees Sixty-three thousand two hundred and forty-six only), including IDC. This proposal is under the process and approval of the Government of India. The project is proposed to be completed by the end of 2028.

Table 13.1: Corridor Details

Corridors	Elevated (Km)	UG (Km)	Total Length (Km)
C-3 Madhavaram to SIPCOT	19.1	26.7	45.8
C-4 Lighthouse to Poonamallee Bypass	16.0	10.1	26.1
C-5 Madhavaram to Sholinganallur	41.2	5.8	47.0
Total Length	76.3	42.6	118.9

Corridors	Elevated (Nos)	UG (Nos)	Total Stations (Nos)
C-3 Madhavaram to SIPCOT	20	30	50
C-4 Lighthouse to Poonamallee Bypass	18	12	30
C-5 Madhavaram to Sholinganallur	42	06	48
Total Stations	80	48	128
Total Project Completion cost INR 63,246 Crores including IDC			

Phase II DPR

- The Government of Tamil Nadu (GOTN) have accorded approval for the implementation of three Metro Rail Corridors for a length of 118.9 Km under Phase II of Chennai Metro Rail Project.
- GOTN have recommended the same to the Central Government for funding under a 50:50 equity share mode and loan assistance from Bilateral /Multilateral agencies.
- The Proposal is approved by MOHUA – Government of India (GOI).

Phase II Funding Details

- Funding for a portion of Chennai Metro Rail in Phase II (i.e. 52.01 Km, i.e. Madhavaram Sholinganallur of Corridor 3 and Madhavaram CMBT of Corridor 5) has been tied up with JICA as a State Sector Project.
- Balance portion of Funding (66.89 Km): The Screening Committee of DEA has accorded in principle approval for funding of the entire Corridor-4 and balance section of Corridor 3 &5 and posed for funding from Multilateral Banks, i.e ADB, AIIB & NDB.
- Multilateral Banks have completed their preliminary Mission programme, and detailed discussions with MDBs are in progress.

Fig.13.4: Chennai Metro Phase II

13.3 General Consultants

- Consortium of Nippon Koei India Pvt. Ltd., Aarvee Associates Architects, Engineers & Consultants Pvt. Ltd. and Balaji Railroad Systems Pvt. Ltd were awarded a General Consultancy contract (GC1) for the 52.01 km JICA Corridor (Corridor-3 & Corridor-5) of the Chennai Metro Phase II project.

- JV of AECOM India Pvt Ltd., Oriental Consultants Global Co Ltd (OCG) & Nippon Koei Co. Ltd (NK) were awarded a General Consultancy contract (GC2) for the balance section of 66.89 km (C4, Part of C3 & C5 Corridors) of Chennai Metro Rail Project under Phase II.

  The Delhi Metro Rail Corporation (DMRC) prepared the project report on the metro system in Chennai. In February 2009, a contract was awarded for the construction of a 4.5 km long viaduct along the Inner Ring Road between Koyambedu and Ashok Nagar and in March 2009, the contract was awarded to a consortium of five companies headed by the French company EGIS for technical consultancy. In August 2010, a contract for supplying rolling stock was awarded to Alstom. In January 2011, a joint venture between Larsen and Toubro (L&T) and Alstom was awarded the design and construction of track works and a depot at Koyambedu. Construction of an elevated section in the first phase began in 2011. Contracts for the construction of underground sections of the first phase were awarded to Gammon India and Mosmetrostroy in February 2011. Subsequently, in June, tenders for the elevated stations of the first phase were awarded to Consolidated Construction Consortium. The contract for the supply of lifts and escalators was awarded to a joint venture of Johnson Lifts and SJEC Corporation. The contract for power supply and overhead electrification was awarded to Siemens, and contracts for automatic fare collection (AFC), tunnel ventilation, and air conditioning were awarded to Nippon Signal, Emirates Trading Agency, and Voltas, respectively. In July 2012, the first tunnel boring machine was launched, and by October that year, eleven machines were commissioned to bore tunnels by three consortia – Afcons-Transtonnelstroy, L&T and SUCG.

13.4 Salient Features

Gauge Standard	1435 mm
Route Length (Between Dead Ends)	
C-3 Madhavaram to SIPCOT:	Elevated-19.1 km Underground -26.7 km
C-4 Lighthouse to Poonamallee Bypass:	Elevated- 16.0 km Underground - 10.1km
C-5 Madhavaram to Sholinganallur:	Elevated - 41.2 km Underground -5.8 km
C-3 Madhavaram to SIPCOT-- Stations:	Elevated -20 Underground- 30
C-4 Lighthouse to Poonamallee Bypas--Stations:	Elevated - 18, Underground -12
C-5 Madhavaram to Sholinganallur – Stations:	Elevated - 41 +1 (At-Grade) Underground-6

Traffic: PHPDT – C3: 2025- 16289 2035 – 22115
– C4: 2025- 11707 2035 -18944
– C5: 2025- 17539 2035 – 24528

Stations: A Elevated 140 x 21.95 m 2-Level Cantilever
B Underground 190 x 21.80 m 2-Level Cut-and-Cover

C UG with Ext.Concourse 150x 21.4 m 2-Level Cut-and-Cover

D Underground 150x21.4 m 3-Level Cut-and-Cover

E Elevated (Split Concourse) 140x32.35 RCC Frame Str.

F Elevated (Split Concourse) 140x37.04 RCC Frame Str.

Design Speed: 80 Kmph

Scheduled Speed: 32 kmph

Headway: **C3** 600 sec.

C 514 sec

C5 450 sec

C4 277 sec

Rolling Stock: 3 Coach Unit, 2DMC + 1 TC 6 Coach Unit, DMC + TC + MC + MC + TC + DMC

Coach Dimension: 22m x 2.9m x 3.9m

Coach: Light-weight Stainless Steel / Aluminium

Axle Load: 16t

Braking System: Regenerative Braking

Propulsion System: 3-phase with VVVF control

Traction: 25 kV AC OHE system

Signaling: Communication-based Train Control System (CBTC) with 90-second Headway,

Telecommunication: Integrated System with Optical Fibre Transmission System, SCADA, CCTV, CVRS

Maintenance Depot: Madhavaram Depot, Poonamallee Bypass Depot

Fig.13.5: Elevated Corridor During Construction

13.5 Salient Features of Phase I Extension of Chennai Metro Rail Project

Phase I of the Chennai Metro Rail Project, with two corridors covering a total length of 45 Km with 32 metro stations, is currently in operation. Corridor-1 is the Blue Line from Washermanpet Metro to Airport, and Corridor-2 is the Green Line from Puratchi Thalaivar Dr. M.G. Ramachandran Central Metro to St. Thomas Mount Metro. The Phase-I extension project from Washermanpet to Wimco Nagar for 9.051 Km has now been completed. The newly inaugurated Phase-I Extension of the Chennai Metro Rail Project covers 9.051 km and consists of 2 Underground and 5 Elevated Metro Stations at the heart of North Chennai. Phase I of the CMRL Project covers a total distance of 54 km with 21 Underground and 18 Elevated Metro Stations. Hon'ble Prime Minister of India inaugurated the Phase 1 extension.

Fig.13.6: Washerman pet Metro Station

Fig.13.7: Hon'ble PM of India Inaugurates Chennai Metro Ph 1 Extension: at washerman pet

Fig.13.8: View During Inauguration of Ph 1 at Washermanpet

Fig.13.9: Internal Views of a Metro Station

Fig.13.10: Completed Elevated Station

The book includes the Chennai Metro, with sections detailing the author's team's involvement in Phase I extension and Phase II, specifically Corridor 5, Package 3, as lead design Checker for Systra, and in the detailed design for part of Package 5 for MMSB, Malaysia. The corridors designed by Systra included elevated and underground corridors, as well as stations, beside the Depot. The design and drawings, which included both corridors and stations —both underground and elevated —and the depot, were prepared by Systra. The Author's team reviewed and approved them. However, for the Tunnel Section, using a TBM was not within the scope of the author's team (except for a small Section in Phase II).

13.6 Geometry parameters of the station

- **Track level structure:** Precast I girder supporting the track resting on precast pier cap.
- **Station concourse:** post-tensioned / cast-in-situ pier cap to support cast-in-situ beams and precast beams at concourse level.

- **Station platform:** RCC cast-in-situ beams & rectangular columns to support the platform only.
- **Depot structure:** RCC cast-in-situ column beam.

Software used for analysis and design is Staad Pro, OASys (Adsec), MS Office, AutoCAD, and MIDAS-Civil.

The list of Design Codes and Standards used includes IRS, IRC, IS, Euro Code, AASTHO LRFD

13.7 Engineering Design

The structural configuration was designed to meet the Functional and Architectural requirements. The planning and design were based on the interface between electrical systems, equipment & machinery, architecture, firefighting, and other critical systems. The elevated stations were designed mainly as single pier-supported structures, i.e., they used a single row of piers at the center of station plans. The structural frame was planned with RCC piers, beams, and a reinforced concrete slab. The concourse slab is supported by Precast/Cast-in-situ RCC beams. The Track Structure consisted of a U-Girder supported on a track-level pier cap. The track and platform structures were separated by expansion gaps. All buildings of the Depot were designed as 3D reinforced concrete frame structures with beams and slabs. Block masonry was used for walls, both for internal and external purposes. Further briefs are added in the following paragraphs.

13.8 Design Methodology

The viaduct consists primarily of a Pre-tensioned U-girder with a circular pier and a pile group of a minimum. 1.0m diameter pile. Design method for Substructure and foundation is as below:

13.8.1 Design method of Pier, Portal beam, Pile cap and Pile for portal piers

For U-Girder spans, dead load reactions are taken from the output using software analysis. SIDL and LL are taken as per the design basis report. Pier, portal, pile cap and pile group are modelled as a space frame using STAAD or ST1 with supports as springs as given in IS:2911, duly modified for seismic case. For calculating the time, cracked inertia is taken as 0.75Ig for pier and pile only as per RDSO guidelines. Where firm soil strata are present, the stiffness of the pile group can be neglected while calculating the time, assuming the piles are firmly embedded in the soil. Based on time, seismic forces are calculated for the pier and foundation wrt the response reduction factor. Design of pier and pile is done using software OASYS for crack width, SLS stresses and ULS capacity. Pile cap is designed using worksheets for beam theory and truss theory. The point of application of longitudinal forces will be at the seismic restrainer level, and the point of application of all the transverse forces shall be as hereunder:

- Centrifugal/Transverse Seismic force due to Track live load is at 1.83m above Rail level
- Transverse Seismic force due to SIDL at the Rail level
- Transverse Seismic force due to DL at the respective center of gravity.

The design of Corbel pier caps is per AASHTO LFRD 207.

13.8.2 Design Method for Stations

Structural Systems for Station Buildings are mainly single pier supported structures, i.e., the whole structure is supported by one row of RCC Piers in the central region of the Station Width. The Structure consists mainly of RCC Columns/Piers, Beams and Slab. RCC Cast-in-situ Post-tensioned Cross Arms have been provided to support Concourse Level, Platform Level and Steel Roofing above completely through RCC Stub Columns starting from Concourse Level. Concourse level Slab is supported on Precast/cast-in-situ RCC Beams. Track structure consists of a U-Girder Supported on Track track-level Pier cap. The Track and Platform Structure are separated by a suitable separation gap. The depot structure is proposed to be designed as an RCC Frame Beam-Slab Structure. The structure will be provided with expansion gaps at appropriate locations. The Columns are provided of sufficient size to control deflection under seismic/wind loads. Ductility detailing shall be done as per codal provisions. The structural configuration is selected to best suit the Functional and Architectural requirements. The depot building is designed as a 3-D RCC frame structure for different load combinations with columns and shear walls. Block Masonry is a filler material with no impact on the behaviour of the RCC frame. The slab panels are designed as per the coefficients in IS 456-2000. The analysis & design of idealised model is carried out using the "Staad Pro" computer program, providing design for beams & columns as per IS 456-2000 and IS 13920. For Seismic analysis slab at each floor has been idealised as a rigid diaphragm, with masses lumped at the joints with 5% damping. Seismic Coefficients are as applicable to Seismic Zone III as per IS 1893-2002. Dynamic analysis shall use Coefficients as applicable to Seismic Zone III as per IS 1893-2002. Dynamic analysis shall be done for the structural performance of these buildings. Following is the code-wise preference for the design of structural elements:

i. Pile & pile cap – IRS, RDSO, IRC
ii. Central Pier –IRS, RDSO
iii. Concourse & Platform beam & slab – IS:456, IS:13920
iv. Track level structure - IRS, RDSO
v. Depot structure – IS:456, IS:13920
vi. Depot track supporting structure – IRS, RDSO

13.8.3 Design Briefs

a) Details of Elevated Stations and Viaducts.

1. Foundations:
    - Bored Cast-in-Situ Piles and Pile Caps
2. Substructure:
    - Concrete Piers, Eccentric Piers, extended flat piers
    - Pre-cast Pre-stressed Pier Cap for U-girder
    - Pre-cast Pier cap for I-girder
    - Portal

3. Super Structure:
    - Deck with I-Girder
    - Track with U-Girders
    - Platforms
    - Station Roof: steel portal with sheeting

b) Details of Underground Stations

1. Launching Shaft & Retrieval Shafts
2. Cut and cover Tunnels
3. Diaphragm Wall and Plunge Columns for Stations
4. D Wall, Temporary Structures, Strut and water system for Cut and cover tunnel
5. Undercroft Slab with provision of Drainage arrangement
6. Platform slab with provision of all MEPF facilities
7. Concourse slab with provision of all MEPF facilities
8. Roof Slab

13.9 Madhavaram Depot

Madhavaram Depot consists of the following Building Structures

1. Administrative Building
2. Workshop Way and Workshop Office
3. Automatic Coach Washing Plant (ACWP)
4. ETU & PWL
5. Inspection way
6. Stabling Yard
7. Stabling office
8. WPT
9. Depot Store
10. Security office

Fig.13.11: Internal View of a Depot Building

13.10 Phase I (Extension)

13.10.1 Details of Elevated Stations and Viaducts.

1. Foundations:
    - Bored Cast-in-Situ Piles and Pile Caps
2. Substructure:
    - Concrete – 1 pier,2 pier group,3 pier group
    - Pier cap
    - Portal Beams
3. Super Structure:
    - PSC I-Girder
    - PSC U Girder
    - Platforms
    - Station Roof: steel portal with sheeting

Fig.13.12: Concourse Level Showing Paid and Unpaid Area

13.10.2 Details of Underground Stations Design Details

1. Launching Shaft & Retrieval Shafts
2. Cut and cover Tunnels with a Temporary structure with a Strut and Waler System
3. Diaphragm Wall (DW) and Plunge Columns for Stations
4. DW, Temporary Structures, Strut and waler system for cut-and-cover tunnel
5. Undercroft Slab with provision of Drainage arrangement
6. Platform slab with provision of all MEPF facilities
7. Concourse slab with provision of all MEPF facilities
8. Roof Slab

Fig.13.13: Underground Corridor Constructed Using

Fig.13.14: Casting Yard Showing Cast U-Girder & Cage in Mould TBM

13.10.3 Wimco Nagar Depot- Buildings

1. Depot Building: G+3 Floors
2. Entry Structure to Depot
3. Workshop
4. Automatic Coach Washing Plant (ACWP)
5. ETU & PWL
6. Inspection way
7. Stabling Yard
8. WPT

13.11 Structural design of Foundations

- The computation of bending moments along the pile length due to forces applied at the pile cap level was based on a space frame model with actual stiffness of piles restrained by springs simulating the soil stiffness. If there were indications of soil liquefaction of the upper strata, the soil springs were idealised.
- The structural design of the pile and pile cap was checked in SLS and ULS conditions.

13.11.1 Soil Dynamic Stiffness – Earthquake loads

- The lateral stiffnesses of soil under dynamic loads such as earthquake loads are Chennai for Elevated Stations 10/12/2020 Report Page 26 of 29 generally higher than static values used for all other static loads.
- Dynamic stiffnesses, calculated using geotechnical tests to evaluate shear and compression velocity, G values, and other parameters.
- These parameters were not provided in the geotechnical reports. However, correlations were made between dynamic stiffness and SPT values, as required.

- In case of a multiple pile system, the spacing between the piles was kept not less than 3 times the diameter of the pile in soil and 2.5 times the diameter when founded on rock.
- Pile cap was designed based on truss analogy for a pile group having up to 5 piles. For other pile groups, bending theory using the space frame model was employed for pile cap design without taking support from the soil above.
- The top of the pile cap was 500mm below the existing/made-up ground, and the weight of the earth cover will be applied on top of the pile cap when necessary. The earth cover on the pile cap, which would have any favourable effect, was neglected. The following limiting values were used in the computation of safe load:
- The result of the investigation was used to adopt the value of the angle of internal friction ϕ and cohesion of soil, c.
- Angle of wall friction δ was considered equal to ϕ degrees.
- Coefficient of earth pressure 'K' was taken as 1.0.
- Maximum overburden pressure at the bottom of the pile for calculation of shaft resistance and bearing resistance was limited to 15 times the diameter of the pile.
- The entire overburden was considered submerged for the calculation of safe load.
- Factor of safety at working load = 2.5.
- Bulk density corresponding to 100% saturation

13.11.2 Details of Foundations

The foundations were designed based on the following:

- 0.8m/1.2m diameter bored cast-in-situ vertical piles in soil/rock were designed for the foundation of piers/columns.
- Open/Pile foundations were primarily adopted for the Access/Entry structure in firm soil/ rocky strata at shallow depth.
- Foundation design was carried out as per IRC: 78, and the capacity of piles in Rock was as per Appendix 5 (Notification no.68) of IRC: 78-2000.
- For pile carrying capacity, the SLS check was considered only.
- The lateral load capacity of the pile was evaluated by soil structure interaction analysis using Winkler's model spring and checked by using the empirical formulae given in IS: 2911 (Part1/ section-2).
- Initial load tests were conducted first to get the safe vertical and lateral loads. The safe load was taken as the least of (a) the load arrived at from the initial load test and (b) the calculated safe load based on a static formula. Initial tests were conducted for 2.5 times the safe vertical load based on a static formula.
- The working load on the pile for vertical and horizontal loads were checked by conducting routine tests during execution.

Fig.13.15: **Erection of Precast U Girder**

Fig.13.16: **Shows top-down Erection of Precast Pier caps (Followed by Erection of U-girders)**

13.12 Construction Stages of an Elevated Metro Station

STAGE 1: Casting of foundation and pier from ground to concourse level.

STAGE 2: Erect and place the central precast segment of the pier cap at concourse level.

STAGE 3: Casting of pier from concourse to platform level.

STAGE 4: Erect and place the central precast segment of the pier cap at track level.

STAGE 5: Launching of the gantry arrangement.

STAGE 6: Lifting of pre-cast segments of track level pier cap.

STAGE 7: Matching of precast segments with already placed central precast segment of pier cap at track level.

STAGE 8: Post-tensioning of PT cables (first stage PT) in the transverse direction to carry the self-weight of the pier cap at track level.

STAGE 9: Release the segments from the gantry hooks.

STAGE 10: Lifting of pre-cast segments of the concourse level pier cap.

STAGE 11: Matching of precast segments with already placed central precast segment of pier cap at concourse level.

STAGE 12: Post-tensioning of PT cables (1st stage pt) in transverse direction to carry the self (1st stage pt) weight of the pier cap at the concourse level.

STAGE 13: Remove the gantry, erect, and place pretensioned beams and cast slab on the concourse level. Stressing of the 2nd stage PT cables at the concourse level pier cap.

STAGE 14: Erect and place post-tensioned beams and cast track slab at track level. Erect and place pretensioned beams and cast a slab at the platform level. Stress 2nd stage pt at track level pier cap.

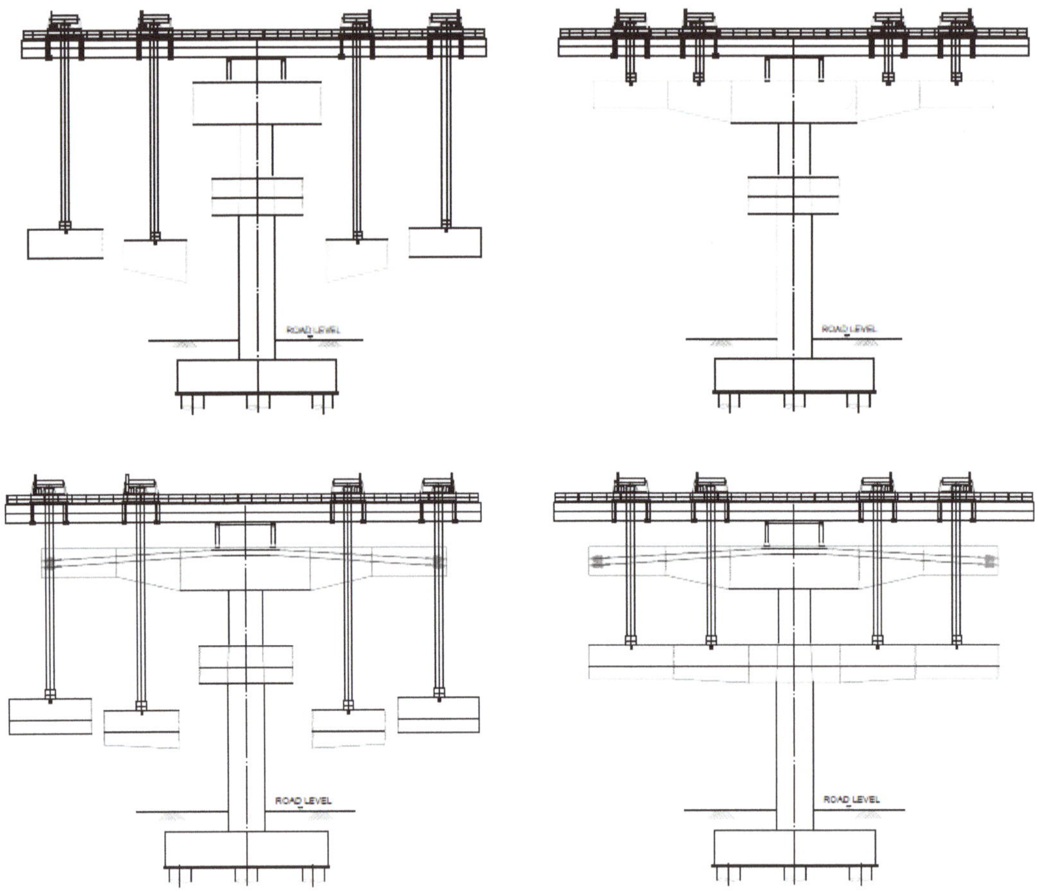

Sketches Showing Erection

Fig.13.17: Photos During Different Stages of Construction

Fig.13.18: Two Tracks with Inspection Pit at Depot

Fig.13.19: Pier & Pier cap ready to receive U-Girder

CHAPTER 14

Bangalore (Bengaluru) Metro

14.1 Background

A brief note on the Bengaluru Metro System is included in Part 2 of the book. Detailed Project Report prepared by **DMRC**. This Section also have details as available from reports by **ADB, RITES**, **CES** & Others including details from **Wikipedia**.

Bengaluru is the capital city of the state of Karnataka in southern India. It is India's fourth most populous city and one of the largest and fastest-growing metropolitan areas in the country, with an estimated population of 12.3 4 million in its urban area as of 2017, up from 8.5 million in 2011. Rapid population growth and urbanisation, a lack of integrated urban planning, and lagging transport network development have contributed to Bengaluru being one of the most congested cities in the world. Peak-hour traffic congestion in the city is worse than in other major cities in India.

Public transport. Compared with other Indian megacities, Bengaluru relies heavily on road transport. Public transport in the city consists of railway, bus, and metro rail. The railway is mainly used for intercity services, although a few morning and evening suburban trains serve commuters living outside the city. Public bus service is the mainstay, mainly operated by the Bangalore Metropolitan Transport Corporation.

The metro rail is the newest form of public transportation in the city. Established in 2005 as a special purpose vehicle jointly held by the Government of India and the State Government of Karnataka, Bangalore Metro Rail Corporation Limited (BMRCL) constructed and operates the Bengaluru Metro. Phase 1 of Bengaluru Metro, which is 42.3 km long, was completed and became fully operational in 2017 **(Fig.14.1).** It comprises the East–West (Purple Line) and North–South (Green Line) lines. Phase 2 will extend the purple and green lines and construct two new corridors (Pink and Yellow lines) with a total length of 72.1 km.

Fig.14.1: Bengaluru Metro During Erection of Elevated Corridor

Bengaluru Metro comprises approximately 225 spans, with span lengths ranging from 21m to 45m. Generally, spans are 28m in length, simply supported, and resting on a pile foundation. The superstructure is a precast box girder, in general, designed to support two tracks with inclined webs, utilising the segmental method of construction.

14.2 Introduction

In 1993, the State of Karnataka established a committee to look into mass rapid transit. This committee had recommended the metro project put forward by SR in 1983 along with the same circular railway. The state then created Bangalore Mass Rapid Transit Ltd **(BMRTL)** in 1994, with terms of reference to seek a public/private partnership for a mass rapid transit project. The government introduced a special city cess dedicated to the anticipated mass rapid transit project. BMRTL commissioned a feasibility study which came up with an elevated, LRT-based, network on six routes. A private consortium led by **United Breweries Group** undertook further development of the project on BOT basis. However, the project hadn't taken off.

Later in 2003, the Government of Karnataka commissioned the **DMRC**, which had carried out a Detailed Project Report (DPR). The study recommended an underground 2-line metro, 60 km and a 50 km line at the Railway Station in Bengaluru. The economic rate of return was forecast at 22.3%. The financial forecast assumes a government subsidy for interest payments and some depreciation, i.e. fare revenue will cover somewhat more than direct operating costs. The Government accepted the report. In the meantime, the Government of Karnataka replaced BMRTL with Bengaluru Metro Rail Corporation Ltd (BMRCL). Bengaluru Metro Rail Corporation Limited **(BMRCL),** a joint venture between the Government of India and the State Government of Karnataka, is the agency responsible for building, operating, and expanding the Metro network.

Bengaluru Metro is the second-longest operational metro network, with an operational length of 76.95 Km, behind Delhi Metro. Upon its inauguration in 2011, it became the first metro system in South

India. In 2016, it became the first underground metro system in South India. In November 2024, the metro had 68 operational metro stations, of which 59 were elevated, eight were underground, and one was at-grade. The trains initially consisted of three coaches but later expanded to six as ridership increased. Power is supplied by a 750V direct current through a third-rail system, like the Kolkata Metro. As of March 2024, the metro system had an average daily ridership of 636,000 passengers. In early December 2024, the ridership was 9.2 lakh, the highest recorded. Metro corridors, including their colour code, are given below and the master plan is included in **Figure 14.2**.

1. **Purple** – 25.7 km – 22 Stations – Baiyappanhalli to Kengri - Line 1
2. **Yellow** 18.8 km – 14 Stations – R.V. Road to Bommasandra – Line 3
3. **Blue** 38 Km - 23 Stations – Silk Board to Kempagowda Intl. Airport – Line 5
4. **Green** 30.5 km – 28 Stations – Nagasandra to Silk Institute—Line 2
5. **Pink** 19 km 19 Stations – Kalena Agrahara to Nagawara – Line 4
6. **Orange** 21 km 21 stations – Hebbal to Sarjapur – Line 6

14.3 Detailed Project Reports (DPR)

DMRC prepared the DPR for Phase 1 of the Bangalore Metro Rail Project and submitted it to BMRCL in May 2003. DMRC also submitted the extension of the N-S line in two parts, one in October 2007 and the other in June 2008.

Phase 1 includes,

i. East-West Corridor from Baiyappana Halli Terminal to Mysore Road Terminal, 18.10 km.
ii. North-South Corridor from Hesaraghatta Cross Station to Puttenahalli Cross – 24.20 km.

Fig.14.2: Bengaluru Metro Plan (Metro News)

14.4 Phases

The Bengaluru metro network was developed in phases.

Phase 1: Phase 1 comprises two lines: the East-West Purple Line, 18.80 km, and the North-South Green Line, 24.20 km (Lines 1 & 2).

Phase 2: Phase 2 consists of Purple Extensions, 8.81 Km + 15.50 Km and Green Lines, 6.29 Km +3.77 Km, Line 3, 18.82 Km, Line 4, 21.25 Km & ORR line, 18.00 Km.

Phase 3: A 33-kilometre line is proposed to extend to cover the eastern side, and a 31.37-kilometre extension to the western side.

A) Phase 1

The foundation stone for the construction of Phase 1 was laid by then-Prime Minister Dr Manmohan Singh in June 2006. Navayuga Engineering Company Limited was awarded the first contract to construct Reach-1 of the east–west corridor (later named the Purple Line) in 2006. Civil construction on the first section of the Purple Line between Baiyyappanahalli and Mahatma Gandhi Road commenced in April 2007.

Detailed Project Reports (DPRs) for a northern extension and a southern extension (from Rashtreeya Vidyalaya Road to Yelachenahalli) of the North-South corridor (later named the Green Line) were submitted in October 2007 and June 2008, respectively. With these extensions, the total route length for Phase 1 became 42.3 km. The objective was to connect the metro to the Outer Ring Road at northern and southern ends, and also cover the industrial areas of Peenya in the north-west, thereby providing better connectivity and increasing ridership. In October 2008, the Government of Karnataka approved this extension.

i) Underground section

Both lines in Phase 1 have tunnel sections in the city centre, which are also the first metro tunnels built in South India. Construction of underground sections in Phase 1 commenced in late 2012. The delay was due to the cancellation of the initial tenders called in early 2008, as the entire DPR had to be revised, and the bids received were too high. A second round of tendering was conducted in late 2009, with the gigantic, majestic interchange station as a separate package. This package was awarded for the tunnel sections in 2011, and construction commenced in 2012.

The tunnels, bored with tunnel boring machines (TBMs), are located approximately 18.3 meters below ground level and have a diameter of 5.5 meters, with a spacing of 5 metres between them. Tunnel boring of the underground section UG1 (on the east-west corridor) was completed in March 2014. Track works and third-rail electrification works were completed on the 4.8 km east–west tunnel of the Purple Line between Cubbon Park and Magadi Road, and the Bangalore Metro Rail Corporation Ltd (BMRCL) began trials in November 2015. The entire stretch of the Purple Line was opened in April 2016.

Tunnel boring of the underground section UG2 (on the north–south corridor) was completed. In TBM Godavari, the cutter head was broken, and spares had to be awaited. Trial runs on the 4.0 km east–west tunnel of the green line between Sampige Road and National College were commenced in March 2017. The entire stretch of the Green Line was opened in June 2017, thus completing Phase 1.

Fig.14.3: Metro Green Line

Fig.14.4: Metro Purple Line train

ii) Opening

After missing deadlines, Namma Metro's first section (Reach-1) finally opened to the public on 20 October 2011. There was an overwhelming response from the public at the commencement of operations. According to BMRCL sources, within the first three days of operations, 169,019 people rode the metro. By the end of the fourth day, approximately 200,000 passengers had already commuted on the Bengaluru Metro. Bengaluru Metro's first 12-day cumulative revenue was ₹1 crore.

The northern section of the Green Line was initially scheduled to be opened by the end of 2012. However, it was delayed and finally opened in March 2014. BMRCL MD Pradeep Singh Kharola stated that about 25,000 passengers travelled on the line on the opening day. In the first month of operations, 7.62 lakh people used the line, averaging 24,605 per day. The first underground section (on the purple line) commenced operations in April 2016, providing connectivity between the east and west of the city. Subsequently, the second underground section, along with the southern reaches, was opened in June 2017. Once the east and west reaches were interconnected with the opening of the underground section of the Purple Line, ridership surged significantly. After the north–south underground section was opened, the network provided connectivity in all four directions, with interchanges between the lines, and this further increased ridership. Ridership continued to increase and was around 4,50,000 daily in September 2019.

B) Phase 2

The State Government accorded approval for the preparation of the detailed project report (DPR) for Phase 2 by DMRC in early 2011. The High-Power Committee (HPC) gave in-principal clearance to proceed with Phase 2 in July 2011, and the Karnataka government provided its approval in early 2012.

Due to delays in DPR preparation, approval from the Central Government for Phase 2 was cleared in August 2013. The Union Cabinet finally announced that it had approved plans in end January 2014.

After an extension of 3.195km to Challaghatta, Phase 2 spanned a total length of 75.29 km, comprising 13.79 km underground, 0.48 km at grade, and 61.37 km elevated. It added 62 stations to the network, of which 12 are underground. Phase 2 involved extending the two Phase 1 lines in both directions, as well as constructing two new lines.

The south end of the Green Line was planned to be extended from Yelachenahalli to Silk Institute along Kanakapura Road, and the north end from Nagasandra to Madavara on Tumkur Road (NH-4). The east end of the Purple Line would be extended from Baiyappanahalli to Whitefield, and the west end from Mysore Road to Kengeri, covering a total of 18.82 km. A fully elevated line from RV Road to Bommasandra would be constructed, passing through Electronic City. The second new line is planned to be 21.25 km long, from Kalena Agrahara to Nagawara. The line is mostly underground, 3.79 km with an elevated section of 6.98 km and a 0.48 km at-grade section. There are 18 stations on the line, of which 12 are underground and six are elevated.

Karnataka Industrial Areas Development Board (KIADB) was responsible for acquiring land for Phase 2. It was estimated that 102.02 hectares of land would be required for Phase 2 (including Phase 2A. Construction work began on the south extension of the Green Line from Yelachenahalli to Silk Institute (6.29 km and the west extension of the Purple Line from Mysore Road to Kengeri (8.81 km) by October 2016. Construction work on the west extension of the Purple Line was awarded in two packages, while the Green Line south extension was awarded in a single package. The construction work for the north extension of the Green Line and the east extension of the Purple Line began in July 2017.

Construction of the yellow line, a new route from Rashtreeya Vidyalaya Road, was awarded in three packages and commenced in November 2017. The first package (6.418 km) includes the entry line to Hebbagodi depot and five Metro stations. The second package was for a 6.38 km stretch from Hosa Road to Bommanahalli. Both packages were awarded to ITD Cementation India. The third package was for a stretch from RV Road to Bommanahalli, 6.34 km long, an elevated section with five stations. Construction of the 7.5 km elevated section of the new pink line between Kalena Agrahara and Tavarekere stations commenced in February 2018.

i) Phase 2A

In September 2016, the government announced that a new 18 km line connecting Silk Board with K.R. Pura will be included in Phase 2 as Phase 2A of the project. The line would run along the eastern half of the Outer Ring Road, with 13 proposed stations. DPR prepared by **BMRCL** was approved by the State in March 2017. The ORR Metro line, also known as the Blue Line, is planned to have an interchange station with the extended Purple Line at K.R. Pura and a connection to the Yellow Line at Silk Board. A second time, as the first tenders were to be quashed, tenders for the ORR Metro line (east) were called in two packages in December 2019, and multiple firms received bids in March 2020. The first package included 2.84 km of ramps for a flyover at Central Silk Board junction and 9.859 km of viaduct with six elevated stations. The second package was for an 8.377 km viaduct with seven elevated stations, a 1.097 km depot line and a 0.30 km pocket track.

ii) Phase 2B

There had been a proposal to build a 33 km high-speed rail line from MG Road to Kempe Gowda International Airport. Initially, it was planned to be executed by an independent SPV, but later, it was decided that BMRC would manage, and a regular metro line with fewer halts would be built instead of the high-speed rail proposed earlier, resulting in less travel time between the city and the airport. As early as February 2012, the Central Government requested that BMRC initiate Phase 2.

Following this, in September 2016, suggestions were invited from the public to choose any one of nine possible extension routes from existing and proposed metro lines to the airport. The proposed extension routes had an average length of 30 km. A 25.9 km extension from Nagawara via Kannur and Bengaluru was the shortest, while the 35.4 km extension from Yeshwanthpur via Yelahanka, Kannur and Bengaluru were the longest. A 25.9 extension of the Kalena Agrahara – Nagawara line via Kannur and Bengaluru to the airport emerged as the most popular choice.

Since Bengaluru International Airport Limited (BIAL) forbade underground construction from the southern side of the airport due to security concerns, it was eliminated. Bengaluru Development Minister announced in May 2017 that the government had finalised the Nagawara-Ramakrishna Hegde Nagar-Jakkur-Yelahanka route to the airport. The new line of 39.8 Km would begin at Krishnarajapura and align through the northern part of ORR, passing Nagawara, Hebbal, and Jakk, heading towards the airport along Ballari Road. The Union Cabinet cleared two much-awaited lines of the Bengaluru Metro's Phase 2A and 2B in April 2021.

C) Phase 3

A 124 km Phase 3 was initially proposed in May 2016. Certain significant sections that were proposed for Phase 3 were included in Phase 2 as Phase 2A and Phase 2B. In March 2020, it was announced that two corridors totalling 44.65 km would be built as fully elevated lines under PPP mode (Kempapura-Jayaprakash Nagara 4th Ph (along ORR-West) and Hosahalli-Kadabagere). In March 2022, during a budget speech by the state government, a new 35 km corridor spanning from Hebbal to Sarjapura was announced. The Chief Minister stated that the Detailed Project Report (DPR) would be prepared for the proposed corridor. Later, in August 2024, the Union Cabinet approved Phase 3, which will add two lines and 31 stations. SECON Pvt. Ltd. has officially started geotechnical soil investigation work on Magadi Road for the upcoming 44.65 km Bangalore Metro Phase 3.

Two corridors have been finalised for implementation in Phase 3 of the Bengaluru Metro network, as under:

i. Corridor 1: JP Nagar 4th Phase to Kempapura
ii. Corridor 2: Hosahalli to Kadabagere

Corridor 1: JP Nagar 4th Phase to Kempapura, considering the centre line of JP Nagar 4th Phase Station, the corridor is 32.15km long. At Kempapura Station, track Integration with under under-construction Phase 2B corridor is planned. Provision already exists in the station and track layout of Phase 2B corridor. The corridor is proposed to be completely

Corridor 2: Hosahalli to Kadabagere, considering the centre line of Hosahalli as 0.00m, this corridor is 12.50 km long, starting from (+)450m and running up to 12050m. The corridor is proposed to be completely elevated. A total of 9 stations has been proposed along the corridor. A common Maintenance depot has been proposed for both corridors on Sunkadakatte State Government land, covering an area of approximately 30 hectares. A total of 22 stations has been proposed on the corridor.

D) Phase 4

Bangalore Metro Phase 4, the BMRCL plans to expand the Bangalore Metro network by adding a proposed line covering 129 km. The plans for the proposed metro lines under Namma Metro Phase 4 deviate from the Comprehensive Mobility Plan (CMP) drafted in 2020. The proposed metro lines extend beyond the Bruhat Bengaluru Mahanagara Palike (BBMP) limits into areas defined by fragmented growth.

According to media reports, the 129-km Metro Phase 4 corridor, which does not align with the CMP, may be rejected by the centre. A rail enthusiast has raised questions about the necessity of metro extensions to satellite towns, given the higher costs and the primary design for intra-city travel, which diverges from conventional infrastructure plans. BMRCL engaged **RITES** to prepare the DPR for two Corridors of Bengaluru Metro from JP Nagar IV Phase to Kempapura along ORR West and from Hosahalli Metro station to Kadabagere on Magadi Road. The study area encompassed the Bangalore Metropolitan Area (BMA) and the International Airport Area Planning Authority (BIAAPA). The Bangalore Metro Phase 4 is expected to boost connectivity in areas such as Bidadi, Harohalli, Attibele, Kunigal Cross, Sarjapur and Varthur. These localities are expected to experience increased demand for residential and commercial properties, resulting in higher property prices.

14.5 Phase 1 lines and sections

Phase 1 includes two corridors: the East-West corridor and the North-South corridor.

i. The East-West corridor is from Baiyappana Halli terminal to Mysore Road terminal and is 18.10 km in length.
ii. The North-South corridor starts from Heasraghatta Cross Station to Puttenahalli Cross and is 24.20 km in length.

The Mysore Road to Kengeri metro line features a fully elevated double-track viaduct spanning about 7.5 kilometres, constructed with prestressed concrete box girders for strength and rapid execution. Spans between piers range from 16 to 31 meters, with emergency walkways and integrated drainage systems to ensure safety and durability. The six metro stations along this stretch were designed with seamless multi-modal integration, providing easy access to buses, autos, and feeder services. Stations included features like ramps, elevators, and tactile flooring for accessibility, durable platform canopies for weather protection, and efficient passenger flow layouts to handle peak commuter volumes while minimising congestion. These structural and design elements combine to offer a safe, efficient, and commuter-friendly metro corridor.

Two lines spanning a length of 42.30 kilometres, of which approximately 8.82 kilometres are underground and approximately 33.48 kilometres are elevated. There are 40 stations in Phase 1, of which seven are underground, two are at grade, and 31 are elevated. A depot has been constructed at Baiyappanhalli for the east-west corridor, and a complete depot with all necessary facilities has also been established in Pennya. There were numerous delays during construction, resulting in several postponements. The difficult geological conditions below ground, with a mix of soft soil, high groundwater levels, hard granite and large boulders, were a major impediment to tunnel boring. A major delay occurred due to a broken cutter head on the TBM Godavari, for which spares had to be ordered from Italy.

14.5.1 Purple Line

The Purple Line runs east to southwest in the Bangalore Metro, connecting Whitefield (Kadugodi) in the east with Challaghatta in the southwest. It takes approximately 1 hour and 29 minutes to traverse the 43.35 km distance at an average speed of 29 km/h. The line is 43.45 km long and has 37 stations. It is elevated on both the east and west sides and has a 4.8 km underground section in the middle. The line passes through prime activity centres of the city like Whitefield, MG Road, Majestic, Vidhana Soudha, and other areas.

The first 6.7-kilometre, 6-station stretch of the Purple Line between Baiyappanahalli in the east and Mahatma Gandhi Road opened in October 2011. This was the first and inaugural section of the Bangalore Metro. The second 6.4-kilometre, 6-station stretch between Mysore Road and Magadi Road opened in November 2015. The underground section, a 4.8 km stretch from Cubbon Park to Bengaluru City (KSR) Railway Station, was opened in April 2016, thus linking the east and west sections.

Fig.14.5: Interchange Majestic – Between Purple & Green Lines

The system runs on standard-gauge tracks. The Metro has been implemented with a 750 V DC Third Rail Traction System and cab signaling with CATC, SCADA, and AFC. The rake interchange link is planned for Majestic Station.

Under Phase 2, the 7.5 km southwest extension opened in August 2021. The eastern extension, which spans 15.2 km to Whitefield, was under construction, and 13.7 km of the extension was opened for service in March 2023. In October 2023, Baiyyappanhalli-KR Pura and Kengeri-Challaghatta were also open to the public. It connects the industrial and suburban areas of the south-west with CBD (MG Road, Trinity Circle) and IT areas of the east.

14.5.2 Green Line

The Green Line is aligned north to south and connects Madavara in the north-west to Silk Institute in the south-west, covering 33.46 kilometres and has 30 stations. It is elevated on both the north and south sides and features a 4.0 km underground section in the middle, like the Purple Line. The

Line passes through industrial areas in the north and through commercial hubs, namely Majestic, Chikpete, City Market and connects large residential catchments in the south. The Green Line opened in March 2014. The stretch connected Sampige Road to the Peenya Industrial Area. The second 2.5-kilometre, 3-station stretch of the Green Line from Peenya Industry to Nagasandra opened in May 2015. The third stretch connecting Sampige Road to Yelachenahalli was inaugurated in June 2017. These stretches, including the underground section, were inaugurated by the late Shri Pranab Mukherjee, then President, in March 2017. Under Phase 2, a 6.29-kilometre stretch from Yelachenahalli to Silk Institute was inaugurated in January 2021, marking the first section of Phase 2 to be put into service. The northern extension, spanning 3.14 km to Madavara, was opened to the public in November 2024.

14.5.3 Yellow Line

The Yellow Line, an elevated corridor connecting Electronic City, spans Rashtreeya Vidyalaya Road to Bommasandra, covering 18.82 kilometres with 16 stations. It begins at an interchange station (Rashtreeya Vidyalaya Road) and passes through two other interchange stations: Jayadeva Hospital station, where it crosses the Pink Line; and Central Silk Board station, where the Blue Line will terminate. The Bangalore Metro network plan is included in **Fig. 14.2**, and **Fig. 14.5** shows the green and purple lines with an interchange station at Majestic.

14.5.4 Pink Line

The Pink Line of the Bangalore Metro Lines connects Kalena Agrahara in the south with Nagawara in the north, aligning from South to North. It covers 21.25 kilometres and has 18 stations. The Pink Line is also under construction and is expected to be operational in 2025–26, forming part of Bengaluru Metro's Phase 2 expansion. Additionally, the Pink Line has two phases, with the first phase expected to be completed by December 2024, while the entire line is anticipated to be operational by around 2026.

14.5.5 Blue Line

The Blue Line, or the ORR-Airport Metro Line, is 58.19 kilometres, consisting of 30 stations. It connects the Central Silk Board with the Kempegowda International Airport. The Blue Line of the Bangalore Metro is also under construction, specifically Phase 2A (Silkboard to KR Puram) and Phase 2B (KR Puram to Airport), with completion expected by June 2025 and June 2026, respectively.

The much-awaited Bangalore Metro is likely to open in two phases. The Kempegowda International Airport (KIA) Terminal – Hebbal Section is expected to open between June and September 2026. It will be followed by the Hebbal – KR Pura section in the next few months in December. Additionally, of the six coaches on the Blue Line Metro train, four will feature luggage racks for the convenience of passengers bound for Kia.

14.5.6 Orange Line

The Orange Line of the Bangalore Metro is part of Phase 3 of the Bangalore Metro and is still under planning, with a length of 80 kilometres. According to the plan, the proposed routes of the Orange Line under Phase 3A connect Kempapura to JP Nagar 4th Ph (along ORR-West). The Orange Line is expected to be completed by 2028, featuring 22 stations.

14.5.7 Red Line

The proposed Red Line of Bangalore Metro (Namma Metro) is to connect Hebbal with Sarjapura. It falls under Phase 3A of Namma Metro and will be a 37 km line with 28 stations.

14.6 Bangalore Metro Lines – Orange Line Properties Acquisition

Bangalore Metro Rail Corporation Ltd (BMRCL) has recognised 700 properties along the Outer Ring Road (ORR) stretch for the construction of the Orange Line under Phase 3. The stretch extends from Kempapura to JP Nagar IV.

At an estimated cost of Rs.1,900 crore, the properties identified for acquisition along the proposed line connect the Blue Line (KIA-Silk Board) and Pink Line (Kalena Agrahara-Nagawara) on either side. The identified properties are for building Metro Stations, entry and exit points, and constructing viaducts in certain areas. The properties recognised for acquisition consist of vacant land and buildings with ground floors and upper floors (+2 and +3), although there are no high-rise buildings along the way. Furthermore, defence authorities and government agencies, such as BMTC, BBMP, and BDA, own these properties. The BMRCL aims to acquire the identified properties in three batches by issuing notifications. Those concerned can expect the first notification within following two months.

The proposed Orange Line will have facilities to interchange with existing Metro Stations and those that are planned, as well as with other modes of transport. The Orange Line will connect with the Blue Line in Kempapura and Hebbal in the northern parts. Furthermore, at Hebbal, the interchange facility will connect with the suburban rail and the Phase 3A line (Sarjapur to Hebbal). On Tumkur Road, the operational Peenya station of the Green Line will be connected, where the Orange Line will go above the Green Line. Also, it will intersect with another Orange Line from Hosahalli station to Kadabagere on Magadi Road.

The BMRCL has proposed building a depot on seventy-five acres of land, which will be used for both lines. On Mysore Road, the Orange Line will be elevated above the Purple Line, which is operational, and the Nayandahalli Metro Flyover. While on Kanakapura Road, the Orange Line will connect with the JP Nagar Metro Station. However, on Bannerghatta Road, the line will terminate at JP Nagar IV Phase with the facility to interchange. As per the original plan, BMRCL aims to construct a road cum Metro-flyover at two points. They are located from Kanakapura Road to Sarakki, approximately 1.3 km, and from Ittamadu to Hosakerehalli, 1.5 km. BMRCL has also proposed to study the possibility of constructing a road cum Metro-flyover along the whole corridor.

14.7 Bangalore Airport Metro Blue Line

Bangalore Airport Metro, featuring 32 metro stations, is under construction. Once operational, the metro line will enhance airport connectivity for residents. Bangalore, the Silicon Valley of India, boasts an efficient public transportation system, which includes the Metro. The Blue Line is a part of the city.

Fig.14.6: Station Platform for an Elevated Corridor

The Blue Line is being constructed in two phases: Phase 2A and Phase 2 B. Phase 2A comprises 13 metro stations between Central Silk Board and KR Puram. Phase 2B is the Bangalore Airport Metro Line, which features 17 metro stations spanning the route between KR Puram and Kempegowda International Airport (KIAL) in Devanahalli.

The Bangalore Airport Line is being constructed in two phases: Phase 2A and Phase 2. Phase 2A is the Outer Ring Road corridor of the Metro Project. The project is under construction, situated between the Central Silk Board and Krishnarajapura. The corridor includes 13 elevated metro stations spanning over 18.236 km. Airport Metro Phase 2B is the Airport Line corridor of the Metro Project. It will be constructed between the Kempegowda International Airport (KIA) Terminal. The metro corridor spans 37 km and comprises 17 metro stations.

Bangalore Airport Metro line is all set to be launched in two phases next year. The first phase of the Bangalore Metro's Blue Line is expected to be operational between June and September 2026. This phase spans the area between the Bangalore Airport Terminal and the Hebbal Section. On the other hand, the second phase of the project is expected to become operational by December. The approximate cost for the Bangalore Airport Metro is INR 10,584 Crore. For this, the central and state governments will provide funding of INR 3973 crore. The funding for the land acquisition is being done by the state government. The remaining amount will be raised through loans. In March 2021, it entered into a loan agreement with JICA. BMRC has raised INR 3643 crore. ADB to fund the construction of the Blue Line (Phases 2A and 2B).

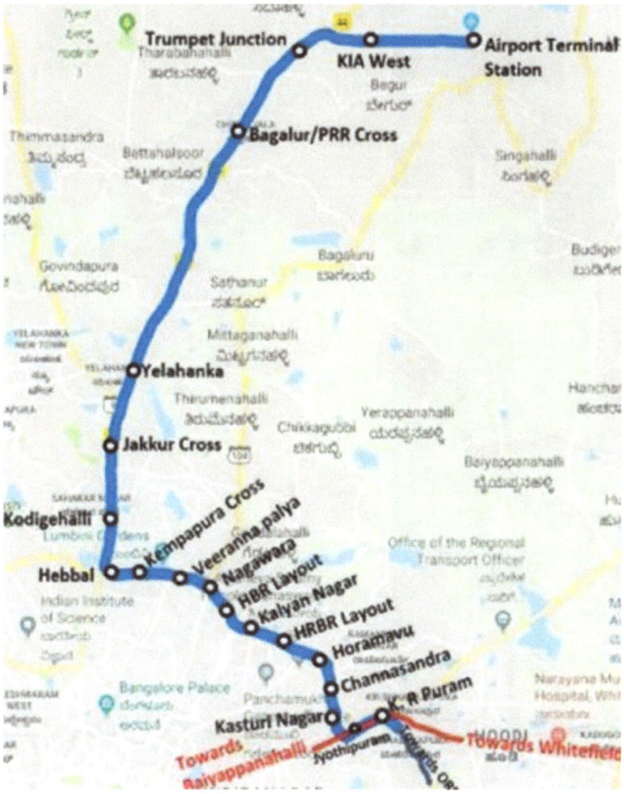

Fig.14.7: Airport Metro Corridor (driverspark.com)

Fig.14.8: Metro Corridor in Ooperation

14.8 Traffic and Transportation

M/s Wibursmith Associates, Bengaluru, did the detailed traffic study. The Transport Demand Model was developed based on detailed household and traffic surveys conducted in 2002-2003 and subsequent studies for extending the N-S line in October & December 2007. The network for all four extensions was developed from the primary database and used in the transport Demand Model. The traffic demand on these four extensions, the proposed RV Road – Electronic City/Bommasandra corridor, and the new corridor between IIM(B) and Nagawara, in place, has been projected.

A study in March 2021 found that the total number of registered vehicles in the Bangalore Metropolitan Region was 100.1 lakh, with two-wheelers constituting the highest number. The number of vehicle registrations in Bengaluru increased from 61.1 lakh to 100.1 lakh from 2015 to 2021. Road network inventory within the study area was carried out for primary and secondary roads, of which 44.9% have a ROW of less than 20 m, about 73.8% of surveyed road network lengths have carriageway widths up to 2 lanes, and about 20% are four lanes or more. The daily traffic counts have been computed for 24 hours at various screen line locations and presented. Road during a typical working day, traffic at different locations varies from 5,634 to 164,930 vehicles, with the highest volume recorded on Yashwant Tumkur Road. Road accidents are lower than in other big cities. The mass rapid transit system is being developed in stages in Bengaluru. Currently, Phase 1 covers a length of 43 km on two lines, i.e. Kengeri to Baiyyappanahalli and Silk Institute to Nagasandra. Both lines are in operation. Phases 2A & 2B are currently under implementation, and Phase 3 is in the planning stage.

Fig.14.9: Elevated Corridor During Construction

14.9 Phase 1 Routes

This phase involved the construction of 2 new lines on standard gauge tracks:

Line-1 – Purple line: Baiyyappanahalli – Mysore Road

- **Length**: 18.10 km
- **Type**: Elevated & Underground (4.8 km)
- **Depot**: Baiyappanahalli
- **Number of Stations**: 17
- **Station Names**: Mysore Road, Deepanjali Nagar, Attiguppe, Vijayanagar, Hosahalli, Magadi Road, City Railway Station, Majestic, Sir M Vesveshwaraya, Vidhana Soudha, Cubbon Park, MG Road, Trinity, Halasuru, Indiranagar, Swami Vivekananda Road and Baiyappanahalli

Line-2 – Green line: Nagasandra – Yelachenahalli

- **Length**: 24.20 km
- **Type**: Elevated & Underground (4 km)
- **Depot**: Peenya

- **Number of Stations:** 24
- **Stations Names:** Nagasandra, Dasarahalli, Jalahalli, Peenya Industry, Peenya, Goraguntepalya, Yeshwanthpur, Sandal Soap Factory, Mahalaxmi, Rajajinagar, Kuvempu Road, Srirampura, Sampige Road, Majestic, Chickpet, KR Market, National College, Lalbagh, Southend Circle, Jayanagar,

14.10 Travel Characteristics and Demand Estimates

The study area encompassed the International Airport, which was divided into several internal and external traffic zones. Classified traffic volume count surveys were conducted at required mid-blocks for 24 hours on a typical fair-weather working day. The traffic at different locations varied from 86034 vehicles at ORR B.T.M. Layout (near Water Tank) to 1,46,724 vehicles at ORR (near Hebbal).

An origin-destination survey of vehicles entering the study area was conducted at predetermined outer cordon locations. It was noted that approximately 47.7% of the trips were work— and business-based, 26.5% were social, and shopping/recreation accounted for about 19.6%. The PT Stop survey was conducted at 25 bus stops (on both sides) to assess passengers' maximum boarding and alighting times at different stops. Pedestrian volume counts were conducted at 13 locations within the study area. Parking surveys were conducted on important days.

A Household Interview Survey was carried out on a sample basis. It is observed from the table that about 72% of the trips were made for work and business purposes. In contrast, the remaining trips were for education, shopping, social, health, and recreation.

A four-stage travel demand model has been used for transport demand forecasting. The maximum peak hour peak direction trips (PHPDT) are shown in **Table.14.1**

Table 14.1: Maximum PHPDT on Corridor 1 & 2 of Phase 3

Corridor	Max. PHPDT			
	2028	**2031**	**2041**	**2051**
JP Nagar 4th Phase to Kempapura	23303	25173	29930	33061
Hosahalli to Kadabagere	18227	19679	23272	25707

14.11 NMV and Pedestrian Facilities

Pedestrian facilities, like a continuous footpath 2m wide, demarcation of pick and drop for PT/IPT, Zebra crossing at intersections, tabletop crossings, relocation of encroachments, and strengthening of the ROW, have been proposed near the station influence area for ease of pedestrian movement. For non-motorised vehicles, facilities like cycle tracks have been planned since land availability near the station for 0.6.3.

The Metro Rail system should be user-friendly, ensuring accessibility for people with disabilities, people travelling with small children or carrying luggage, as well as people with temporary mobility problems and elderly people, differently abled and Elderly Persons and others.

14.12 Intermodal Integration

The intermodal integration should be formulated to facilitate traffic dispersal and circulation facilities based on.

Minimising pedestrian/vehicle conflicts and effective passenger interchange with feeder modes.

Facilitating passenger interchange with other transit systems

Circulation area with designated space for embarking and disembarking for vehicular traffic and NMT.

The availability of total carriageway and footpath widths is required to cater to the proposed traffic volumes.

Fig.14.10: View of Bangalore-Hosur Metro Corridor (drivespark.com)

14.13 Last Mile Connectivity

BMRCL has proved parking spaces through park & pay with a uniform rate through contractors as per the fare chart shown below

Class of Vehicles	PARKING FEE		
	For the First Four hours or part thereof.	For every sub- seatuent hour or part thereof	Parking charges for the day
Cars/Jeep	Rs. 30/-	Rs. 10/-	Rs.60/-
Motor Cycles/ Two wheelers	Rs. 15/-	Rs.5/-	Rs. 30/-
LCVs	Rs.75/-	Rs.25/-	Rs.150/-
Bicycles	Re. 1/- per hour subject to a maximum of Rs.10/- per day.		

Two-Wheeler Parking is available in 9 Locations on the Purple Line and in 12 Locations on the Green Line Four-Wheeler Parking is available in 7 Locations on the Purple Line and in 10 Locations on the Green Line

BMTC is running Metro Feeder Bus services covering 17 Metro Stations at a frequency of 15 to 30 minutes, with 1981 trips and 159 Schedules.

Fig.14.11: Vehicles for Last mile connectivity

Bounce provides Last-Mile Connectivity at all metro stations. Bounce operates Keyless Scooters, which help customers to commute easily and are also affordable.

Bicycles are deployed at all the Metro Stations. They can be used by customers for last-mile connectivity.

14.14 Operational Highlights February 2019

(BMRCL)

Operational efficiency achieved in the month : 99.98%

Ridership for the month : 1,06,19,238

Revenue earned during the month : Rs. 28,45,18,381/-

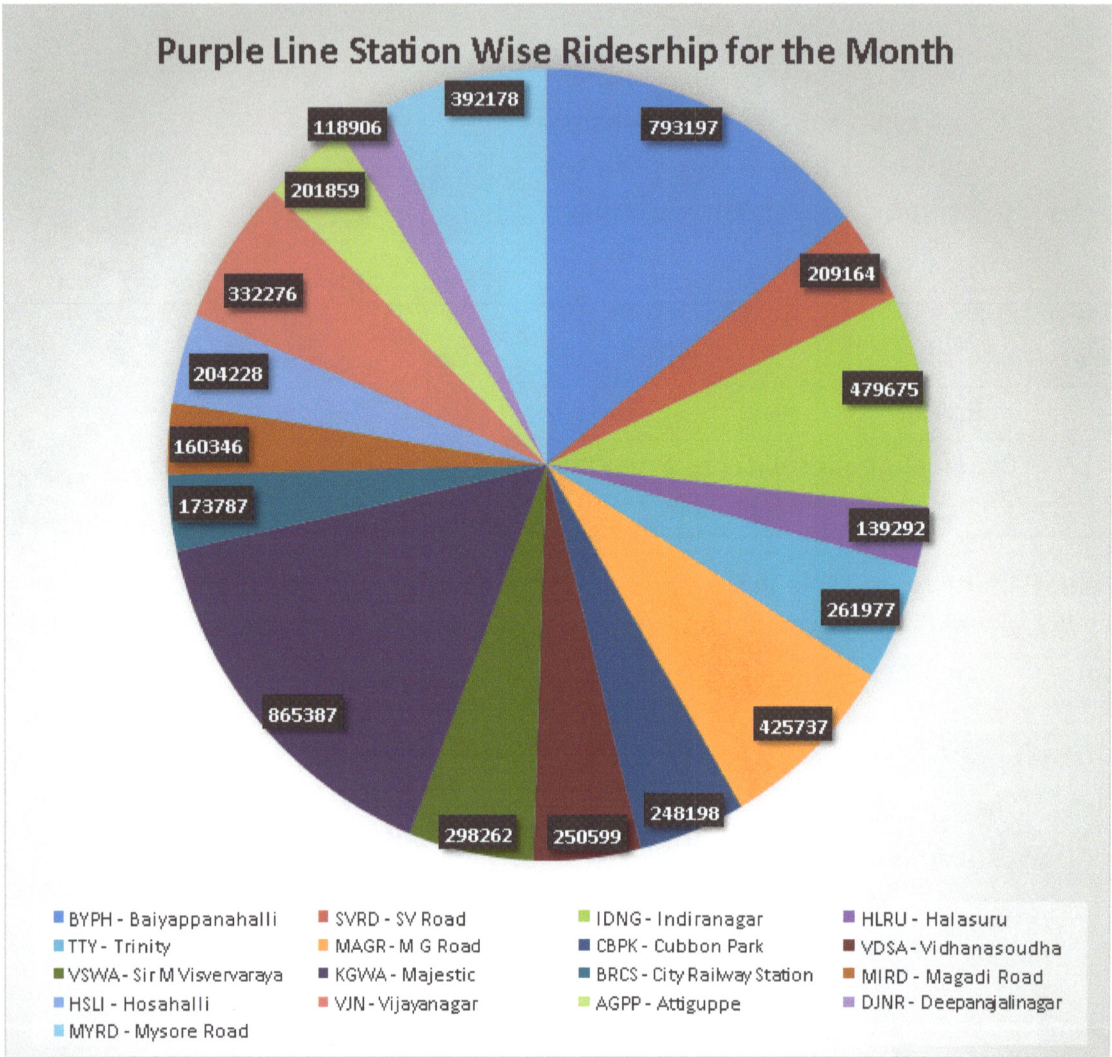

Fig.14.12: Station-wise Ridership- Purple Line (BMRCL)

14.15 Design Data

Materials Concrete

Instantaneous modulus:

For the Central module, E is taken as per clause 6.2.3.1 of IS 456:2000

For the End module, E is taken as per clause 5.2.2.1 of IRS – CBC

- For f_{ck} = 60 MPa E = 360000 MPa
- For f_{ck} = 50 MPa E = 340000 MPa
- For f_{ck} = 45 MPa E = 32,500 MPa
- For $_{fck}$ = 40 MPa E = 31,000 MPa
- For f_{ck} = 35 MPa E = 29,500 MPa

Where f_{ck} is the Characteristic Compressive Strength of 150mm Cube at 28 days.

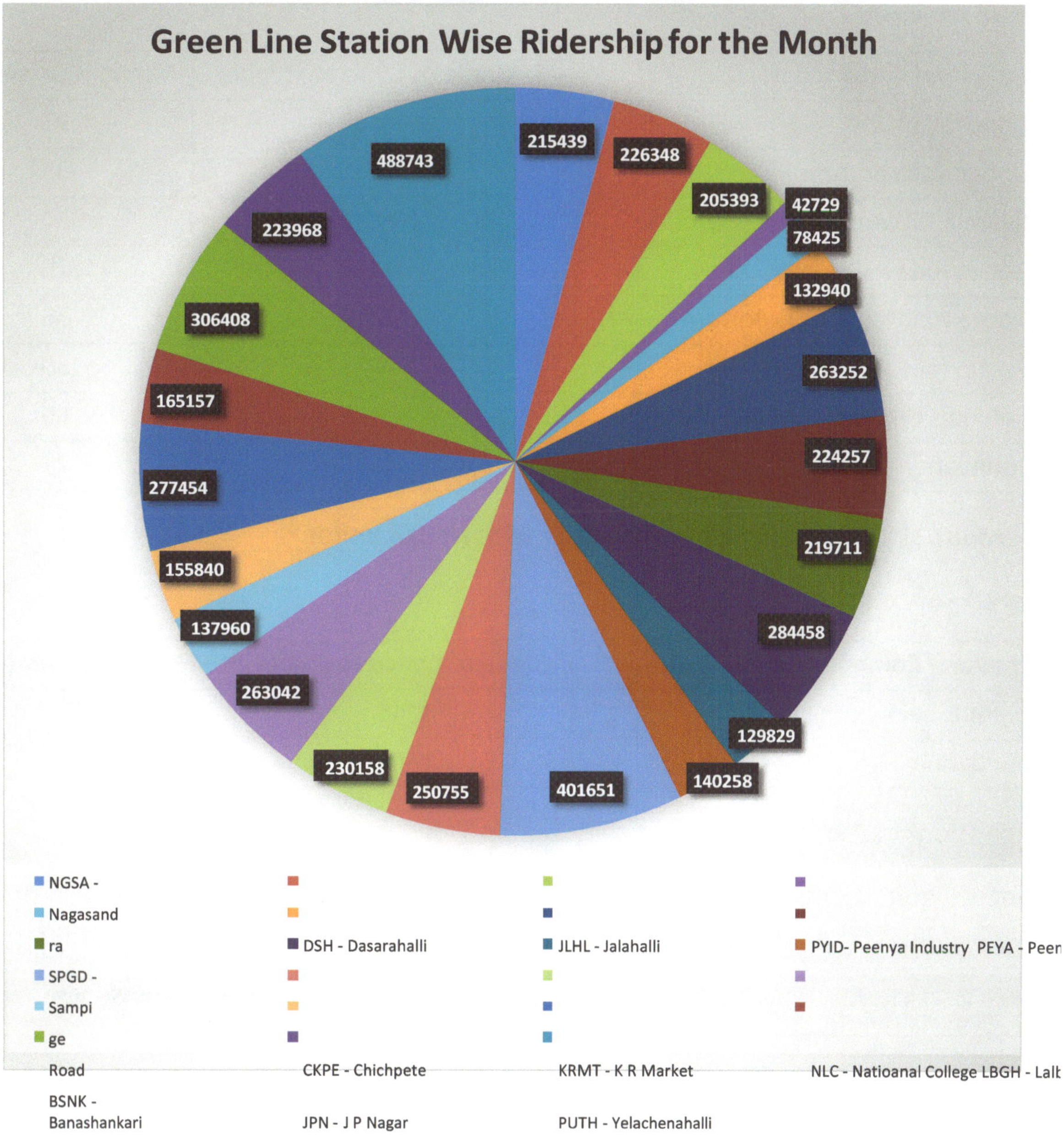

Fig.14.13: Station-wise Ridership- Green Line (BMRCL)

Long-time modulus

- Long-time modulus is calculated through time-dependent analysis of structures for the effects of creep, shrinkages.
- In the absence of detailed study of time-dependent effects, long-term modulus may be assumed equal to 50% of instantaneous modulus.

Modular Ratio: Modular ratio for all concrete grades shall be taken as for tensile reinforcement = $280/f_{ck}$

For Compression reinforcement = $420/f_{ck}$

The minimum grade of concrete is as follows

Table 14.2: Grade of Concrete

Sl. No.	Structural Components	Grade of Concrete (cube)	Nominal Cover (mm)
1.	Upper slabs Top	M35	25
2.	Upper slabs Bottom	M35	35
3.	Beam & Columns above concourse	M35	50
4.	Above road level slabs Top	M35	25
5.	Above road level slabs Bottom	M35	35
6.	Above road level Beams	M35	50
7.	Columns below concourse	M45	50
8.	Foundation	M35	75

For above-ground structural elements in contact with Nallah water

Minimum grade of concrete

Sl. No.	Structural Components	Grade of Concrete (cube)	Nominal Cover (mm)
1.	Columns below concourse	M45	65
2.	Foundation	M45	90

Thermal Expansion = 1.17×10^{-5} /°C

Fire Resistance Requirements =As per the National Building Code, a 3-hour fire rating is recommended for assembly buildings. The fire resistance design of the structure conforms to IS456:2000

Table 14.3: Cover for Fire Resistance Nominal Cover to meet Specified Period of Fire Resistance as Per IS456:2000 –Table

Fire Resistance	Nominal Cover						
	Beams		Slabs		Ribs		Columns
	Simply Supported	Continuous	Simply Supported	Continuous	Simply Supported	Continuous	
h	mm	mm	mm	mm	mm	mm	mm
0.5	20	20	20	20	20	20	40
1	20	20	20	20	20	20	40
1.5	20	20	25	20	35	20	40
2	40	30	35	25	45	35	40
3	**60**	**40**	**45**	**35**	**55**	**45**	**40**
4	70	50	55	45	65	55	40

Prestressing Steel

Pre-tensioned strands for precast I-girder entire spans at the casting yard conform to IS:14268, Class 2, Low Relaxation, uncoated, stress-relieved strands.

Young's Modulus E 195,000 MPa.

Breaking Strength & 0.2% Proof Strength, Jacking Force

- **Breaking** strength of strand = **260.7** kN
- **0.2%** Proof Load = **234.6** kN
- **0.1%** proof stress (85% of minimum ultimate tensile strength) = **221.6**kN
- **Jacking force** 90 per cent of 0.1% Proof stress 0.9 * 0.85 = 76.5% ultimate tensile strength, taken ultimately at 75% ultimate strength = **195.5**kN

Incidentally, as per IRSCBC: 1997 cl. 16.8.1, the maximum seating force in the strand is limited to 75% of the characteristic strength of steel for pre-tensioned tendons. The jacking force in the tendon permitted up to 80% of the characteristic strength of the strand.

Limit jacking force to -75% of the ultimate tensile strength of the strand for pre-tensioned Steel. Post-tensioning tendons.

Post-tensioned strands are used for precast I girders and segmental Box girders at the casting yard.

The prestressing steel conforms to IS 14268, class 2 Low Relaxation, uncoated Prestressing Units: 12K15, 19K15 (longitudinal units)

Breaking Strength & 0.2% Proof Strength, Jacking Force

- **Breaking** strength of strand = **260.7** kN
- **0.2%** Proof Load = **234.6** kN
- **0.1%** proof stress (shall be taken as equal to 85% of minimum ultimate tensile strength) = **221.6**kN
- **Jacking force** in the strand is limited to 90 per cent of 0.1% Proof stress= 0.9 * 0.85 = 76.5% ultimate tensile strength, taken ultimately at 75% ultimate strength = **195.5**kN

As per IRSCBC: 1997, the maximum seating force in the strand is limited to 70% of the characteristic strength of steel for post-tensioned tendons. The jacking force in the tendon permitted up to 80% of the characteristic strength of the strand.

Jacking force limited to 75% of the ultimate tensile strength of the strand for post-tensioned steel

Sheathing: As per the technical specifications, Anchorage set-in 6 mm

HDPE ducts with Friction (wobble) 0.003 per m

For cables with plan curvature, the above value will be enhanced by 10%

Friction (curvature) 0.20 rad-1

Structural Steel

Two types of structural steel are proposed as per the following standards:

IS: 4923 Hollow steel sections for structural use

IS: 2062 Steel for General Purposes Grade B-Designation 410-B

The hollow steel sections would be square (SHS) or rectangular (RHS). Traditional rolled sections, such as plates, angles, channels, and joists, would also be used where necessary.

internally threaded bolt sleeves (hot-dipped galvanised 300 gm/sqm) manufactured from IS: 2062 Grade B mild steel shall affect the base connections and connections with concrete. The sleeve for hexagon-head bolt M20 Class 8.8 as per IS: 1364 (Part 1) with galvanised spring washer.

The connections within the steel structure would be affected essentially by directly welding members with/ without gusset plates. The minimum metal thickness for SHS/RHS sections for main chord members and bracings should be 4mm, as applicable for steel tubes in clause 6.3 of IS: 806.

For high-strength steel, structural steel conforming to Grade Fe540HT as per IS: 8500 is to be used.

Young's Modulus: E= 200,000 MPa Tensile Strength / Yield Strength:

For Hollow steel sections (conforming to IS:4923), Tensile strength shall be **450**MPa for hollow steel sections having a Yield strength of **310MPa**

Structural Steel (Conforming to IS: 2062):

Tensile strength **410** MPa.

Yield strength

250 Mpa (for t<**20**mm),

240Mpa (for 20mm< t < 40mm)

Thermal Expansion Coefficient: □ = 1.2x10-5 /°C

Reinforcement Steel (Rebars)

Only thermo-mechanically treated reinforcement bars of grade 500D (min.) conforming to IS: 1786

Young's Modulus: **E= 200,000 MPa**

Yield Stress: **f_y = 500MPa**.

Time-Dependent Characteristics of Materials

Long-term losses in prestressing (pre-tensioning and post-tensioning) and evolution of material

Characteristics to be calculated using:

- Formula and parameters, given in IRS-CBC
- Or using the CEB-FIP model

For the calculation of long-term effects following the IRS-CBC model, the following Humidity ratio = 70%

Shrinkage coefficient: Automatically and depending on the type of concrete and the humidity ratio

Creep coefficient computed in FIP formulae and depending on stress on concrete at each stage of construction and service, time, modulus of concrete, and humidity ratio

Clearances for Road Traffic: As per IRC guidelines

Clearances for Rolling Stock:

Vertical & Horizontal clearances for rolling stock as per the Latest SOD for Meter Gauge for BMRCL Specification

The distance between the rail's top and the plinth's top is 219mm.

The top of the rail to the top of the finished platform level is 1090mm.

Seismic Design

The seismic analysis will be conducted as per IS: 1893 (Part 1, Section 2). The project site falls in Zone II—ductile detailing per IS13920 & IS4326; while the horizontal design would be as per clause 6.4.2 of IS: 1893 (Part 1- 2002). Response reduction factor (R) = 5.0 for a framed portion (concourse area) and ductile detailing as per IS 4326 and IS 13920. The value of horizontal design seismic acceleration coefficient is not to be less than 0.09 for a framed structure.

Vertical Seismic coefficient

As per clause 7.12.2.2 of IS: 1893 (Part 1-2002), all horizontal projections are to be designed and checked for stability for 5 times the design vertical seismic coefficient.

As per clause 6.4.5 of IS: 1893 (Part 1-2002), the vertical seismic coefficient is to be taken as 2/3rd of the design horizontal seismic coefficient.

Percentage of Live Load considered in seismic weight calculation

Table 14.4: Live Load used for Seismic Analysis

Type of live load	Longitudinal Direction	Transverse Direction
Train Load	0%	50%
Imposed Load ≤ . kN/ ^{2}	25%	25%
Imposed Load > 3.0 kN/m^2	50%	50%

Percentage of Train Live Load considered in combination with the earthquake

In both seismic conditions (longitudinal and transverse), when calculating the stresses due to the vertical effect of the live load, 50% of the design train live load must be considered during the earthquake.

As per IRS Bridge Rules, Correction Slip No. 22, dated January 17, 1994, in transverse/ longitudinal seismic conditions, only 50% of the gross tractive effort/braking force was considered.

The design seismic force resultant at all components were computed as follows:

$r_1 \pm 0.3\ r_2 \pm 0.3\ r_3$

$\pm 0.3\ r_1 \pm r_2 \pm 0.3\ r_3$

$\pm 0.3\ r_1 \pm 0.3\ r_2 \pm r_3$

Where

r_1 = Force resultant due to complete design seismic force along X direction

r_2 = Force resultant due to complete design seismic force along Y direction

r_3 = Force resultant due to complete design seismic force along Z direction

Rail Structure Interaction (RSI)

Unit Load Cases - Parameters

For most of the verifications, a general longitudinal uniform load of 1.6T/m for two tracks was used to check structural members.

Rail-Structure Interactions (RSI) studies analyse the normal stress variation in the long-welded rail generated by:

- Different expansions between the supporting structure and the rail.
- Rotations at deck ends due to deck flexure.
- Acceleration/braking horizontal loads.

Note: The RSI studies are designed to investigate the interaction between rail and deck under acceleration/braking,

Ttemperature variations of the deck/rail, and end rotation. Therefore, the effect of RSI was included in the LL, LF, OT, and DT load cases.

In the RSI studies, the following Parameters/loads are to be considered.

– Vertical train load

- Braking & acceleration loads

  *Standard BMRCL Rail type.

- Ballast-less track. The force/displacement bilinear laws must be compatible with the project's fastening system, which is a continuous welded rail.

- Distance between the top of the rail and the soffit of the deck at the support section
- Creep & shrinkage of concrete not considered (i.e. assumed finished at the track lying phase).
- Thermal effects on rail and deck

For tracks on curvature, the radial force due to the temperature effect in LWR is taken into account.

Stress Limits

Maximum allowable stress in rails: +/- 92 Mpa

DeflectionVertical Deflections at Mid-Span of Track Beam:

under Live Load + dynamic impact: **L/1800** (L =span length). The deflection calculated is the sum of the deflections due to the longitudinal flexure of the beam and the transverse flexure of the bottom slab of the U in the case of the U-beam.

As per IRS-CBC §11.3.4, deck deflection must be checked under the Basic load combination: - DL+SIDL+LL.

Deflection for Station Building: as per IS: 456-2000

Vertical Live Load

The Train Live Load was based on the axle configuration of **Standard Gauge**

Coach length- 21.80m, **Two bogies**, **Axle loads**-160 Kn, **Number of Coaches/Train**-6 **Maximum Speed** – 80/Kmph

Maximum BM and SF are based on one car each (Six Coaches) on Both Tracks, supported by a single slab. Condition

One track is loaded, and both tracks are loaded

All axle loads = 160kN

Horizontal Train Live Load (Braking and Traction)

- Braking load is 18% of the unfactored vertical loads
- Traction loads are 20% of the unfactored vertical loads.

Since a single girder supports both tracks, the tractive force of one track and the braking force of another will be taken in the same direction to produce the worst loading condition. As per IRS Bridge Rules, correction slip no.22 dated 17 / 1 / 1994, in transverse/ longitudinal seismic conditions, only 50% of gross tractive effort/braking force may be considered.

Coefficient of Dynamic Impact: As per the latest IRS Bridge Rules code.

Superimposed Dead Loads (SIDL)

For the **platform slab**, the assumptions are

Floor finishes: 3.6 KN/m2 uniform load.

Suspension load: 2.0 KN/m2 uniform loads. (False ceiling, plumbing & electrical equipment, Escalator Pits, etc. This load is applicable wherever necessary.

Light partition wall: 1.0 KN/m2 uniform load.

For the concourse area: Floor finishes 3.6 KN/m2 uniform loads.

- Load due to additional toilet fill:3.0 KN/m2 uniform load.
- Suspension load:2.0 KN/m2 uniform loads and Others as per actual

Fig.14.14: Elevated Metro Station

SIDL for two tracks

Table 14.5: SIDL used for Two Tracks

Cables	0.7 kN/m
Cable troughs with cover	7.4 kN/m
Cable trays	0.1 kN/m
Concrete plinths for rails + Rail + Pad	31.0kN/m
Miscellaneous (OCS, signnaling)	4.0 kN/m
Handrail	0.8 kN/m

Live load on Platform, Concourse level, Staircases and Foot Bridge

Table 14.6: Live load used for Platform & Concourse Levels

Location	Distributed Load (kN/m2)	Concentrated Load kN
Public/Staff Room		
Concourse Floor	5.0	4.5
Staircase Area/Foot bridge	5.0	4.5
Platform	5.0	4.5
Office Accommodation	5.0	4.5
Shop	5.0	4.5
Technical Room		
Heavy Equipment* Room (ASS, UPS, Electrical Switch)	15.0**	40.0
Other Technical* Room	10.0**	20.0

*To be checked for actual load and its location.

As shown in the table above, floor slabs were designed for the distributed load. Slabs should also be checked in critical locations using the concentrated loads listed in the above table for possible local internal forces exceeding those produced by the normally distributed loads, e.g., cantilevers, punching shear in slabs, and all other locations sensitive to concentrated loads.

Dispersal of loads at a spread-to-depth ratio of 1 horizontally to 1 vertically through concrete and 0.75 horizontally to 1 vertically for other surfacing.

Derailment load (DR): As per the latest IRS Bridge Rules code.

Overall Temperature changes and shrinkage effects:

The seasonal temperature variation, although more pronounced than its daily variation, is not critical for structural design, as its impact becomes insignificant due to creep. Uniform temperature variation is considered a long-term loading. The cracked moment of inertia, which shall be at least 0.75 x gross moment of inertia, can be used in the analysis of reinforced concrete structures.

The total shrinkage strain of concrete = 0.0003 (Cl. 6.2.4 of IS:456-2000)

Coefficient of thermal expansion of concrete = 1.17E-05 per 0C

Equivalent temperature for shrinkage = – 25.64 0C

Differential Temperature (DT)

The differential temperature considered, as defined in IRC: 6-2000, Clause 215, Closed section, shall also be checked for a difference in temperature between the outside and inside of the box girder of 5 0C.

Wind Loads

The wind loads will be taken as IS: 875-1987 Part 3. Wind loads and Seismic Forces are not considered to act simultaneously.

The value of wind load: Wind velocity at height $V_z = V_b$.

$K_1.K_2.K_3$ Where: V_b (m/s): Regional essential wind speed, taken 33 m/s for Delhi.

K_1: Probability factor, considered 1.05

$K_{2:}$ Terrain height and structure size factor, considered 0.84

$K_{3:}$ Topography factor, considered 1.00 □V_z = 33 * 1.05 * 0.84* 1.0 = 29.106 m/sec and, corresponding wind pressure, Pz = 0.6 V_z2 = 0.6*(51.79)2 N/m2 =0.508 Kn/m2

Differential Settlement

* 10mm – Long-term settlement

* 6mm – Short-term settlement

Construction Loads: As per the construction methodology and as per actual

Collision load (CL)

As per IS875 (Part 5) -1987 clause 6.1.2, a concentrated load of 630kN at 1.20m above the road level to be considered.

Load Combinations Methodology& Load Combinations

The central module of the station is to be analysed for load combinations as per IS 456 and IRS CBC. All the worst combinations are considered for design. Load combinations as per the codal provision of IS 456. The load factor of the Train live load and track SIDL is to be considered as per the IRS CBC code. Load Combinations for the design of the end module of the station as per IRS-CBC are presented in Table 1.

Service Limit State (SLS) Loading Combinations

Fig.14.15: Platform of UG Metro

Fig.14.16: Elevated Metro in operation

Reinforced Concrete Structure

The design criteria for checking permissible stress for SLS Load Combinations are not covered in IS 456; therefore, refer to IRS-CBC-1997.

As per IRS-CBC-1997, Table 11, permissible stress in concrete & reinforcement as follows: -

- Permissible compression in Concrete = $0.5f_{ck}$

Prestressed Concrete Structure

Table 14.7: Details used for Prestressed Concrete Design Precast pre-tensioned/post-tensioned I Girder Simple Spans (SLS Check)

No.	Load Combination	Allowable compressive stress	Reference	Allowable Tensile stress	Reference
At transfer and/or Construction stage					
	Cl. Cl. 16.4.2.2(b) of 1 DL + DS + App. IRS CBC 1997 of	16.4.2.4(b) PR 0.5 f_{ci} but < 0.4 IRS CBC	f_{ck} 1 N/mm2		1997
During Service					
2	SLS G I	0.4 f_{ck}	Cl. 16.4.2.2 (a) of IRS CBC 1997	No tension anywhere	-
3	SLS G II	0.4 f_{ck}	Cl. 16.4.2.2 (a) of IRS CBC 1997	No tension anywhere	-
4	SLS G III	0.4 f_{ck}	Cl. 16.4.2.2 (a) of IRS CBC 1997	No tension anywhere	-

Precast Segmental Box Girder Simple Spans (SLS Check)

No.	Load Combination	Allowable compressive stress	Reference	Allowable Tensile stress	Reference
At the transfer and/or Construction stage					
1	DL + DS + App. PR	0.5 f_{ci} but < 0.4 f_{ck}	Cl. 16.4.2.2(b) of IRS CBC 1997	No tension anywhere	-
2	Group 1+ 50% EL	0.5 f_{ci} but < 0.4 f_{ck}	Cl. 16.4.2.2(b) of IRS CBC 1997	No tension anywhere	-
During Service					
3	SLS G I	0.4 f_{ck}	Cl. 16.4.2.2 (a) of IRS CBC 1997	No tension anywhere	Note 2 under Table 11, IRS CBC 1997 and cl 17.3.3
4	SLS G II	0.4 f_{ck}	Cl. 16.4.2.2 (a) of IRS CBC 1997	No tension anywhere	Note 2 under Table 11, IRS CBC 1997 and cl 17.3.3
5	SLS G III	0.4 f_{ck}	Cl. 16.4.2.2 (a) of IRS CBC 1997	No tension anywhere	Note 2 under Table 11, IRS CBC 1997 and cl 17.3.3

Fig.14.17: Metro Yellow Line

Fig.14.18: Blue Line during Construction

Load Combinations for the design of the end module of the station as per IRS-CBC (Table 1)

Table 14.8: Load Combinations

Limit State	Loads	Symbol	G I	G II (EQ)			G II (WIND)		G III	G V
			Cracking + Stress	G II a Stress	G II b Stress	G II c Stress	G II a Stress	G II b Stress	Stress	Stress
SLS COMBINATIONS	Dead Loads	DL	1.00	1.00	1.00		1.00	1.00	1.00	1.00
	Shrinkage & Creep	SC	1.00	1.00	1.00		1.00	1.00	1.00	1.00
	Prestressing	PR	1.00	1.00	1.00		1.00	1.00	1.00	1.00
	Super Imposed Loads	SIDL	1.20	1.20	1.20		1.20	1.20	1.20	1.20
	Earthquake (**)	EQ		1.00	1.00					
	Overall T(2)	OT							1.00	
	Differential Temperature	DT							0.80	
	Differential settlement	DS	1.00							
	Live Load	LL	1.10		0.30			1.00(2)	1.00(2)	
	Derailment Loads	DR								1.00
	Wind Load						1.00	1.00		

Limit State	Loads	Symbol	GI	G II (EQ)			G II (WIND)		G III	G V
				GII a	GII b	GII c	GII a	GII b		
ULS COMBINATIONS	Dead Loads	DL	1.25	1.25	1.25	0.90	1.25	1.25	Not to be considered, per IRS: CBC Cl. 10.3.1	1.25
	Prestressing	PR	1.15/0.87(*)	1.15/0.87(*)	1.15/0.87(*)	1.15/0.87(*)	1.15/0.87(*)	1.15/0.87(*)		1.15/0.87(*)
	Super Imposed Loads	SIDL	2.00	1.50	1.50	0.80	2.00	2.00		2.00
	Earthquake (**)	EQ		1.50	1.20	1.50				
	Overall T(2)	OT								
	Differential Temperature	DT								
	Differential settlement	DS	1.00							
	Live Load	LL	1.75		0.30			1.40(2)		
	Derailment Loads	DR								1.75
	Wind Load						1.60	1.25		

(*) 1.15/0.87: According to IRS CBC ART. 11.3.3. When the Prestressing pressure increases the Section Capacity vs. Shear, Pressure is multiplied by 0.87. When the Prestressing pressure decreases, the Section Capacity vs. Shear pressure is multiplied by 1.15.

(**) It should be noted that the temperature load case is never combined with seismic loading.

Notes:

1. Long Welded Rail loads are naturally included in the OT load case.
2. 50% of LL and LFP will be used in Group II b combinations.
3. The Structure also must be checked for the following construction stage combination:

1.0 DL+1.2 SIDL+0.8 EQ+1.0 PR

Ultimate Limit State (ULS) Loading Combinations: as per IS456:2000, Section 5.

Rail Stresses

The rails are checked in RSI calculations under various forces, including the LWR effect. Rail Check	**Loads**	**Combination 1**	**Combination 2**
	TW	1.0	1.0
	OT	1.0	
	DT		1.0
	LF	1.0	1.0

(TW, OT, DT and LF are all issued from RSI calculations)

Fatigue Check

The fatigue phenomenon needs to be analysed only for those structural elements subjected to the repetition of significant stress variation under traffic load for steel Structure

Fig.14.19: Overhead Launching Girder Moving Forward

Fig.14.20: Span-by Span Construction

14.15 Civil-Engineering

Considering the traffic demand and availability of right-of-way, the Metro rail system, catering to the design PHPDT, is proposed for adoption on the Bangalore Metro corridors. Broad parameters, like other metros in Bengaluru, as envisaged, are

1. Traffic Handling Capacity (PHPDFT): 60,000 with six-car trains with 2 mins headway
2. Alignment-Gauge: Standard Gauge (1435 mm)
3. Maximum gradient-: 4%
4. Maximum radius of curvature: Elevated 120 m & UG 200 m
5. Minimum radius of Curvature in Depot: 100 m
6. Minimum radius at Station Location: 1000 m
7. Design Speed: 90 Kmph
8. Mam. Axle Load: 15 t
9. Min. Track Centre: 4.2 m
10. Rail Level from Ground at Corridor: 7.5 m /8.5m
11. Rail level from the Ground at the Station: 12.5 m

14.16 Design Parameters

DMRC recommended a similar system to Delhi for Bengaluru, with certain modifications to the design norms, considering the city's specific needs and with the aim of standardising parameters for other metropolitan and major cities in the country.

For the elevated section of the Bangalore Metro, a box-shaped deck, like that adopted for the Delhi Metro, was recommended. It has an overall top width of 9.519m and a track center of 4.2 m to accommodate both tracks. The section has side walkways at the coach's floor level for passenger emergency evacuation.

The Bengaluru Metro system has modern, lightweight stainless steel rolling stock. Trains are air-conditioned and consist of 6 coaches from the beginning. The maximum acceleration (1.0 m/sec2) and maximum deceleration (1.1 m/sec2) parameters. It is proposed that the system has a standard gauge.

14.17 Design Principles of the Underground Section (RDSO)

The proposed structure of the Underground station may be a rigid box section with permanent walls serving as an external wall support system, and a beam-slab & column forming the internal structural framing. The roof slab may support the soil and vehicular surcharge, while the passenger and plant loads may be carried by the concourse slab. The track and platform loads may be supported by the base slab. The permanent walls may resist the lateral earth and hydrostatic pressures, in addition to the surcharge and structural loads.

In the design of Underground station structures, the following factors should be considered: -

a. Method of construction, including temporary works and construction sequence.
b. Ground/structure interaction, including the effects of temporary works.

c. Ground pressure, shear force and bending moment distribution during construction and in the long-term.
d. Short- and long-term ground and groundwater response.
e. Other static load changes, such as excavation, surcharge, traffic loadings and the like.
f. Long-term surface water level changes.
g. Dynamic (such as seismic or vibratory plant) loads and displacements.
h. Safe evacuation of passengers in case of accident/derailment/fire, etc.

For the purposes of assessing ground and groundwater pressures, the underground station structures can be considered effectively impermeable rigid box structures subject to earth pressure. The Design shall minimise the effects (such as movement and distortion of the ground).

Fig.14.21: UG Platform of Purple Line Station (24 coaches.com)

For the purposes of assessing ground and groundwater pressures, the underground station structures may be effectively treated as impermeable, rigid box structures subject to earth pressure.

Fig.14.22: Track Laid in Tunnel Section

The Design shall minimise the effects (such as movement, distortion of the ground and the like) on all Existing Building Structures (EBS) that may be affected by the Works. Wherever necessary, additional support will be provided for these EBS. The design of all cut-and-cover structures may consider, but not be limited to the following: -

a. The variation in ground conditions along the alignment; The geological/ hydrogeological features and their variations, including rock joint orientation and spacing, etc.
b. The variation in engineering properties of soil or rock within the influence of the proposed Works.
c. All dewatering and groundwater cut-off systems required to maintain dry and stable conditions within all excavations required for these works.
d. Any ground treatment before, during or after construction of the works (e.g., groundwater recharge), which is required to stabilise the ground and existing building structures (EBS) in
e. Methods by which the completed structure shall be secured against flotation. (Any temporary dewatering system shall not be turned off unless and until provisions have been made to ensure that the structure will not be subjected to leakage or flotation when the groundwater returns to the design levels).
f. Differential groundwater pressures.
g. Methods of waterproofing the completed structure.

Fig.14.23: Dr. B.R. Ambedkar Metro Station, Vidhana Sodha

h. Any difficulties that are envisaged at the site with respect to access, clearances, working space and obstruction to excavation.
i. Maintenance of traffic flows along/ on roads, including access to adjoining properties and roads.
j. Control of heave, swell, piping and instability of the excavations.
k. The noise levels produced during the construction and subsequent operation of trains.
l. The depth of construction required.
m. The effects of vibration and vibration-induced movements – e.g., earthquake.
n. To minimise adjacent ground and EBS movement and distortion.

Fig.14.24: Integrated Completed Elevated Metro Station

Fig.14.25: Metro Platform at Elevated Station (timesnownews.com)

Fig.14.26: Majestic Station Platform near Entrance

Fig.14.27: Elevated -UG Hybrid Station at Whitefield.

Fig.14.28: Purple line- UG Metro Station Platform

Fig.14.29: Metro during Operation (24 Coaches.com)

14.18 Underground Section

BMRCL initially floated tenders for the construction of the 13.9 km underground section in four packages during June 2017. As the tenders had to be cancelled due to high quotations and the second round of tendering resulted in tenders being awarded to firms during the March–June period, 2019, in which (L&T) won two tunnelling construction works and piling for stations began in May 2019. L&T, using a TBM, began in July 2020. Tunnel boring work by the other two contractors (viz. Afcons & ITD-Cem) commenced in 2021.

14.19 Underground Construction Using Tunnel and Instrumentation

After facing years of delays, Bengaluru Metro's longest underground section is making steady progress. Three under-construction stations and a Tunnel Boring Machine (TBM). MG Road, Bengaluru's most famous high street, is getting a new metro station more than 62 feet below the surface. Part of the Pink Line, this station will be integrated with the existing Purple Line, bringing the CBD closer to the city's southern and northeastern suburbs. The 21.26-km Pink Line has a 13.76-km underground section. It has 18 stations, including 12 undergrounds. Of the three stations, civil work is nearly complete at MG Road and Shivajinagar. Work on installing air-conditioning systems. Tracks have also been laid on both tunnels, and preparations are underway to commence work on electrical, signalling, and other systems.

Civil work has resumed at Lakkasandra, the third station en route to Langford Town. The gigantic machine bored 461 metres in September and is expected to break through in late October. Encardio-rite was awarded the monitoring contract for both the underground lines, i.e. UG-1 and UG-2, and the majestic station. The scope of work broadly consisted of:

- Pre- and post-construction condition survey for buildings within the zone of influence.
- Supply, installation of geotechnical and structural instruments used for safety monitoring.
- Manual and automatic monitoring, optical stabilised visual monitoring.
- Programming and commissioning of automatic systems where required.
- Daily, weekly and monthly reporting with data evaluation & interpretation.
- Calibration of data loggers and sensors.

Instrumentation used included:

- Inclinometer system for diaphragm walls and subsurface monitoring around stations and tunnels.
- Inclinometer cum Magnetic extensometer for subsurface lateral movement as well as settlement monitoring.
- Piezometers (vibrating wire, casagrande and standpipe) for monitoring groundwater level and pore pressures.
- Center hole load cell for anchors.
- Strain gauges for strut monitoring for pile testing and diaphragm walls.
- Borehole extensometer for subsurface settlement and ground settlement points for surface settlement monitoring.
- Crack meters, tilt meters, bi-reflex targets, mini prism targets and building settlement points for structure monitoring.
- Portable readout units and automatic data loggers.

Package UG-1 and UG-2 required special settlement monitoring as the subsurface strata consisted of rock and soil saturated with water, which became very critical during tunnelling.

Coastal TTS JV is the Contractor with OCB Rites SYSTRA as Consultants for **Package UG-1**. Encardio-rite was awarded the subcontract for the instrumentation and monitoring of the underground section of the North–South corridor (3.4 km), spanning from the ramp of KIMS Hospital to Malleshwaram Mantra Mall Road, with City Market Station and Chickpet Station in between.

CEC SOMA CICI JV is the Contractor for **Package UG-2,** and the Consultant is OCB Rites SYSTRA. Encardio-rite was associated with CEC-SOMA-CICI JV for the instrumentation of a 4.4 km section starting from the ramp of Chennaswami Cricket Stadium and Magadi Road Ramp. The stations between the ramps are Cubban Park, Vidhan Sudha, Vishveshwarya College and City Railway Station. The alignment had several historical monuments that needed special attention and critical safety monitoring.

For Majestic Station Contractor was GYT-Coastal JV. Encardio-rite was associated with GYT-Coastal JV for the Majestic Station project. Both UG-1 and UG-2 cross each other at Majestic Station. The construction of the crossover double-layer crossing station at the heart of the city was very challenging and required critical monitoring.

Tunnel boring of the underground section UG2 (on the north-south corridor) was completed in September 2016. The tunnel faced a major delay when the cutter head of TBM Godavari broke, and spares had to be awaited. Trial runs on the 4.0 km east-west tunnel of the green line between Sampige Road and National College commenced at the end of March 2017. The entire stretch of the Green Line was opened, completing Phase 1

14.20 Design Basis of Elevated Section

i) General

The concrete structures of the project are proposed to be designed in accordance with the Indian Code IS: 456-2000, which is considered mandatory for concrete structures in the country. However, for some technical issues that are not adequately covered in IS codes, reference has been made to the IRS code or other internationally accepted codes of practice, as listed.

ii) Methodology for analysis and design

The design should model the station using accepted computer software, in accordance with usual practice, and the software used must be approved by the relevant authority. The primary software used to present the results of analysis and design is to be approved by the appropriate Authority. In addition, the designer should verify the results obtained from the primary by a verification software. However, the critical designs must be supported by manual checks. Dynamic analysis is performed to obtain the design seismic force by the response spectrum method as per IS:1893 (Part-1)-2000. Analysis of the framed structure is carried out considering fixed support at the top of the pile cap. The interaction between soil structures was considered in the design of the pile foundation. The structural design of the central module of the station conformed to the codal provisions of IS: 456,

IS: 4326, and IS: 13920. Structural design of the end module of the structure conformed to the codal provisions of

i. Foundation, Substructure and track beam as per IRS-CBC,
ii. Platform beam as per IS 456, IS 4326 and IS 13920. The design of the water-retaining structure shall conform to the codal provisions of IS 3370. Curtailment of reinforcement should not be done in the column. Dynamic analysis must be conducted for all structures. The appropriate grade of concrete and nominal cover to reinforcement is governed by the following main considerations:

    i. Durability of Concrete
    ii. Fire Resistance Corrosion Protection to the Reinforcement
    iii. Bar Size (Minimum diameter shall not be less than 10mm)
    iv. Nominal maximum aggregate size

The stations were broadly divided into two parts, i.e. stations with a concourse and stations without a concourse. The station, with a concourse (approximately 70.0m) located in the middle, is separated from the station arms without a concourse (approximately 35.0m on either side) by expansion joints. The station features a concourse that includes a substation, a pump house, a water tank, and other service buildings on the ground floor, as well as ticketing facilities and other amenities at the concourse level. A steel roof portal is also erected on the platform.

Based on the detailed Geotechnical investigation, which describes the test conducted and the parameters recommended for design, submitted to BMRCL for comments/approval prior to the start of foundation design. A total of 36 boreholes were drilled along this proposed corridor: 26 boreholes along the proposed alignment for Corridor 1, and the remaining 10 borings along the proposed alignment for Corridor 2. Boreholes have been drilled to a maximum depth of 30.00 m each along the proposed Metro alignment.

Fig.14.30: Elevated Corridor in Operation

iii) Pile Capacity

a. Minimum 1.0m diameter bored cast-in-situ vertical piles in soil/rock have been designed for the foundation of piers/columns in the main station building. Minimum number of piles is not less than: (i) Central module of station – three and (ii) End module of station – four.
b. Open foundation has been contemplated primarily for the Access/Entry structure in firm soil/ rocky strata at shallow depth.
c. For pile carrying capacity, the SLS check is considered.
d. The lateral load capacity of the pile is evaluated either by using empirical formulae given in IS: 2911 (Part-1/ section- 2) or by soil structure interaction analysis using Winkler's Spring model.
e. The capacity of piles in rock shall be taken as per IRC:78-2000.
f. The calculation should be carried out for rock socketing as per the latest IRC:78 (including all amendments), including the details on socketed piles in rock.
g. Initial load tests (not on working pile) shall be conducted first for the determination of safe vertical and lateral loads. The safe load was taken as the least of (i) the load arrived at from the initial load test and (ii) the calculated safe load based on the static formula. An initial test was conducted for a load of 2.5 times the safe vertical load, as determined by the static formula.
h. The working load on the pile for vertical and horizontal loads is checked by conducting routine tests during construction.
i. In case of a multiple pile system, spacing of piles was not less than 3 times the diameter of the pile in soil and 2.5 times the diameter when founded on rock.
j. The top of the pile cap was kept about a minimum of 500mm (approx. as per site condition) below the existing ground level, and the weight of the earth cover is applied on top of the pile cap when unfavourable.

The limiting values were not exceeded for the computation of safe load:

The result of the sub-structure investigation is used to determine the values of the angle of internal friction ϕ and the cohesion of the soil. Angle of wall friction δ is equal to ϕ deg., and the Coefficient of earth pressure 'K' is taken as 1.0.

The reinforcement in the pile cap was provided based on the higher value assessed from both theories, namely, the truss analogy and bending theory. Detailing of reinforcement was done according to the method that provides the highest reinforcement. For other pile groups, the bending theory is applied to the design of the pile cap.

The structural design of the pile and pile cap had to be checked in SLS and ULS conditions. Ductile detailing for the central module of the pile conforms to the codal provision of IRC112-2011.

Foundations Stiffness – Static Stiffness Soil static mechanical characteristics and stiffness for lateral loads were taken from IS: 2911 (Part -I/Section 2 Appendix

Soil Dynamic Stiffness – The lateral stiffness of soil under very dynamic loads, such as earthquake loads, is generally higher than static values used for all other static loads. Dynamic stiffnesses were calculated using specific geotechnical tests to evaluate shear and compression. STATION PLANNING

14.21 Types of Elevated Corridor

The choice of Superstructure for viaduct Structures has been made with consideration for the ease of constructability and maximum standardisation of the formwork for a wide span range. Accordingly, like other metros in India, the type of superstructures considered is,

i. Precast segmental box girder with post-tensioning.
ii. Precast pre-tensioned U-Girder superstructure.

Precast segmental box girder with post-tensioning consists of precast segmental construction with post-tensioning and joints glued, preferred technique in fast-track projects. The pre-stressing cables are placed in the conduits inside the structural concrete, which are grouted with non-shrink cement slurry. The match cast joints at the interface of two segments are provided with shear keys for maintaining the correct line and level of the superstructure.

Precast, pre-tensioned U-girder superstructure with a single unit of U-two Girders for each span and two such units to cater for the two tracks. The U Girders are pre-tensioned, pre-cast, transported to the site and erected on the spans using road cranes. As units are single units per span, they are faster to erect compared to segmental box construction. The Pier cap will have to accommodate two U-girders; it must be wider and made as a precast, pre-stressed unit. This feature also facilitates faster construction, using road cranes. Since these units are erected as a single unit per span, the speed of construction of the Superstructure becomes significantly faster compared to segmental construction. Since the Pier cap will have to accommodate two U-girders, it must be wider, and it is made as a precast, pre-stressed unit. This feature also facilitates faster construction. The internal width of a U Girder is generally kept at 4m, and hence this type of superstructure can be adopted for spans on curves with a radius of only up to 300m.

14.22 Structures for Elevated Stations

The locations of stations are either in the middle of the road or partially on the service road, or off-road, as given hereunder.

a. The elevated alignment generally passes on the median of the road, and the stations are also proposed above the road, with entries planned from both sides of the road beyond the existing service road. The proposed stations have two side platforms, and access to the platforms is through staircases, escalators and elevators housed in the paid area of the concourse.

b. The Traffic Integration facility at stations includes approach roads to the stations, circulation facilities, pedestrian ways, an elevated approach for Metro and non-Metro commuters, and adequate halting/ temporary areas for various modes likely to be in Station areas.
c. Connecting bridge at Concourse level has been planned for crossing the road for the use of non-Metro commuters, also through the unpaid areas.
d. Number of stations have been planned for commercial development for an area of about 360 sq. m at the concourse level and off-road entry structure. The area at ground level is planned for use in intermodal transit and parking.
e. A 6m wide service road has been provided around the stations.

Station structures are designed as a combination of Precast and In-situ concrete.

14.23 Station Planning

The stations were planned based on the following:

a. Station planning is dependent on the peak hour traffic load for each station in the design year.
b. The station is planned for 6 6-car trains.
c. The total evacuation time for the movement of all passengers in an emergency from the platform level to the landing at the next level does not exceed 5.5 minutes in elevated stations, considering that the stations are open and the risk is much less.
d. The station planning complies with the "Guidelines and space Standards for Barrier Free Built Environment for Disabled and Elderly persons" published by the Ministry of Urban Affairs and Employment, India in 1998.
e. Seven designs have been proposed for various stations, forming the basis for planning all the Stations.

Type A: 135 m X 20.50 m (Typical)

Type B: 135 m X 22.30 m (with common concourse for corridors 1 & 2)

Type C: 135 m X 20.50 m (Three to accommodate Flyover and with provision for property development- one & three floors)

Type D: 135 m X 37.75 m (With two sides and one island platform)

Type E: 77.2 m X 30.30 m (With one side and one island platform connecting Depot)

Important Issues for Station Planning

a. NMV and Pedestrian Facilities
b. Accessibility for Disabled People
c. Intermodal Integration with Existing Modes
d. Feeder Bus System
e. Parking Area

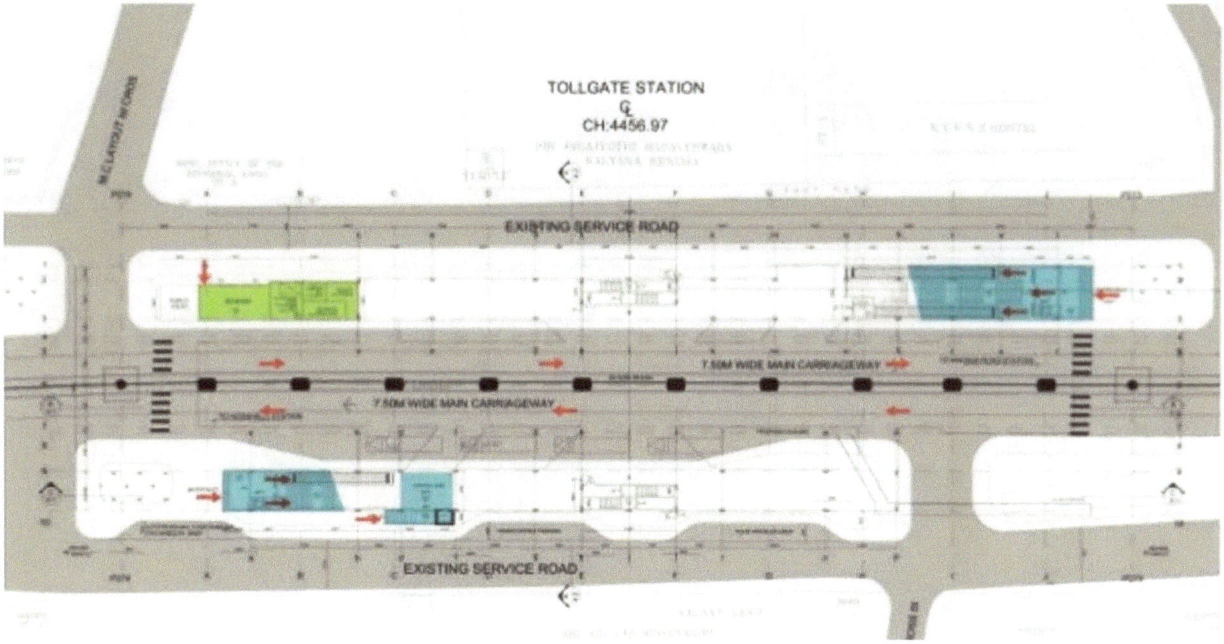

Fig.14.31: Typical Road Level Plan of Elevated Station

The plan should include intermodal integration proposals to facilitate traffic, dispersal and circulation facilities based on

i. Minimising pedestrian/vehicle conflicts and effective passenger interchange with feeder modes,
ii. Facilitating passenger interchange with other transit systems,
iii. Circulation area with a designated space for integration with NMT for last-mile connection

The feeder buses have been proposed as high-quality, ultramodern, and customer-oriented, capable of delivering fast, comfortable, and cost-effective urban mobility. Easy-to-board, attractive, and environmentally friendly minibuses are proposed for the feeder system.

The system should be user-friendly, ensuring accessibility for people with disabilities, those travelling with small children, individuals with temporary mobility issues, and the elderly. The plan needs to provide space for seamless movement around metro stations.

Typical station Plan at concourse, Longitudinal Section of the Station and Cross sections are included in Figures nos.14.8, 14.9 and 14.10.

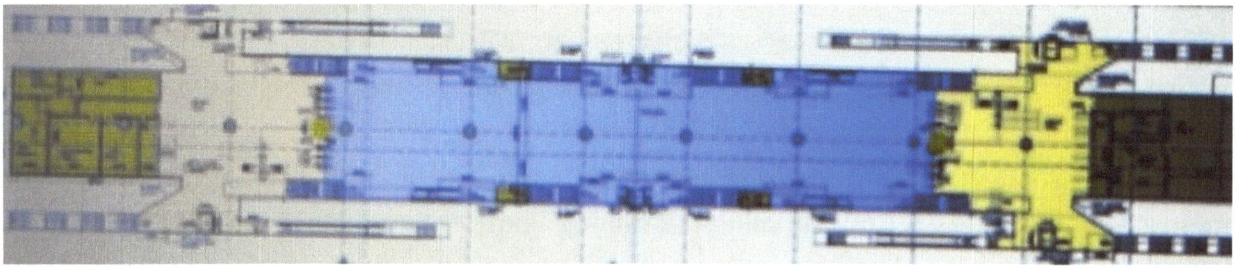

Fig.14.32: Typical Concourse Plan

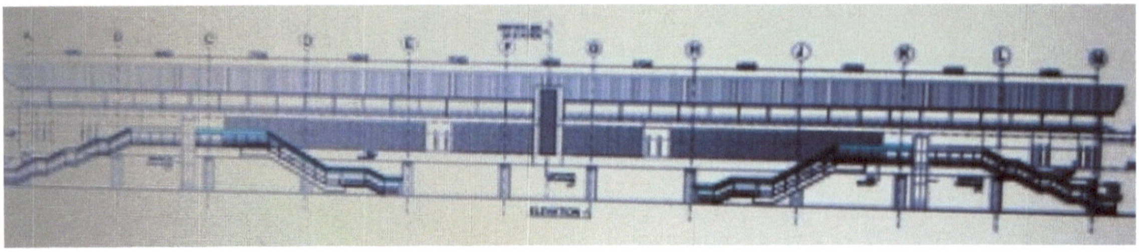

Fig.14.33: Typical Longitudinal Section

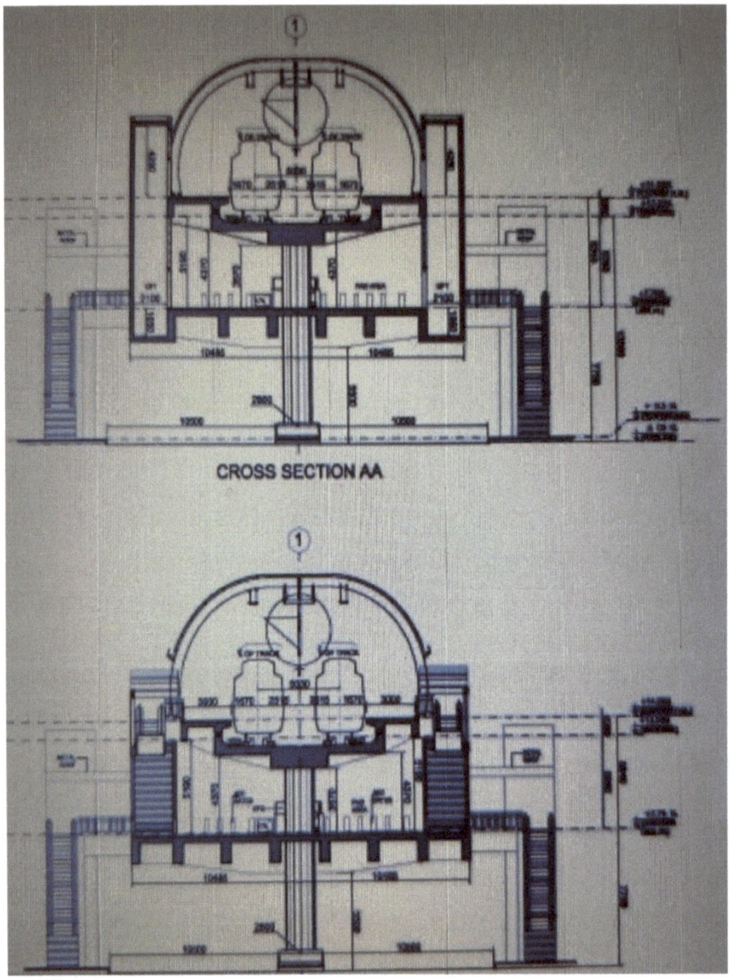

Fig.14.34: Sections at two Locations of Station

14.24 Construction

The construction approach used in the Bengaluru Metro is like those used in other metro construction projects. The construction method employed incorporates both in-situ and segmental construction techniques. The materials used are as per the Contracts and Design Documents. Concrete was produced using a Batching Plant, independent for each contract, and/ or procured from batching plants operating outside the construction sites. For concrete supplied by an independent supplier, a quality/supervising engineer was deployed at the batching plant to ensure the quality and adherence to design specifications for the mix design, as well as the quality of the ingredients, water, and admixtures.

Storage of materials, both at the site and the independent plant, was strictly in accordance with the contract.

Fig.14.35: Erection of Precast Segments Using Overhead Launching Girder

Fig.14.36: Segmentl Construction Using Underslung Launching Girder

For the precast sections, pre-casting yards were established at the site in accordance with design norms. The contractor had to submit the casting yard plan, both for match-casting precast prestressed box sections and/or for prestressed beams, pretensioned or post-tensioned, to the client through the site engineer of the authorised organisation for approval. The casting method was in line with the methods used in other metropolitan areas.

Transportation of precast units to the site was done using an appropriate vehicle and at a time approved by the Site-in-charge/ Client's representative. Construction Photographs are included in the following.

14.24.1 Photographs

Photographs showing different aspects of precast elements.

Fig.14.37: Lifting of Segments in Place

Fig.14.38: Elevated Corridor with Pre-cast Girders

Fig.14.39: Erection of Segments (Segmental Box, Girders & U-Girders)

Fig.14.40: Box Segment and Precast Beam Sections

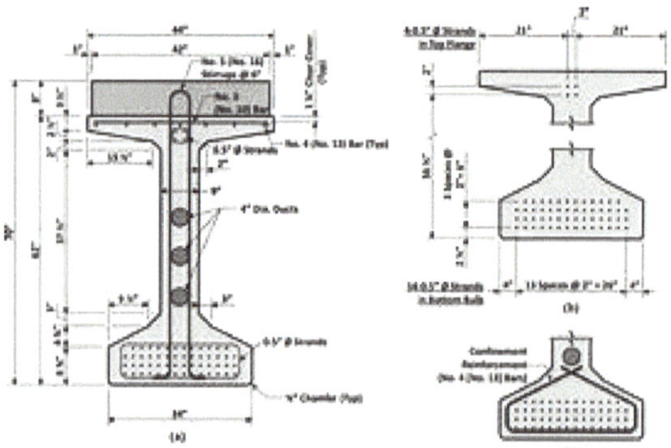

Fig.14.41: Sections of Pre-cast Prestressed Girders

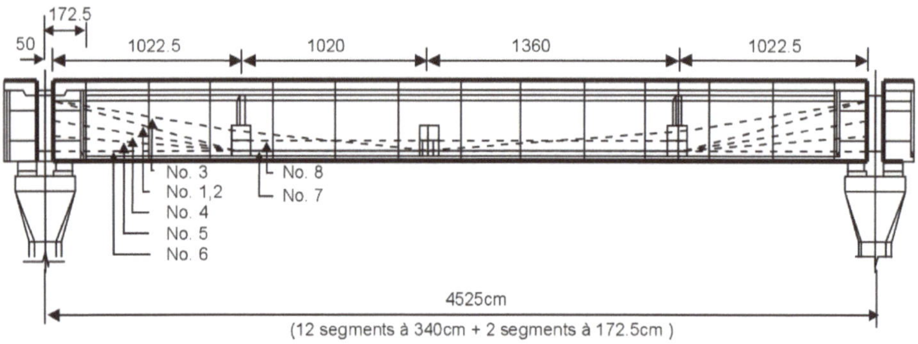

Fig.14.42: Tensioning of Precast Segments (Section)

Fig.14.43: Precast Retaining Walls

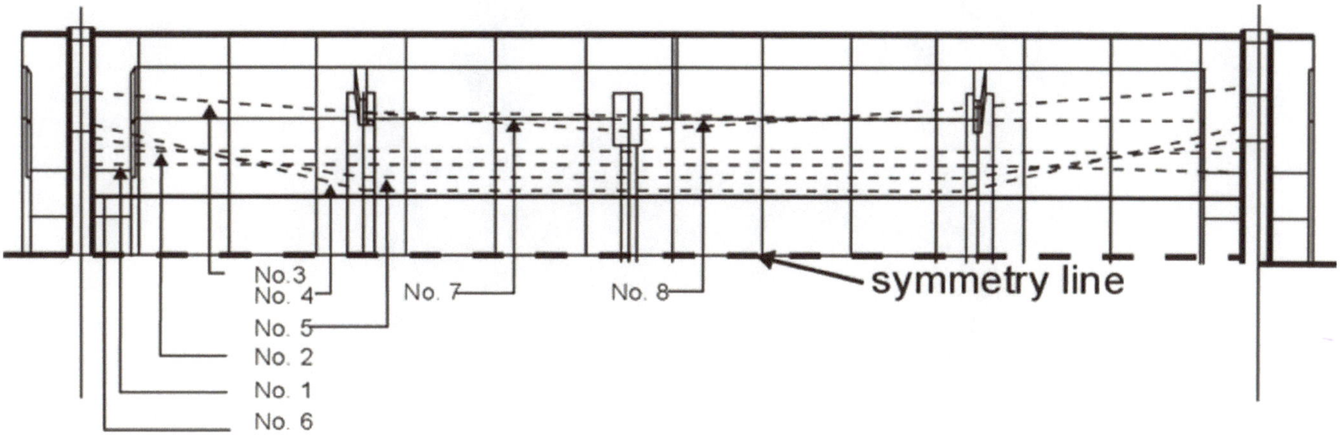

Fig.14.44: Tensioning of Precast Segments (Plan)

Fig.14.45: Box Formwork (bridgeformwork.en.cn)

Fig.14.46: Lifting of Box Section

Fig.14.47: Stacking of Precast Segments (structurea.net)

Fig.14.48: Lifting of U-Girder ((themetrorailguy.com)

Fig.14.49: Tunnel sections (itd cementation)

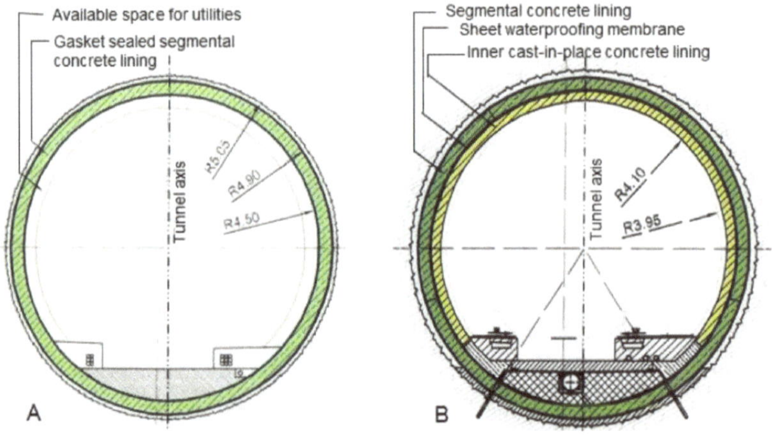

Fig.14.50: Tunnel sections (researchgate.com)

14.24.2 Photographs of Other Than Precast Elements

Fig.14.51: Corridor Approach to Station

Fig.14.52: Erecting & Placing of Roof Truss

Fig.14.53: View of Erected Trusses

Fig.14.54: Foundation during Construcrion

Fig.14.55: TBM in Position before Boring (undergroundexperts.info)

Fig.14.56: Tallest Station under Construction (bangaloremirror.indiatimes.com)

Fig.14.57: Pillar & Girders during Construction (metrorailnews.in)

14.25 Quality and Safety

Despite strict contract conditions in Indian Metro Projects, both for internal and external funding, it is far away from reaching the Goal. All parties involved in the projects, besides the Clients, GC, DDC, manufacturers, suppliers, Contractors, Subcontractors, Labour Contractors, and the supervising engineers and inspectors, need to adhere strictly to quality and safety. Durability and Safety are prime aspects to achieve the cost control and success of the project.

More details are available in a couple of metro projects discussed earlier. A few important issues are included for ready reference.

i) Quality

The factors influencing durability include,

* Environment.
* The cover to embedded steel.
* The type and quality of constituent materials.
* The cement content and water/cement ratio of the concrete.
* Workmanship, to obtain full compaction and efficient curing; and
* The shape and size of the member.

ii) Safety

Fig.14. 58: Derailment guard installed across the network

It may not be out of place to mention, even today, 57% of people who die are unskilled workers. 24% of fatalities occurred with main contractors, and 69% of fatalities belong to the sub-contractor. 36% of fatalities occurred due to either a fall from height or a fall of materials

The risk of injury is not confined to difficult operations, but also to relatively simple activities.

It is not because there are inherent hazards, but because safety methods are not properly in place and are often left to individual workers or operators. Accidents often occur due to a lack of foresight, inadequate or insufficient planning, carelessness, or a deficiency in assessing risk and implementing mitigation measures.

The importance of Quality and Safety as a cost-controlling measure is often overlooked.

- Defects or failures at construction sites result in huge costs and delays, particularly accidents involving personal injuries or fatalities.
- It is now well established that Quality and Safety are critical items on all construction projects for multiple reasons:
  - Protecting the welfare of employees.
  - Providing a safe work environment.
  - Controlling construction costs.
- The objective is to provide a safe work environment to the workmen as the Primary Welfare measure. To achieve this, Clients, GCs, and contractors must deploy experts in safety, in addition to the Supervising Engineers.
- Providing Safe equipment and encouraging workmen to adopt safe work procedures.
- Controlling/eliminating all possible hazards, including Noise, Dust, Heat and Cold.
- Ensuring basic amenities like drinking water, toilets, canteens, rest places, etc.
- Ensuring all the statutory benefits applicable to construction workmen.
- Qualified Personnel

Fig.14.59: Batching Plant

Fig.14.60: Erection of Segments in a busy Road

Supervising Engineers are required to ensure the Contractor's

* Inspection & Testing
* Concrete Batching Plant
* Men & Machinery
* Deployment of Qualified Personnel
* Deployment of appropriate machinery

Supervisors also need to ensure Strength, Water Permeability, Crack Control, and Cover to Reinforcement during production and pouring. In addition, the following items need careful attention.

Periodic Calibration (Monthly)

Equipment Maintenance Program

Concrete material testing before batching (including grading, moisture content, etc.)

Concrete testing before dispatch

Timely sampling of concrete test specimens at the point of discharge

BMRCL, in early 2014, had two road-cum-rail rescue vehicles that could be used to perform evacuations or reload derailed trains back onto the track.

The trains are equipped with derailment prevention equipment, and the tracks are equipped with concrete barriers to prevent trains from leaving the viaduct. The support pillars are earthquake-proof and are designed to have a lifespan of at least 100 years. Trains are equipped with sensors to detect impending collisions and have automatic braking systems to prevent speed limits from being exceeded. In case of train stoppages midway between stations during emergencies, pavements beside tracks are provided on elevated viaducts as well as in underground sections.

Since its opening, suicides on Metro tracks have been a prevalent issue, raising concerns over the absence of platform screen doors —a safety feature present in other comparably lengthy Indian metro systems, such as the Mumbai Metro, Kolkata Metro, and Chennai Metro. All lines from the Pink Line onwards are equipped with platform-screen gates in underground stations. In August 2024, BMRCL decided to retrofit all existing metro stations with PSDs.

Yellow tactile tiles are used at all stations to guide the visually impaired. The tiles start at the ramp and lead to the staircases and lifts. Disabled and elderly passengers can avail a wheelchair at all metro stations.

14.26 Track Structure

Track on Metro Systems is subjected to intensive usage with very little time for day-to-day maintenance. The track structure selected for the Metro must be long-lasting and require minimal or no maintenance, while ensuring the highest levels of safety, reliability, and comfort, with minimal noise and vibrations. Keeping the above in mind, two types of track structures are proposed for this corridor, like those in other Indian metros. The normal ballasted track is planned inside the Depot. The ballast-less track is planned on Viaducts, as the regular cleaning and replacement of ballast at such locations is not possible.

The track is laid with 1 in 20 canted rails, including on the turnouts, and the wheel profile of Rolling Stock should be compatible with the rail Cant and rail profile. Turnback and CBTC systems are designed to operate at an operational frequency of 2 minutes to accommodate a future increase in PHPDT.

14.27 Power Supply and Traction

Like other corridors, the 750 V DC Third Rail traction system is proposed for Bengaluru Metro phase 3. It is suggested to operate with a peak period headway of 3.5/4.0 minutes for Corridor 1 and 4.0 for Corridor 2.

A communication-based Train Control (CBTC) signalling system provides an adequate safety level of CENELEC SIL 4 (Safety, Integrity Level) and permits an operational

14.27.1 Sources of Power Supply

Bengaluru has 66kV and 220kV power transmission and distribution networks to cater to various types of demand near the proposed Corridors. Considering the reliability requirements and the complete length of the corridors, three (03) Receiving Substations are proposed to provide power supply for traction and auxiliary services.

14.27.2 Standby Power Supply

Auxiliary substations (ASS) are envisaged to be provided at each station for stepping down the 33kV supply to 415V for auxiliary applications. The demand for power at each elevated station is about 200 kW at the beginning and may reach 300 kW in the horizon year. Each elevated station shall be provided with an Auxiliary Substation with two numbers. 33kV/415V, 3 phase, 500 auxiliary transformers to cater to depot cum workshop load.

14.27.3 Solar Energy Harnessing System

Considering the futuristic technology and potential for solar power generation, the provision of grid-connected solar Power is on the rooftops of stations, sheds, and buildings. Based on the city's solar radiation intensity, the peak solar power generation in the corridor is expected to be about 200 kWp for each elevated station and 500 kWp for the Depot.

14.28 Rolling Stock

Rolling Stock is proposed for the Bangalore Metro Phase 3 corridors, like Bangalore Metro Phase 2A and 2 B. The rolling stock is planned as a 3 / 6 coach train with a stainless-steel body, measuring 21 m x 2.88 m x 3.848 m, with a capacity of 975 passengers for the 3-coach car and 1,904 for the 6-coach car. The system is designed for a maximum speed of 90 km/h and 80 km/h. for max operating speed. Rolling stocks are also planned for a 3-phase driving system with VVVF control and regenerative braking.

14.29 Signaling-System

In September 2009, the consortium led by Alstom Project India Limited was awarded a contract to supply a control and signalling system for the first phase of the project. The consortium was led by Alstom and comprises Alstom Transport SA, Thales Group Portugal SA, and Sumitomo Corporation.

Alstom provided the design, manufacture, supply, installation, testing, and commissioning of the train control and signalling system, and Thales provided the design, installation, testing, and commissioning of the telecommunication system for Phase 1 of the metro system. It included the Urbalis 200 Automatic Train Control system, which would ensure optimal safety, flexible operations and heightened passenger comfort.

The integrated control centre at Baiyyappanahalli has direct communication with trains and stations, and is equipped with CCTV, providing visual and audio service information. Passengers have emergency voice communication with train staff.

The Communication Train Control (CBTC) Signalling system provides adequate safety, and the level of CENELEC SIL 4 (Safety Integrity Level) permits an operational headway of 90 seconds with continuous automatic train control. CBTC Technology suits UTO, (Unattended Train, Operation) / DTO (Driverless Train Operation). Considering the above, the CBTC system is recommended for Phase 3.

14.30 Telecommunication System

The telecommunication system is the backbone for signaling and provides telecommunication services to meet operational and administrative requirements. The proposed telecom system and transmission media need the following sub-systems:

GE (Gigabit Ethernet) based Transmission System Telephone Exchange.

Mobile Radio Communication System Public Address System (PAS) Centralized Clock System

Passenger Information Display System (PIDS) Close Circuit Television (CCTV)

Central Digital Recording System (CDRS)

Central Communication Fault Reporting System (CCFRS)

14.31 Fare Collection

Mass Rapid Transit Systems proposes to handle many passengers for Bengaluru Metro-Phase-3. Ticket issues and fare collection are vital to the system's efficient and proper operation. The automatic fare collection system and the proposed ticketing system are simple, easy to use/operate, easy on accounting facilities, capable of issuing single/multiple journey tickets.

14.32 Ventilation and Air Conditioning System

The proposed corridors of Metro Phase 3 have been planned as wholly elevated. The Stations' air conditioning and ventilation requirements primarily serve ancillary spaces, such as staff rooms and equipment rooms. It is essential to maintain an acceptable environment for the operation and maintenance of equipment by properly controlling temperature, pressure, humidity, and potential gas accumulation.

14.33 Maintenance Depot

To cater to the stabling and maintenance of trains to be put into commercial operations up to 2041, the Bangalore Metro Rail Corporation Limited (BMRCL) is preparing five new depots to augment its existing three depots. The Baiyappanahalli depot, being completely remodelled with two levels, is unique as it will cater to two Metro lines operating in opposite directions. The present Location of the depot will remain, and the remodelled depot will be the first on the network to have two levels. It will cater only to the 16 trains that will run on Outer Ring Road (Phase 2A) and 21 trains on Airport Lines (Phase 2 B). Both run in opposite directions, and this depot is in the middle of both.

n the remodelled depot, one level will be built below ground, while the other will be at grade level. The underground section will have 21 stabling lines, while the grade-level section will have 20 lines. BMRCL is expecting to increase its fleet to a total of 159 trains by 2026-28 in phases. Five new depots are being built. The Shettygere depot, situated near the airport, is currently under construction. It will have ten stabling lines at grade, with 12 elevated lines to be built in future.

At the upcoming Kothanur depot, which has 20 stabling lines, construction for the Pink Line is at the midway point. At the Anjanapura (North-South Green Line) depot, construction is halfway through. Challaghatta depot, with 38 stabling lines, is the most expensive of all depots. At Hebbagodi depot, work is generally complete and is slotted for the opening in December this year.

Fig.14.61: Hebbagodi Depot (oneindia.com)

Fig.14.62: Baiyappanahalli Depot (sawarajyamag.com)

At present, the Kadugodi (Whitefield) depot takes care of the Purple Line, and the Peenya depot of the Green Line. Most operations at Baiyappanahalli Line have been shifted to Kadugodi since the former is being remodelled. The Operations Control Centre (OCC) will remain at Baiyappanahalli.

The maintenance depot at Sunkadakatte for Corridors 1 and 2 of Phase 3. According to the operation plan for the year 2051, Corridor 1 will require 41 rakes, and Corridor 2 will require 16 Rakes. Accordingly, rolling stock maintenance facilities are proposed to be created for 57 rakes of both corridors. To facilitate seamless operation from either end of the corridors, it is suggested that a stabling facility for 40 rakes at the Sunkadakatte depot and eight rakes at Peenya be built, and the balance at terminal stations.

Fig.14.63: Yellow line Depot (english.jagaran.com)

Fig.14.64: Kothanur Depot (metrorailnews.in)

The Bengaluru Depot design parameters are like those included in the Kolkata and Delhi Metro depot plans. The maintenance depot is to have The infrastructure to maintain the rakes with necessary facilities viz stabling lines, scheduled inspection lines, a workshop for overhaul, unscheduled maintenance including significant repairs, wheel profiling, heavy interior/ under frame/ roof cleaning, etc. for the rolling stock operational on the corridor as well as maintenance facilities for Civil track, buildings, water supply; Electrical Traction, E&M; Signaling & Telecommunication; Automatic Fare Collection etc.

14.34 Environmental

(Rites)

Land use along the corridor consists of built-up areas and vegetation. The city's elevation ranges from 850 to 950 meters above sea level, and it is in Seismic Zone II. Bengaluru city has a tropical savanna climate with distinct wet and dry seasons. Maximum rainfall occurs between June and November. According to the Air Quality Monitoring data by KSPCB, PM10 and PM2.5 exceed the permissible limits.

14.34.1 Impacts

The project yields benefit in terms of improved accessibility, reduced fuel consumption, lower vehicle operating costs, shorter travel times, enhanced quality of life, and decreased road accidents. A huge number of trees are likely to be cut /transplanted during construction along the proposed corridors and depot. The dust is produced due to excavation, loading & unloading, and transportation of construction materials

Mitigation, Management Plan and Monitoring Plan measures and management plan for Water and Energy management, Compensatory afforestation, Management of construction material and waste, Energy efficiency, Construction Safety Management, Labour Camp and Workplace facilities, Air Pollution Control, Noise pollution and Vibration control, Rainwater Harvesting, Management Plan for Depot, Solid Waste Disposal, Training to engineers and managers from the Implementing Agency. Environmental Monitoring is proposed during the construction and operation phases of the project. The Environmental Management Cell must supervise and coordinate the monitoring and implementation of EMP.

14.35 Disaster Management

(Rites)

It is a crisis that results in massive damage to life and property, uproots the physical and psychological fabric of the affected communities and outstrips the capacity of the local community to cope with the situation. Disasters are situations that cause acute distress to passengers, employees, and bystanders and may be triggered by external factors. As per the disaster management act, 2005"disaster" means a catastrophe, mishap, calamity or grave occurrence in any area, arising from natural or manmade causes, or by accident or negligence which results in substantial loss of life or human suffering or damage to, and destruction of, property, or damage to, or degradation of, environment, and is of such a nature or magnitude as to be beyond the coping capacity of the community of the affected area.

14.35.1 Measures

The inherent characteristics of the metro system make it an ideal target for terrorists and miscreants. Metro systems are typically open and dynamic systems that carry thousands of commuters. Moreover, the high cost of infrastructure, its economic impacts on society, and its status as the lifeline of the city with high news value pose a greater threat to its security. Security is a relatively new challenge in the context of public transport. It addresses problems caused intentionally, differing from safety, which addresses problems caused accidentally. Security problems or threats are caused by individuals whose actions aim to undermine or disrupt the public transport system and/or harm passengers or staff. Providing an excellent and reliable security system is a prerequisite for the metro system.

CHAPTER 15

Jaipur Metro

15.1 Introduction

Jaipur Metro, launched in June 2015, is operated by the Jaipur Metro Rail Corporation (JMRC). It currently runs along the Pink Line, covering 11.97 kilometres from Mansarovar to Badi Chaupar, passing major stations such as Civil Line, Chandpole, and Sindi Camp. The Jaipur Metro route length will expand with the upcoming orange line, which is currently in the planning phase and will connect Sitapura to Ambabari, thereby improving connectivity in the north and south. Jaipur Metro is fast, convenient and pocket-friendly. Properties near metro stations have seen a 15-20% rise in value due to improved connectivity and increased demand. The metro boosts real estate growth and urban development, making it a smart investment in Jaipur's growing infrastructure. Jaipur Metro offers fast and convenient travel across the city, helping people avoid traffic and save time. With expanding routes and smart services, it is becoming a key part of life. The updated metro timing of Jaipur gives an exact time for people to travel easily.

Metro Link to Satellite Town: Plans are underway to connect nearby towns, including Bassi, Bagru, Chomu, and Kotputli, to the Jaipur Metro. The government is also studying new routes, such as Badi Chaupar to Galta Ji, to improve access for more people. DPR for Jaipur Mero prepared by **DMRC**. Inputs are taken from **Ayesa, Spain, Wikipidea** & Others.

15.2 Important features of the Metro

* It runs on solar power, supporting green transportation

* The pink line passes through Jaipur's walled city, a UNESCO World Heritage Site.

* The project was completed in record time, just 4 and a half years for Phase 1.

* Jaipur Metro offers different types of tickets that suit different kinds of passengers, whether one is a daily commuter or tourist, or travelling in a group. One can choose a token for a single ride, a smart card for regular travel or a tourist card for unlimited rides. Group tickets are also available for 21 or more people, with discounts.

15.3 Route Map

The Jaipur metro route map offers a detailed view of the current Pink Line, which spans 11 stations from Mansarovar to Badi Chaupar. The map shows station names, nearby landmarks and upcoming lines. Commuters can use this to commute easily and locate key facilities, such as lifts, escalators, ticket counters, and more. The Jaipur metro schedule is clearly marked, helping passengers save time and understand the routes.

15.3.1 Phases

The **Jaipur Metro** is the sixth metro in operation in India. Construction on the most elevated part of the first line, called **Phase 1A,** comprised a 9.63 km route and nine stations. Construction of the first line started in November 2010 and was completed in 2014. Line 1, which runs from Mansarovar to Chandpole Bazaar, began its commercial service in June 2015. The Jaipur Metro Rail system is India's sixth metro rail, and is the first Metro in India to run on a triple-storey elevated road and metro track. **Phase 1-B**, which is from Chandpole to Badi Chaupar, began operation in September 2020.

The Jaipur Metro Rail Corporation Ltd. was created in January 2010. Among the Rapid Transit systems of India, it has been recorded as the fastest to conduct the trial run after starting construction, when it commenced trial runs in Jaipur in September 2013, flagged off by the then Chief Minister of Rajasthan. Physical construction work in February 2011. JMRC consulted the DMRC on rapid transit operation and construction techniques.

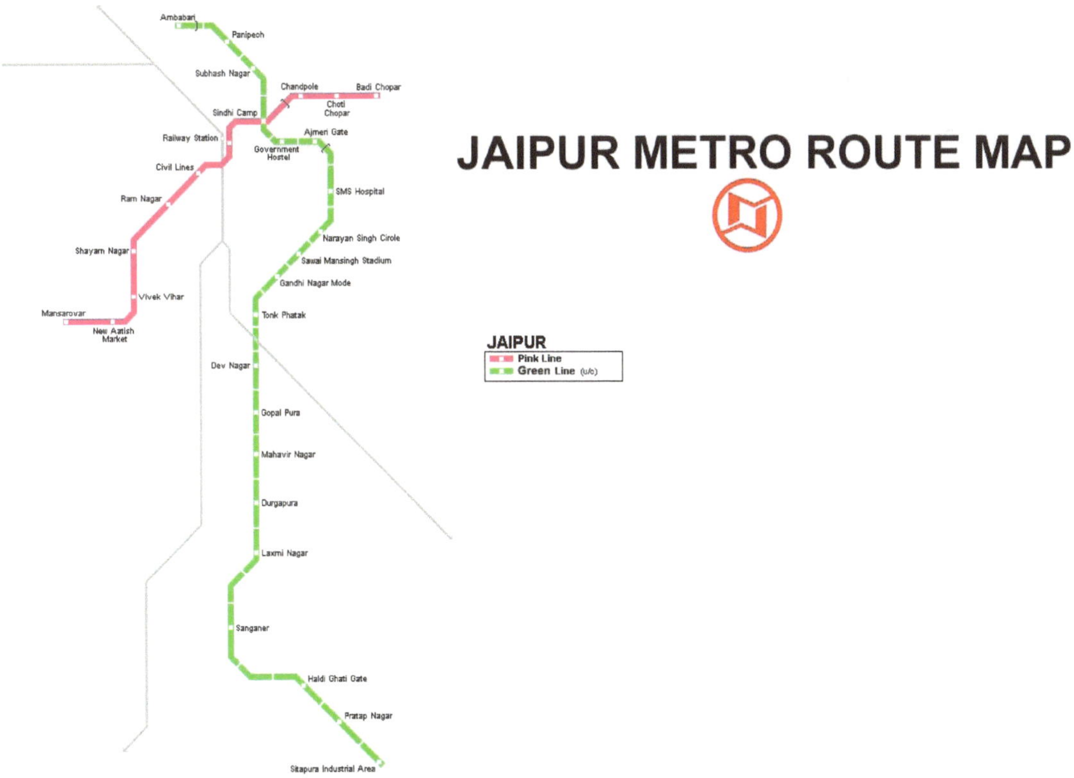

Fig.15.1: Jaipur Metro Network Plan (Yometro)

The first line was opened by the then Chief Minister of Rajasthan. Phase 1-B of the project started operating in September 2020.

Fig.15.2: Jaipur Metro

The Jaipur was built in 2 phases. Phase 1 consists of the Pink Line, and Phase 2 consists of the Orange Line. The implementation of *Phase* 1A of the project, including civil works, permanent way, depot, traction and power supply, etc., was being managed by DMRC. Phase I-A completed 9 stations, and the corridor, of which 0.95 kilometres is underground, and 9.13 kilometres is elevated. The remainder of Phase 1-B, which is 2.34 kilometres long and includes 2 stations, was completed in September 2020. Phase 2, 23.099 kilometres in length, was scheduled for completion by 2021.

15.3.2 Pink Line

Fig.15.3: Exterior of the Railway Station - Pink line Phase 1A

The first route of Jaipur Metro (East-West Corridor) connects Mansarovar to Badi Chaupar via Civil Lines and Chandpole. In Phase I-A, the metro was operating between Mansarovar to Chandpole. The construction of Phase 1 B between Chandpole and Badi Chaupar was completed and started operation in June 2015, with a colour code of Pink. The depot for this line is located at Mansarovar.

Phase 1A

The Jaipur Metro was planned to be built in phases. Phase-1A, 9.63 km, was completed in 2014, and Phase-1B in 2020. Phase II was scheduled for 2020.

The Pink Line was completed with the completion of Phase 1B, which is 2.34 kilometres long and features 2 stations, in September 2020. The tunnel excavation work between Chandpole and Badi Chaupar was completed in August 2017, and civil work was completed by October 2019.

Phase 1B

The Pink Line was realised with the completion of Phase 1B (2.34 km, two stations) in September 2020. The tunnel excavation between Chandpole and Badi Chaupar was completed in August 2017, and civil work was completed by October 2019. It was opened to the public in September 2020.

Phase 1C

For Phase 1C, JMRC planned to extend the Pink Line from Badi Chaupar to Transport Nagar and prepared a detailed project report (DPR) for Phase 1C, incorporating public suggestions. The total length of the project is 3.412 km, comprising 2.452 km underground and 0.960 km elevated. There is one underground station at Ramganj Chaupar and one elevated station at Transport Nagar. This phase missed its deadline. Once constructed, the estimated ridership along the entire stretch, from Mansarovar to Transport Nagar, is expected to exceed 1.38 lakh.

Phase 1D

For Phase 1D, Jaipur Metro Rail Corporation (JMRC) planned to extend the Pink Line from Mansarovar to Ajmer Road. For this purpose, the corporation has prepared a detailed project report (DPR) of Phase 1D, incorporating public suggestions.

According to the DPR, the total length of the project will be 1.312 km, elevated, with 1 elevated station at Ajmer Road and a 0.357 km loop line.

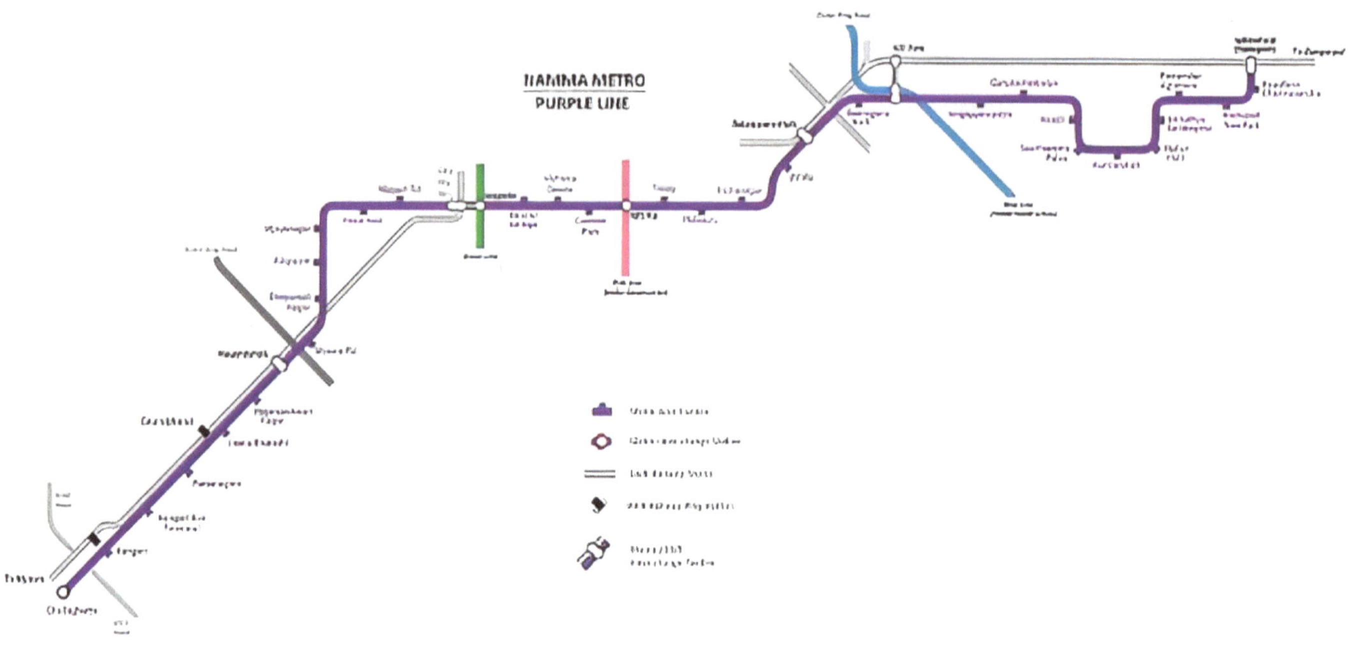

Fig.15. 4: Jaipur Metro Route Map

15.4 Briefs on Line 1B

Delhi Metro Rail Corporation (DMRC) has published a tender for the Engagement of a Detailed Design Consultant (DDC) for the civil, Architectural, and Building works of Two Underground Stations (Choti Chouper and Badi Chouper), as well as the shaft and Tunnels of Phase 1 B.

The project was awarded to M/s Ayesa Ingenieria y Arquitectura S.A.U., and the contract agreement between DMRC & M/s Ayesa Ingenieria y Arquitectura S.A.U. was signed on 2014.

Phase 1B was executed by Jaipur Metro Rail Corporation Ltd. (JMRC) as an underground extension from Chandpole to Badi Chaupar in the heart of Jaipur city. The alignment covered approximately 2.349 km of fully underground section, constructed using standard gauge twin bored tunnels with Tunnel Boring Machine (TBM) technology. The civil contract was awarded to CEC–CICI JV.

The primary objective of Phase 1B was to extend the Phase 1A corridor (Mansarovar to Chandpole, 9.63 km) into Jaipur's dense, heritage-rich Walled City. This extension provided last-mile connectivity to the historic and commercial core, while reducing road congestion and preserving environmental quality in the heritage zone. Civil works commenced in 2013, tunnelling was completed in 2017, trial runs were conducted in 2019, and commercial operations began in September 2020.

The construction faced significant challenges due to extremely narrow lanes, high-density heritage structures, and sensitive archaeological sites within the Walled City, requiring careful TBM operation and continuous structural monitoring. Despite these difficulties, the project successfully delivered enhanced connectivity, enabling seamless metro travel across 12 km from Mansarovar to Badi Chaupar.

The underground alignment safeguarded Jaipur's historic Pink City architecture reduced vehicular traffic in the Walled City, and improved environmental conditions through lower emissions and better air quality. Additionally, it provided easier access to major landmarks such as Hawa Mahal, City Palace, and Jantar Mantar, thereby boosting tourism and contributing to the local economy.

Fig.15.5: A few Photographs from Line 1B

15.5 Orange Line

The authorities are planning phase 2 (North-South Corridor). The Orange Line will be 23.1 km in length, and serve 20 stations (5 underground, 15 elevated).

The Jaipur Metro Rail Corporation Ltd. (JMRC), created in January 2010, operates the Metro in Jaipur. DMRC was the consultant to JMRC for transit operation and construction activities.

When complete, the Orange Line will be 23.099 kilometres and serve 20 stations. It will connect the Sitapura Industrial Area in the South to Ambabari in the North via Ajmeri Gate and MI Road. The tracks will be elevated between Sitapura and Ajmeri Gate and then go underground.

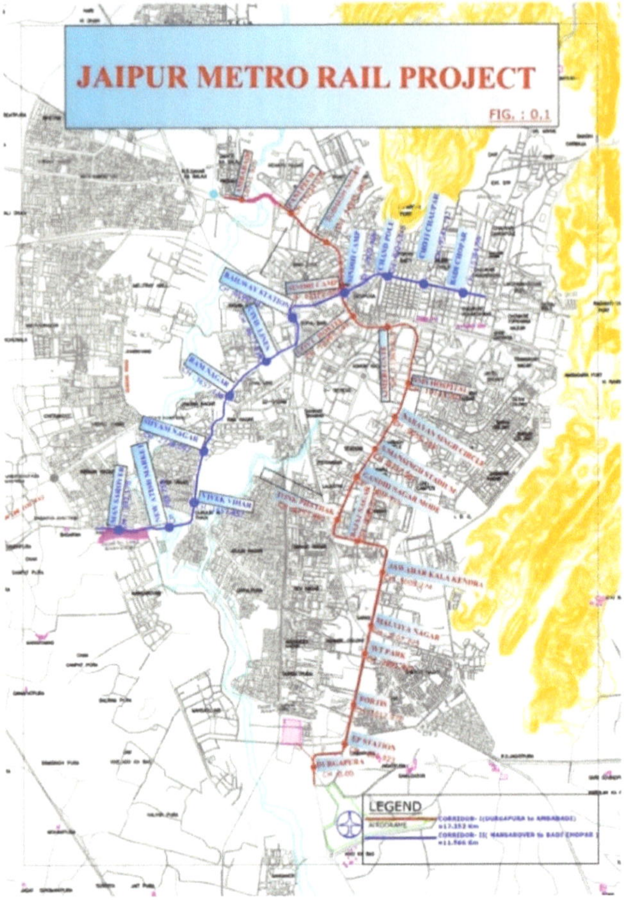

Fig.15.6: Jaipur Metro network

Fig.15.7: Jaipur Metro running on a double storey elevated track

Fig.15.8: Internal View of a Railway Station - Pink line

Fig.15.9: Metro in Operation (1)

15.6 Finances

The estimated cost of the east-west corridor of the Jaipur Metro is ₹3,149. The state government would fund ₹600 crore, while other state urban development and housing departments would bear the rest. Phase 2 is expected to cost ₹6,583 crore, for which the government is mulling over the PPP mode.

FIg.15.10: Jaipur Metro Phase 2

15.7 Basic Features

i) Phase 1 (East-West Corridor)

Total Length 12.067 kms

- Elevated from Mansarovar to Chandpole - 9.278 kms
- Underground from Chandpole to Badi Chaupar – 2.789 kms
- 8 elevated and 3 underground stations

Fig.15.11: Metro in Operation (2)

ii) Phase 2 (North-South Corridor)

Total length - 23. 099 kms

- Elevated Section from Ambabari to Pani Petch, and Ajmeri Gate to Sitapura – length 18.004 kms
- Underground from Pani Petch to Ajmeri Gate – length 5.095 kms
- 15 elevated and 5 underground stations

Financial (Price Level April 2011)

Total Cost 9732 crore
East West Corridor (Phase-I) 3149Cr
North South Corridor (Phase- II) 6583 Cr.
Projected FIRR
Phase – I8.24%
Phase – II 7.37% 3
Economic IRR 18.6%

15.8 Brief on Phase 1B

Delhi Metro Rail Corporation (DMRC) has published a tender for the Engagement of Detailed Design Consultancy (DDC) services for Civil, Architectural, and Building works of Two Underground Stations (Choti Chouper and Badi Chouper), as well as Shafts and tunnels, for Phase 1B of the Jaipur Metro Project

Commercial Operation Date Phase I A: 1-7-2013 Phase – IB: 31-3-2017 Phase – II: 31-3-2017

Fig.15.12 & 15.13: Typical Internal and External Views of Stations

The project was awarded to **M/s Ayesa Ingenieria y Arquitectura S.A.U.** vide. Agreement with DMRC signed on 19/06/2014. Jaipur Metro Rail Corporation Ltd. (JMRC) executed the Jaipur Metro Rail Project Phase 1B as an underground extension from Chandpole to Badi Chaupar in the heart of Jaipur city. The alignment covered approximately 2.349 km of fully underground section, constructed using standard gauge twin bored tunnels with Tunnel Boring Machine (TBM) technology. The civil contract was awarded to CEC–CICI JV, while the Detailed Design Consultant (DDC) was M/s Ayesa Ingenieria y Arquitectura S.A.U. This stretch included two underground stations, Chhoti Chaupar and Badi Chaupar. The original contract value was INR 3,20,00,000, which was later revised to INR 3,97,46,731, with an intended project duration of 36 months up to 20 April 2017. However, the project faced delays, and the actual completion date was 23 September 2020.

The primary objective of Phase 1B was to extend the Phase 1A corridor (Mansarovar to Chandpole, 9.63 km) into Jaipur's dense, heritage-rich Walled City. This extension provided last-mile connectivity to the historic and commercial core, while reducing road congestion and preserving environmental quality in the heritage zone. Civil works commenced in 2013, tunnelling was completed in 2017, and trial runs were conducted in In 2019, commercial operations began in September 2020.

The construction faced significant challenges due to extremely narrow lanes, high-density heritage structures, and sensitive archaeological sites within the Walled City, requiring careful TBM operation and continuous structural monitoring. Despite these difficulties, the project successfully delivered enhanced connectivity, enabling seamless Metro travels across 12 km from Mansarovar to Badi Chaupar.

The underground alignment safeguarded Jaipur's historic Pink City architecture, reducing vehicular traffic in the Walled City and improved environmental conditions through lower emissions and better air quality. Additionally, it provided easier access to major landmarks such as Hawa Mahal, City Palace, and Jantar Mantar, thereby boosting Tourism and contributing to the local economy.

15.9 Salient Features

Gauge Standard

The Standard Gauge (1435mm) is primarily used for metro railways worldwide. Over the past decade, several new metro systems have been constructed in various cities across India. And most of these metros have gone in for Standard Gauge, even though the national gauge for India is Broad Gauge.

Total Route Length Total - 35.166 kms

Elevated – 27.282 kms

Underground – 7.884 kms

Rolling Stock Coach Width – 2.9 m Axle Load – 16 tons 4

Coach Unit Capacity – 1034 passengers

Traction 25 KV – OHE

Signalling ATO

Stations' Capacity for 6 coach units

Land (Stns. & Depots) North South Corridor East West Corridor 37 Ha. 19 Ha.

Land for property development 50 Ha.

15.10 Important Tasks

i) Processes for rapid implementation,

i. Smooth land acquisition
ii. Entire Row for Phase IA acquired
iii. 22 hectares acquired for Property Development
iv. Effective Settlement Committees
v. Quick redressal of legal issues
vi. Support of urban local bodies
vii. Cost optimisation and minimum cost escalation
viii. Kickstart the project with a unique project structure and attract partners for the second phase
ix. Parallel processing of civil works, rolling stock, and S&T works

ii) PPP Consultants engaged through the IIPDF Scheme of the GoI

- General & Technical Consultant – Mott MacDonald
- Financial Consultant – Deloitte Touche Tohmatsu
- Legal Consultant – Luthra & Luthra Law Offices
    - RFQ issued for PPP Concessionaire
    - Four Applicants qualified (out of five applications):
- Reliance Infrastructure Limited ▫ Gammon Infrastructure – Iridium Concessions
- Soma – ICF
- Essar Projects – Samsung CNT

iii) RFP - included Annexures

- Draft Manual of Standards & Specifications (MSS) prepared
- Financial Model
- Technical Schedules, other than MSS, prepared with assistance
- Draft Concession Agreement (DCA)

iv) Government of India support

- Metro Railways (Operation & Maintenance) Act, 2002 and Metro Railways (Construction of Works) Act, 1978, extended to Jaipur.

- Support under the IIPDF Scheme for Project Development
- Initial in-principal approval given to the project by the GoI
- Amended project structure being considered for approval

v) Streamlined Land Acquisition

- Most of the alignment on the medians of major roads: minimal land acquisition
- No major hurdles
- Robust R & R Mechanism

* Other parcels for parking and circulation are being regularly identified and transferred.

15.11 Civil Engineering

The design parameters related to the Metro system described herein have been determined based on a detailed evaluation, experience, and internationally accepted practices. Various alternatives were considered for most of these parameters, but the best-suited ones have been adopted for the system. Horizontal curves are provided with the following parameters: Underground, Elevated Description Section Desirable Minimum 300 m 200 m Absolute minimum 200 m 120 m Minimum curve radius at stations 1000 m 1000 m Maximum permissible Cant (Ca) 110mm Maximum cant deficiency (Cd) 85 mm 85 mm

15.12 Elevated Sections

The viaducts carrying the tracks will have a minimum vertical clearance of 5.5 m above road level. To meet this requirement with the 'Box' shaped pre-stressed concrete girders, the rail level will be approximately 9.8 m above road level. However, at stations located above the central median, the rail level will be 12.5 m above road level, with a concourse at the mezzanine. These levels, however, will vary slightly depending on the location of the stations.

The track centre on the elevated section is maintained at 4.2m throughout the corridor to standardise the superstructure, except at a few locations as detailed below: On curves with radii below 300m but up to 120m at scissors crossings, and in Underground sections, the track centre ranges from 4.2m to 4.5m.

15.13 Underground Sections

The rail level at the midsection in the tunnelling portion shall be maintained at 12.0 m below the ground level, ensuring a cover of 6m is available over the tunnels. At stations, the desirable depth of rail below ground level is 12.5m; the Track centre in the underground section to be constructed by Tunnel Boring Machine (TBM) is 13.05m to accommodate a 10m wide island platform. The track centre in the underground section to be constructed by the cut-and-cover method is 4.5m. Gradients Normally, the stations shall be on a level stretch. In limiting cases station may be on a grade of 0.1%.

Between stations, the grades generally may not exceed 3.0%. However, where existing road gradients are steeper than 2%, gradients of up to 4% (compensated) can be provided in short stretches on the main line for Switch Over Ramps.

15.14 Design Speed

The maximum sectional speed will be 80 km/h. However, the applied cant and length of transition will be decided in relation to normal speeds at various locations, as determined by simulation studies of alignment, vertical profile and station locations. Geometrical design norms are based on international practices adopted for similar metro systems with a standard gauge, on the assumption that the maximum permissible speed on the section is limited to 80 km/h. Horizontal and vertical alignment are largely dictated by the road's geometry, followed by the alignment.

15.15 Viaduct Structure

The proposed viaduct structure for the Jaipur Metro is a Pre-cast segmental box girder, carrying two tracks supported on a single pier located on the median of tenured below the viaduct structure. Ine foundation shall be a pile foundation at most locations. Open foundations are possible at certain isolated locations. The superstructure shall be pre-cast segmental construction, which will cause minimal inconvenience to the road users during the execution stage. Station Locations & Planning Stations have been strategically located to serve major passenger destinations and facilitate convenient integration with other modes of transportation. However, effort has also been made to propose station locations, such that inter-station distances are as uniform as possible and the average spacing of stations is as close to 1 km as possible.

15.16 Stations

A total of 11 stations has been planned along this corridor, out of which 8 are elevated and 3 are underground stations. Alignment of the East-West Corridor includes a section on Viaduct Structure The proposed viaduct structure for the Jaipur Metro is a Pre-cast segmental box girder, carrying two tracks supported on a single pier located on the median of the road. Road clearance of 66 m.

Stations have been divided into two distinct areas, namely public and non-public. The public area is further subdivided into an unpaid and a paid area. Provisions for escalators are made at all stations in the paid area from the beginning. Provisions in civil structures at stations are being made to provide lifts for disabled passengers in the future. Integration facilities at Metro stations include approach roads to the stations, circulation facilities, and pedestrian ways and circulation areas for various modes likely to come to important stations, including feeder buses. Parking for private vehicles has not been proposed due to the scarcity of land along the alignment.

Fig.15.14: Choti Chauper Station During Construction (themetrorailguy.com)

15.17 Geo Technical Investigations

Geotechnical investigation work at the site was carried out to determine the existing subsoil strata, the proposed type & depth of foundations, and the safe bearing capacity of the foundations required for the proposed two Metro Corridors in Jaipur City, based on the results of 35 boreholes. Core drilling was carried out by using a rotary-type boring machine with diamond bits of N size. The casing of 100/150 mm diameter was advanced up to the firm strata as per IS: 1892 (1979). The description of bore logs for bore holes drilled as per IS-5313. The topsoil is generally a silty sand with gravel, varying in thickness.

Weathered/hard rock at a depth of 13 meters was encountered near Ajmer Puilya on the East-West Corridor. The rock is a metamorphic type of quartz.

15.18 Utilities

The proposed Metro alignment passes along major arterial roads of the city's road network, which serve institutional, commercial and residential areas. Many surface and subsurface utility services, such as sewers, water mains, stormwater drains, telephone cables, electric poles, and traffic signals, exist along the proposed alignment. Details of the existing utility services along the proposed alignment have been collected from the concerned authorities.

System

15.19 Power Requirements

A power supply is required for the operation of the Metro system, including the running of trains, station services (such as lighting, lifts, escalators, tunnel ventilation systems, signalling, telecom, firefighting, etc.), and workshops, depots & and other maintenance infrastructure within the premises of the Metro system. The major component of the power supply is the traction requirements for elevated sections, as well as the auxiliary requirements for the Underground section. The power requirements of a metro system are determined by the peak-hour demands of power for traction and auxiliary applications.

15.19.1 Sources of Power Supply

The high-voltage power supply network of Jaipur city comprises a 220kV and 132kV network to cater to various types of demand in the vicinity of the proposed corridor. 220/132 kV sub-stations are located along the alignment of Corridors. Keeping in view the reliability requirements, two input sources of 132 kV Voltage level are normally considered for each corridor. According to the construction sequence, the revenue operation of the elevated sections of the two corridors will commence before the Underground sections are completed. The intersection of the two corridors will be at Sindhi Camp station (the Underground station of the N-S Corridor). Therefore, to achieve the desired reliability, two receiving substations (132/33/25 kV) are proposed to be set up for the E-W Corridor at each location. Based on the discussions with Rajasthan Rajya Vidyut Prasaran Nigam Limited (RVPN), it is proposed to avail power supply for traction as well as auxiliary services from the following grid substations at 132kV voltage through cable feeders.

15.19.2 Auxiliary Supply Arrangements for Stations & Depot

Auxiliary sub-stations (ASS) are envisioned to be provided at each station (three ASSs for Underground stations and one ASS for elevated stations) for stepping down the 33 kV supply to 415 V for auxiliary applications. A separate ASS is required at the depot. The ASS will be located at the mezzanine or platform level inside a room. The auxiliary load requirements have been assessed at 200kW for elevated/at-grade stations, which is likely to increase to up to 300kW by 2031. For underground stations, the requirements are 2000kW, which is likely to increase to up to 2500kW by 2031. To meet the requirement for auxiliary power, two dry type cast resin transformers of 500 kVA capacity are proposed to be installed at the elevated stations (one transformer as a standby) and one transformer of 1.6 MVA at each underground ASS. For Property Development within the station's footprint, a provision to add a third transformer later may be kept at the elevated station.

15.19.3 Standby Diesel Generator (DG) Sets

In the unlikely event of simultaneous tripping of all input power sources or grid failure, the power supply to stations, as well as to trains, will be interrupted. It is, therefore, proposed to provide a standby DG set of 200 kVA capacity at the elevated stations and two sets of 1000/750 kVA at the Underground stations.

15.20 Rolling stock

The Metro uses 1,435 mm standard gauge rolling stock. Trains are maintained at the Mansarovar Depot for the Pink Line. BEML was awarded a ₹318 crore contract for 10 four-car trains for Phase 1. As traffic increases, the Jaipur Metro plans to lengthen the trains to 6 coaches.

Rolling stocks are indigenous and manufactured at a factory in Bangalore. The four-car trains have a capacity of 1506 commuters per train, accommodating 50 seated and 292 standing passengers in each coach. These trains will feature CCTV cameras both inside and outside the coaches, power supply connections within the coaches to charge mobile devices and laptops, humidity control, microprocessor-controlled disc brakes, and be capable of maintaining an average speed of 32 km/h. Trains on the metro can operate at a maximum speed of 80 km/h.

Trains on the metro operate at a maximum speed of 80 kilometres per hour. Maximum speed is limited to 42 km/h on curves

15.21 Traction System

Considering the ultimate traffic requirements, standardisation, and other technoeconomic factors, the 25 KV traction system is likely the best trade-off and is therefore proposed for adoption on the Jaipur Metro System.

15.22 Signaling and Train Control

Metro carries many passengers at a very close headway, requiring a very high level of safety enforcement and reliability. At the same time, heavy investment in infrastructure and rolling stock necessitates optimisation of its capacity to provide the best services to the public. These requirements of the metro are planned to be achieved by adopting ATP (Automatic Train Protection) and ATS (Automatic Train Supervision) signaling systems. Automatic Train Operation (ATO) will be added in future.

The Jaipur Metro utilises cab signalling and a centralised automatic train control system, comprising automatic train signalling modules. An integrated system comprising optical fibre cable, on-train radio, CCTV, and a centralised clock and public address system is used for telecommunication during train operation.

15.23 Telecommunication

The telecommunication system serves as the communication backbone for signalling systems and other systems, such as SCADA and AFC, and provides telecommunication services to meet the operational and administrative requirements of the metro network. The telecommunication facilities proposed help meet the requirements for 1. Supplementing the signalling system for efficient train operation. 2. Exchange of managerial information 3. Crisis management during emergencies 4.

Passenger information system. The proposed telecom system will cater to the following requirements: Train Traffic Control

15.24 Ventilation and Air-conditioning

The underground stations of the Metro Corridor are built in a confined space. Many passengers occupy the concourse halls and platforms, especially during peak hours. The platform and concourse areas have limited access from outside and lack natural ventilation. It is therefore essential to provide forced ventilation in the stations and inside the tunnel for: Supplying fresh air for the physiological needs of passengers.

15.25 Ticketing & Recharge

Jaipur Metro commuters have three choices for ticket purchase. The RFID tokens are valid only for a single journey on the day of purchase, and the value depends on the distance travelled, with fares for a single trip ranging from ₹5 to ₹15. Fares are calculated based on the origin and destination stations using a fare chart and depend on the peak rush in the Metro. Travel cards are available for longer durations and are most convenient for frequent commuters. Tourist cards can be used for unlimited travel on the Jaipur Metro network over short periods. Two kinds of tourist cards are valid for one and three days, respectively

15.26 Operations

A concession for the operation and maintenance of Stage 1 and Stage 2 is to be let on a PPP basis. Trains operate at a frequency of 10 to 15 minutes between 6:45 and 21:00, depending on whether it is a peak or off-peak time. Trains are operating within the network, typically travelling at speeds of up to 40 km/h and stopping at each station for about 20–40 seconds. Automated station announcements are recorded in Hindi and English. Many stations offer services such as Parking lots, ATMs, and mobile recharge. The metro also features a sophisticated fire alarm system for emergency warnings, and fire-retardant materials are used in trains and on the premises of stations. The first coach of every train is reserved for women.

15.27 Environmental Impact

A detailed Environmental Impact Assessment Study has been carried out along the proposed alignment. Comprehensive environmental baseline data were collected. Both positive and negative impacts of the project were assessed. The project has positive environmental impacts, including reduced traffic congestion, savings in travel time, decreased air and noise pollution, lower fuel consumption, and fewer road accidents. However, the project has some negative impacts, especially during its implementation of the project. An important environmental consideration of this project is that neither any forest area nor any plants/ trees of endangered species exist along the proposed alignment, though 1158 trees will need to be uprooted. A few residential/commercial structures are affected. To minimise the negative environmental impacts, an Environmental Management Plan has been drawn up.

Mansarovar Depot includes i) All minor inspections of East-West corridor, ii) All minor repairs iii) Lifting of coaches for replacement of heavy components. iv) Stabling of all the trains.

15.28 Maintenance Depots

The depot planning is based on the following assumptions: i) ii) iii) iv) v) There is no connectivity between the North-South corridor and East-West corridor to transfer trains from one line to another, else it would have reduced depot requirement in the East-West corridor. Enough space is available at Bambana Nala and Mansarovar terminals for the establishment of Depot-cum-workshop and a satellite depot, respectively. Road transport shall be available for transporting heavy equipment for IOH/POH and heavy repairs from the Mansarovar depot to Sitapura depot-cum-workshop and vice versa. All inspection and workshop lines are designed to accommodate 6-car trains. SBLs at terminal stations shall accommodate 6-car trains, and SBLs within depots shall accommodate a composite fleet of 4-car and 6-car trains. It is proposed to establish a depot-cum-workshop at Sitapura Indl. Area and a small depot at Mansarovar. Design of Depot is like other Indian Metros.

15.29 Security

The security of Jaipur Metro has been entrusted to the Rajasthan Police. The latest security equipment has been provided at all Metro Stations. Closed-circuit cameras from Indigo Vision are used to monitor trains and stations, and additional metal detectors, X-ray baggage inspection systems, and dog squads are also deployed, which are used to secure the system. Each underground station has about 45 to 50 cameras installed, while the elevated stations have about 16 to 20 cameras each.

15.30 Bottlenecks

- Processes involved in seeking the GoI support
- Funds from the Government of India for Phase I
- Safety Trials for Phase I A
- Availability of Skilled Manpower for O&M of Phase - IA
- Financial Issues ▫ Market Borrowings & Debt Servicing ▫ Consolidation of the Dedicated Metro Fund
    - Self-Sustainability
- Engaging PPP Concessionaire for Phase II
- Integration of EPC Phase with PPP Phase
- Urban Integration Issues

- Multimodal Integration
- Transit-Oriented Development
- Heritage-related issues

- Public Sensitisation
    - Increased Ridership
- Timely Implementation and avoiding cost overrun

Fig.15.15: The Metro runs on a double-storey elevated track (postoast.com)

15.31 Accidents

There were a few accidents at the Metro project site, injuring labourers and passersby, which attracted criticism for a lack of security measures. In one incident, a 5.5m long wall collapsed, killing two men.

15.32 Issue of Heritage Structures

Phase I-B is not in accordance with Jaipur's archaeological laws, which state that any kind of digging/ tunnelling work in the vicinity of heritage sites is not allowed. According to the archaeological laws, whoever destroys, injures, mutilates, defaces, alters, removes, disperses, misuses, imperils or allows to fall into decay a protected monument, or removes from a protected monument any sculpture, carving, image, bas-relief, inscription or other like object, shall be punishable with imprisonment for a term which may extend to six months with a fine which may extend to five thousand rupees or with both.

15.33 Construction Time Photographs

Fig.15.16: Mansarovar Depot (themetrorailguy.com)

Fig.15.17: Corridor during Erection (livemint.com)

Fig.15.18: Elevated Metro during Operation (3)

Fig.15.19: Elevated Metro during Operation (4)

Fig.15.20: Tunnel Section (TOI)

Fig.15.21: Overhead Launching Girder in Operation (ITD Cementation)

Fig.15.22: Erection at Second Level – ITD Cementation

Fig.15.23: Lifting of Segment in Position

Fig.15.24: Erection at Second Level (2)

CHAPTER 16

Safety during Construction of Metro Rails in the Indian Context

Adherence to safety norms in Metro Construction is a critical activity. As such, in all the Metros constructed since the end seventies, safety aspects are part of the contract. For JICA-funded projects, safety is a key component of the SHE (Safety, Health, Environment) item.

While going through the Metro Rail Project's details on the internet, the author located a safety study report named 'Analysis of Factors influencing Safety in Metro Rail Construction and Safety Perspective of Metro Rail Construction Employees' (by?). The author believes it is appropriate to include this for the benefit of all concerned. The study was conducted on the Chennai and Hyderabad Metros, with reference to the Kolkata, Delhi, Kochi, and Bangalore metros.

The following paragraphs, presented in a slide format, expand on the factors, Problems, Challenges, and Importance.

Why metro, problems, Significance, Organisation of Study, Objectives, Questionnaire, Conclusion and Recommendation.

16.1 Factors

- Factor 1-perspective on the safety of on-site construction workers
- Factor 2 is the safety-related factor specific to an actual construction site, on-site management-related variables, such as the safety system, the safety organisation, and the safety strategy
- Factor 3: the problems in government supervision during the design and construction process
- Factor 4: These items express the lack of practical market restrictions on safety in the construction industry
- Factor 5 items indicate the unpredictability of tasks in metro construction
- Factor 6 items indicate the requirement during an emergency

16.2 Collection of data by questionnaire

- Statistical analysis is used for analysing data
- Sites visited – Chennai metro & Hyderabad metro

Steps in data analysis

1. Classification of questionnaire questions
2. Classification of questionnaire form based on
    i. Education qualification
    ii. Job position
    iii. The company they are working for.

16.4 General problems in metro rail construction in India

Lack of financial resources, cost control & venture capital.

- Technical failure.
- Lack of planning and a cost estimate.
- Lack of financial resources, cost control and venture capital
- Improper implementation of project management processes and training of key project staff
- Public acceptance/opposition due to the social, economic, political & environmental impacts.
- Ignorance of health and safety considerations, as well as the absence of activating health and safety acts
- Lack of efficiency and effectiveness of the Project Management process

16.5 Challenges in Metro Rail Construction

- Completion of the project within the estimated time.
- Completion of the project within the estimated budget.
- Safety during execution.
- Regular funding of the project.
- Construction of a viaduct over a busy road junction.
- Planning and scheduling of the project.
- Construction that minimises environmental impact and respects public safety.
- Land acquisition for metro construction.
- Resource management.
- Other technical and non-technical challenges.
- Corruption-free construction.

16.6 Why metro

Many metro projects are going on in the major cities of India

- A huge investment is required for the construction
- Many projects are about to come
- Many people work on each metro rail project

- Metro rail construction work is very risky
- Its future of India

16.7 Importance of safety in metro rail construction

Construction of the metro rail requires a huge amount of manpower

- Most of the time, skilled manpower is required because the work is risky and requires extra attention during execution.
- While laying viaducts and overhanging components of viaducts, many employees, skilled and unskilled workers together under huge machines and hang concrete components
- Thus, safety is extremely important during such a situation, hence safety in the metro project

16.8 The problem

Any accidents at the site may cause various serious problems, like

- Loss of lives
- Loss of money
- Loss of market value
- Thus, it is very important to have effective safety management in construction
- Where to put more effort to improve effectiveness

16.9 Conclusion

Top management in the metro construction industry is very serious about safety aspects in metro rail.

- B. Tech/ engineering graduates are found to be most sincere about safety rules, regulations, and policies.
- Post-graduates and diploma holders, compared to graduates, are found to be less aware and sincere about safety
- People with inter/10th and other qualifications are very little aware of safety aspects and safety importance

16.10 Recommendation

Lower job position employees need to be taken care of more, while being educated about safety

- Employees who are less qualified to be taught more and should be well trained because they who found weak in the management chain
- The importance of emergency equipment and mock drills should be explained well, and all employees should be trained for emergencies at the site
- A frequently unannounced mock drill shall be made

**** The content of Chapter 16 is a copy of "Analysis of Factors Influencing Safety in Metro Rail Construction and Safety Perspective of Metro Rail Construction employees"**

Acknowledgement

The Author would like to express his sincere gratitude and thanks to Professor SS Chakraborty, Former CMD, CES, for encouraging and giving free hand to get involved in Transportation and the Mass Transit System globally, and Dr AK Chatterjee, a well-known Expert from Kolkata, for suggesting a book on Kolkata Metro concerning the first Metro in the Country, North-South Metro, particularly when very few persons involved in the project are alive. My sincere thanks to Mr Rajesh Prasad, former Director Operations & Advisor, IRCON and Akshay Sharma, Director-Projects, PITCMRL, for their contribution.

I convey my sincere thanks to Mr GR Madan, former GM, Kolkata Metro & Director-MRTS for MMR DA, Neeraj Jain, former CE, Biswanath Dewanjee, Former ED, KMRCL. I am grateful to my colleagues and former MTP experts, namely, BK Ghosh, MK Bhattacharjee, Chandurkar, and BG Kale; system experts, Satyen Bhowmik & V Narayan; and civil/Structural experts. My sincere thanks to my colleagues and fellow professionals who helped me from time to time, by providing data to the best of their knowledge, namely G Ranganath Babu, C Sankarlingam, KC Thapliwal, M Pramanik, S Chatterjee, Soumya Ray, Makarand Khare and Amal Bhattacharya. Thanks to Jitendra Rastogi, Mukul K Lodhi, and Partha P Banejee. My sincere thanks to my colleague Tanmoy Chakraborty, who provided continuous assistance during my tenure at E-W Metro and throughout the writing of this book. Also, thanks go to Auyon Roy for his coordination with the Publisher for publication.

The author drew upon the Detailed Project Reports by DMRC and CES-led consortia, Reports by GC & IC on Mumbai Metro-Line 3, along with reports by RITES, BMRCL and Wikipedia. And the aurhor used the work experience of CES as part of GC, DDC and of RUPL (Author's team) as LDC.

References

1. B. C. Roy, 2004. Planning of Transportation System in an Indian Metropolis
2. B.C. Roy, Kolkata Metro, including East-West Metro
3. Dr BC Roy & Dr N Som, 2000, Behaviour of Braced Cut in Soft Soil Condition, Calcutta Metro
4. Dr. BC Roy & A. Basu, Evolution of Metro Rail Projects in India: Design and Construction Practices
5. Biswanath Dewanjee, First Tunnel across Major River Hooghly for East-West Metro Kolkata
6. PRK Murthy, Metro Rail System in MMR: Planning, Implementation, and Challenges
7. Dr BC Roy & S Wagh; Underground Transportation Tunnel: Cut–and–Cover Approach
8. Dr BC Roy & S Wagh; Globalisation of Structural Design and Construction in Developing Countries
9. Dr BC Roy, December 2003, Thane MRTS-Station Area and Way Structure
10. Dr BC Roy, September 2004, Planning of Transportation System in Indian Metropolis
11. BK Ghosh & Dr BC Roy, August 2006, Kolkata Metro – Planning and Design (North-South Metro)
12. V Narayan, S Bhowmik and Dr BC Roy, August 2006, Kolkata Metro-Design and Construction (North-South Metro)
13. Dr BC Roy, 2006, Top-down Construction of Four-Level Integrated Intersection Planning, Design & Construction
14. Dr BC Roy, 2008, Transport Infrastructure in the Indian Urban Environment
15. Dr BC Roy, 2009, Mass Rapid Transport in Indian Urban Areas
16. Dr BC Roy, 2011, Planning Aspects of East-West Metro and Depot
17. Dr BC Roy, 2013, Kolkata Transit System including East-West Metro
18. Dr BC Roy, 2013, Keynote Address on East-West Metro
19. Dr BC Roy, 2014, Mass Transit System & Evolution of Kolkata Transport System
20. Dr BC Roy, 2015, Urbanisation and Growth in the Indian Transportation Scenario
21. Dr BC Roy, 2016, Urbanisation in India: Mass Transit System
22. Dr BC Roy +1, 2023, Nairobi MRTS Conceptualisation: Transit-Oriented Development for Sustainable Transport
23. Dr BC Roy, Dr R Ekambaram, 2021, Performance Assessment on Influence Zone of Existing Structures for Metro Construction
24. Dr BC Roy & Biswanath Dewanjee, Maiden Endeavour for creating a Transportation Tunnel Below Major River in India
25. Hemant Sharma, BC Roy & B Dewanjee, Kolkata Metro, including the East-West Metro
26. Amal Bhattacharya, 2024, Influence of Process Performance During the Construction Stage- A Comparative Case Study of Kolkata East-West Metro and Pune Metro Line 3.
27. Amal Bhattacharya & VN Heggade, 2004, Fast Track Construction of Kolkata Metro-EWE 1.

28. Harsimran Singh & Soumya Roy, A Unique Metro Station in Kolkata- (Line 3) AM Shete, Dr SP Patil, Prof. SS Sathe, 2020, Seismic Analysis of Pune Metro Tunnel
29. Sandeep Bharadwaj, 2021, Analysis of Chawri Bazar Section of Delhi Metro Tunnel
30. B&S Engineering Consultants, Elevated Viaduct Yamuna River Near Kalindi Kunj-Botanical Garden, Ph 3, Delhi Metro
31. B&S Engineering Consultants, Elevated Viaduct & Six Stations-Karkarduma – Gokulpuri Corridor, Line 7, Delhi Metro
32. B&S Engineering Consultants, Elevated Viaduct & Ramp on H Nizamuddin - Mayur Vihar Corridor, Line 7, Delhi Metro
33. B&S Engineering Consultants, Viaduct for Automatic Walkway/Travellators at Dhaula Kuan, Punjabi Bag and Rajouri Garden, Delhi Metro, Phase 3.
34. Standardisation and Indigenisation of Metro Railways, Systems and Subsystems, 2013, Ministry of Urban Development, GOI
35. Design Basis Report for Elevated Station, 2019, CMRL
36. Design Basis Report for Viaduct,2016, CMRL
37. Detailed Project Report for Pune Metro,2015, DMRC
38. Construction Specifications for Phase 4, Civil Works, DMRC
39. Madan Magdum and Bilavari Karkure: Finite Element Modelling and Design of Underground Metro System
40. Rajesh Prasad - Soil Engineering Challenges in Implementation of Metrowerks in the City of Kolkata
41. Rajesh Prasad – Geotechnical and GI Data for Metro Line 3, Kolkata in the City of Kolkata
42. Rajesh Prasad – Design Standard and Planning for Metro Line 3, Kolkata
43. Mumbai Metro Planning Group (MMPG) Mumbai: Metro Study- Seventh Rail Corridor, December 1997
44. CES-led Consortium Detailed Project Report: Thane MRTS, 2001
45. CES-led Consortium: Detailed Project Report of Andheri- Ghakopar Metro Corridor
46. Wikipedia: Metros in Kolkata, Delhi, Mumbai, Bengaluru and others
47. DMRC: Detailed Project Reports for Metros in Delhi, Bengaluru, Pune, and Others
48. R.S. Grover- Former Member, Electrical – Presentation on Kolkata Metro
49. DMRC Detailed Project Report Ph 4-Delhi Metro
50. DMRC Signalling and Telecommunication Report by Abhisek Garg
51. First Section of Delhi Metro Phase III Operational with Ticketing System from Thales (thalesgroup.com)
52. Ministry of Urban Development, GOI - Traction Systems, General Power Supply Arrangements and Energy Efficient Systems for Metro Railways
53. Ministry of Urban Development, GOI - Standardisation of Signalling and Train Control Systems for Metro Railways.
54. Interim Consultancy Report for Mumbai Metro Line 3.

55. Presentation on Kolkata Metro UG 1 – AFCONS; Atkins
56. O. Vora, A. More & E. Thinakaran – Mumbai Metro Line 3: Initial Drive of Tunnelling Activities
57. Amalendu Gope, Arnab Sur and Ravi Ranjan Kumar – Challenges for NATM Design of Grant Road Station in Mumbai Metro Line 3
58. Vinay K Pandey and SC Mishra
59. SK Gupta, MG Khare & GVR Raju – Geotechnical Challenges in Design and Construction of Cut-and-Cover Tunnel- Mumbai Metro Line 3
60. SK Gupta, MG Khare & JD Calentano – Design and Construction of Bored Tunnels for Mumbai Metro Line 3
61. Dr MG Khare – Instrumentation, Monitoring & Reporting

About the Author

Dr BC Roy

Dr BC Roy (Bidhan Chandra Roy), FNAE, PhD (Engg), MTech (Structures). He is a fellow of Major institutions worldwide, such as the Institution of Structural Engineers (UK), Indian National Academy of Engineers (INAE), International Association for Bridge & Structural Engineering (IABSE), Institution of Civil Engineers (UK), American Institution of Civil Engineers, Institution of India and many others. Dr over 50 years of extensive experience in providing engineering design consultancy services in India & abroad, spanning the transport infrastructure sector, including bridges, Roads, highways, flyovers, railways, metro rail, Industrial, residential, commercial & sports facilities. Dr Roy was the Team Leader for an International Award-Winning Nivedita Setu. He took a lead role for First (N-S) Metro in India, which was the first metro across a major river in the country. DPRs for Thane MRTS & Andheri-Ghatkopar Metro, and DDC & LDC of metros in Delhi, Mumbai, Chennai, Bangalore, Gurgaon and Ahmedabad. He led a team for the integrated transport system in Nairobi, Bangalore, Thiruvananthapuram, Indore, Rajkot, and other. During his long career, he had been involved in developing several innovative structures and several high-profile projects. This is his fourth book. He is a former Board Member of CES, Dy CE of CPPL, Transport Advisor of AECOM, and on the Board of Directors of RUPL and RUICPL.

www.ingramcontent.com/pod-product-compliance
Lightning Source LLC
LaVergne TN
LVHW070943210926
848482LV00002B/84
9798902076278